Problems in Chemistry

UNDERGRADUATE CHEMISTRY

A Series of Textbooks

edited by

J. J. Lagowski

Department of Chemistry
The University of Texas at Austin

Volume 1: Modern Inorganic Chemistry, *J. J. Lagowski*

Volume 2: Modern Chemical Analysis and Instrumentation, *Harold F. Walton and Jorge Reyes*

Volume 3: Problems in Chemistry, Second Edition, *Henry O. Daley, Jr. and Robert F. O'Malley*

Volume 4: Principles of Colloid and Surface Chemistry, *Paul C. Hiemenz*

Volume 5: Principles of Solution and Solubility, *Kozo Shinoda, translated in collaboration with Paul Becher*

Volume 6: Physical Chemistry: A Step-by-Step Approach, *M. K. Kemp*

Volume 7: Numerical Methods in Chemistry, *K. Jeffrey Johnson*

Volume 8: Polymer Chemistry: An Introduction, *Raymond B. Seymour and Charles E. Carraher, Jr.*

Volume 9: Principles of Colloid and Surface Chemistry, Second Edition, Revised and Expanded, *Paul C. Hiemenz*

Volume 10: Problems in Chemistry, Second Edition, Revised and Expanded, *Henry O. Daley, Jr. and Robert F. O'Malley*

Volume 11: Polymer Chemistry: An Introduction, Second Edition, *Raymond B. Seymour and Charles E. Carraher, Jr.*

Problems in Chemistry

Second Edition, Revised and Expanded

Henry O. Daley, Jr.
Bridgewater State College
Bridgewater, Massachusetts

Robert F. O'Malley
Boston College
Chestnut Hill, Massachusetts

Marcel Dekker, Inc. New York and Basel

Library of Congress Cataloging in Publication Data

Daley, Henry O., Jr.
Problems in chemistry.

Includes index.
1. Chemistry--Problems, exercises, etc. I. O'Malley, Robert F. II. Title.
QD42.D26 1988 540'.76 87-33165
ISBN 0-8247-7826-X

MARCEL DEKKER, INC.
270 Madison Avenue, New York, New York 10016

Current printing (last digit):
10 9 8 7 6 5 4 3 2 1

PRINTED IN THE UNITED STATES OF AMERICA

TO THE INSTRUCTOR

There have been many changes in the teaching of freshman chemistry since the publication of the last edition of this book—in particular, the full use of the SI system of units, the introduction of the unit factor method of solving problems, the use of the terms *atomic* and *molecular mass* in place of *weight*, and the elimination of the term *gram-atom*. All of these changes are reflected in this new edition.

In addition, approximately 200 new sample problems and supplementary problems have been added to the chapters. The section on the use of the slide rule has been replaced with one on the use of calculators and computers. Included in this section is a discussion of the problems associated with significant figures while using computers and calculators and those problems that can arise in switching between $\ln_e$ and the exponent e. A series of useful programs written in general BASIC is included; they can be used on any microcomputer. These programs illustrate the use of computers in solving quadratic equations, approximation methods in solving equations, and nonweighted linear least-squares analysis of data.

As with the first edition, we have chosen our problems with two fundamental concepts in mind: 1) to show students that much of what they learn in freshman chemistry is being used by chemists in their careers, and 2) to encourage students to read the chemical literature. It is important for a student to understand that what is done in the laboratory and what is covered in lectures are related. Whether the course is intended for chemistry, biology, or nonscience majors, students should learn that the knowledge they are acquiring in the lecture part of the course is developed or verified by laboratory experiments.

The interpretation of the results of laboratory experiments involves a chemist in one of the most basic everyday activities associated with chemistry, that is, problem solving. Chemistry is by its very nature a problem-solving-oriented course. All too often, this aspect of chemistry is overlooked by students. Most students require practice in solving problems. Today, students are particularly weak in solving word problems. The more problems a student solves, the more proficient he or she becomes in this aspect of chemistry.

It is the purpose of this book not only to provide the beginning chemistry student with a relatively large number of problems that illustrate chemical principles, but also to facilitate his acquisition of problem-solving skills by helping him see that the skills he acquires even at the introductory level are relevant to the activities of practicing chemists. The hope is that the student will come to see that he is doing what is being done, not that he is merely repeating what has been done. To this end, the problems in the present text are based, wherever possible, on experimental results obtained by chemists, and the corresponding references are cited. The problems and techniques for solving them are intended to reflect the real features of one important aspect of chemistry.

Besides showing that problem solving even at the introductory level is directly relevant to the practice of chemistry, there are other aims implicit in the general approach taken in this book. One of the most important is to illustrate the use of significant figures in a practical way. It is one thing to learn that significant figures are important because one must learn them in order to pass a test. It is another to learn from example that they are constantly used by chemists and therefore are important for learning chemistry. Consistent citation of references should also provide an indirect introduction to the chemical literature (the instructor, of course, also has the option of using these references to introduce students directly to the literature). Finally, it will be no small gain if the student comes to appreciate that chemical knowledge represents the accumulation of results from the activities of a great many people, each one making a contribution, however small, to the enterprise called chemistry.

A brief word concerning the use of this book. Although it is intended primarily as a self-study supplement and answers to problems are given in an appendix, it can also be used by the instructor as a source of test problems and as a supplementary text. The first 18 chapters contain about 860 problems plus an additional 240 or so example problems with solutions worked out.

The problems at the end of these chapters are divided into two sections to facilitate the choosing of representative problems on the material covered in these chapters. Each section contains a complete selection of problems arranged in the same order as that of the corresponding material in the chapter. An additional 200 problems in Chapter 19 are grouped to correspond to standard organization of the descriptive portion of the introductory course. Since most of the earlier literature did not use the modern metric system of units, the original data are retained in the problems, whereas the answers are given in SI units.

In preparing this new edition, one thing became obvious in the literature, that is, many of the experiments that gave useful data for problems are rarely carried out today. Instrumentation is the keyword for most laboratory experiments of the present. The mass spectrometer is used to determine molecular mass rather than freezing point depression. In addition, even though data such as percent composition may be listed for a new compound, the data used to calculate the percent composition and the method used are rarely given. Despite this, we have found sufficient material to introduce new problems in just about every chapter except Chapter 19. We chose not to add any new additional problems to this chapter. Some new and interesting compounds such as AgF_3 have been prepared and data were available for use in the chapter on empirical formulas.

Although there are many people who have been helpful in developing this second edition, we would like to thank in particular Dr. Vahe Marganian of Bridgewater State College for the many discussions we have had with him. We would also like to thank Rosemary Daley, the wife of one of the authors, for the proofreading she has done, and the very cooperative and fine editors from Marcel Dekker, including Sandra Beberman, Henry Boehm, and especially Patricia Brecht, without whose help this revision could not have been accomplished.

Henry O. Daley, Jr.
Robert F. O'Malley

TO THE STUDENT

Solving problems by mathematical manipulation is an essential, daily experience of chemists. Students who seek even an elementary knowledge of chemistry must learn to solve a variety of problems. If for no other reason, then, a knowledge of the chemical principles and the mathematical manipulations necessary to solve typical chemical problems is important.

Students almost universally believe that they must *understand* the principles of chemistry before they can apply them to the solution of problems. Many students fail to realize that it is also necessary to *remember* statements of principles, definitions of terms, and many facts of a descriptive nature. The importance of a thorough understanding of the laws and theories of chemistry is not to be underestimated, but the need to *commit* to *memory* a certain body of information is equally important. A student may understand the concepts of an atom, molecule, mole, equivalent mass, and so on as he reads about them in the text, but this understanding is not enough if the student cannot recall the precise definition of the terms when he attempts to apply them to the solution of a problem.

Before trying to solve problems, *study* the relevant sections of the text and class notes. *Memorize* the precise meaning of important terms and statements of principles. The first step toward the successful solution of a problem is to read very carefully the statement of the problem. Next, reflect on the meaning of the terms and *determine exactly what is asked for by the problem.* (This point cannot be overemphasized. Probably the greatest source of confusion in any kind of problem solving stems from the failure to determine what is asked for. Obviously, we need to "understand" the problem. But we often forget that understanding a problem is above all a matter of knowing what kind of answer we are looking for.) Once you have determined both what the problem asks for and what principles are required to provide the appropriate kind of answer, you can proceed in an orderly fashion to perform the necessary mathematical manipulations.

At this point, the most important advice is the following: *Be neat*. Fewer errors and time saved are the rewards of a neat, orderly approach to the solution of a problem. Write as legibly as you can and leave space between the steps.

Example

A compound containing 5.28% silicon and 94.66% iodine was prepared at M.I.T. Calculate the empirical formula of the compound.
[W. C. Schumb and D. W. Breck, *J. Am. Chem. Soc.*, *74*, 1758 (1952)]

$$\frac{5.28}{28.1} = 0.188 \text{ mol Si}/100 \text{ g}$$

$$\frac{\text{g}}{\text{g/mol}} = \text{g} \times \frac{\text{mol}}{\text{g}} = \text{mol}$$

$$\frac{94.7}{127} = 0.745 \text{ mol I}/100 \text{ g}$$

$$\frac{0.745}{0.188} = 3.97$$

$$\frac{\text{mol I}}{\text{mol Si}}$$

$\therefore$ SiI_4 is the empirical formula.

If the proper procedure for solving the problem does not occur to you, read the pertinent sections in your text and notes again. Reconsider very carefully the definition of terms and the statements of principles. Very often, the difficulty is due to an inexact definition or failure to consider all that is implied in a definition or principle.

Although students study to acquire knowledge of a subject for many different reasons, they must all face the practical requirements of passing examinations. To prepare for examinations they must *learn* or memorize definitions and principles, and they must *practice* the application of what they have learned by answering questions and solving problems similar to those that appear on examinations. This book is intended to assist the student in his *active preparation* for examinations in chemistry.

CONTENTS

1
PROPERTIES AND THEIR MEASUREMENTS

1-1 PROPERTIES

The 105 elements and the immense number of known compounds are distinguished from one another by their properties, which are the characteristics by which chemists identify the various elements and compounds. *Chemical properties* are those which are observed when substances are changed into totally different substances. *Physical properties* can be observed without the occurrence of chemical reactions.

1-2 EXTENSIVE AND INTENSIVE PROPERTIES

Extensive properties depend on the particular sample present; examples are volume, weight, and area. Intensive properties are properties that are independent of the size and shape of a particular sample of a substance; examples are color, odor, melting point, and vapor pressure.

1-3 UNITS OF MEASUREMENT

Scientific knowledge is incomplete as long as it is merely descriptive or qualitative. Quantitative knowledge, expressed as measurements, is the immediate goal of scientific investigations. Among the fundamental units of any system of measurement, the units of length, mass, and time are the most basic. In scientific work, these quantities are expressed as units of the metric system.

TABLE 1-1 Base Units of the Metric System

Physical quantity	Unit	Symbol
Length	Meter (metre)	m
Mass	Kilogram	kg
Time	Second	s
Temperature	Kelvin	K
Amount of substance	Mole	mol
Electric current	Ampere	A
Luminous intensity	Candela	cd

1-4 THE METRIC SYSTEM

The modern version of the metric system, developed in 1960, is known as the International System of Units (SI). For a brief history and use of the English and metric systems, see the National Bureau of Standards special publication 304A, 1970. The defined values of the fundamental units are those adopted by the National Bureau of Standards [*Nat'l. Bur. Std. (U.S.) Tech. News Bull.*, *47*(10) (1963) or *J. Chem. Educ.*, *48*, 569 (1971)]. This system was established by international agreement to allow science, industry, and commerce to work with one another.

There are seven base units in the metric system. They are given in Table 1-1. All other physical units are derived from these units. Of these units, the first six are used in Freshman chemistry. For now we will define the first three units only. The rest will be defined as needed in a particular chapter.

The fundamental unit of length is the *meter*. It is defined as the distance that light travels in a vacuum in 1/299 792 458 of a second. (Note: in this system spaces, not commas, are used to separate parts of large numbers.)

Because the metric system is based on the decimal system, it is easy to use. The smaller and larger multiples of each fundamental unit are related by powers of 10. Other units of length are defined in terms of the meter. Table 1-2 lists some of the common prefixes used in the metric system.

TABLE 1-2 Prefixes Used in the Metric System

Prefix	Symbol	Meaning	Example
Kilo	k	1000 x	Kilometer (km) = 1000 m
Deci	d	1/10	Decimeter (dm) = 0.1 m
Centi	c	1/100	Centimeter (cm) = 0.01 m
Milli	m	1/1000	Millimeter (mm) = 0.001 m
Micro	μ	1/1 000 000	Micrometer (μm) = 0.000001 m
Nano	n	1/1 000 000 000	Nanometer (nm) = 0.000000001 m

The unit of mass is the *kilogram* (kg), the mass of the international kilogram at Sevres, France, which is a platinum-iridium cylinder. Smaller units are listed with their symbols.

Gram (g) $= 10^{-3}$ kg

Milligram (mg) $= 10^{-6}$ kg

Nanogram (ng) $= 10^{-12}$ kg

The microgram (10^{-9} g) is also called the gamma (γ). Chemists refer to solutions containing one gamma of solute per liter of solution, for example.

The fundamental unit of time is the *second*. It is defined as 9 192 631 770 cycles of the radiation associated with a specific transition of the cesium atom.

The unit of volume is the cubic meter (m^3). Volumes of liquids are measured in a unit derived from this called the liter (L). A liter is equal to a cubic decimeter (dm^3). It is the same as 1000 cm^3 (commonly, cc). Another common unit used in the laboratory to measure liquids is the milliliter (mL). One liter contains 1000 mL. A mL is thus equal to 1 cc.

1-5 DENSITY AND SPECIFIC GRAVITY

Density is defined as the mass per unit volume. It may be expressed as

$$d = \frac{m}{V}$$

where d is density, m is mass, and V is volume. Since the fundamental unit of mass is the kilogram, while that of volume is the cubic meter, density should be expressed as kg/m^3. Since this leads to large numbers, it is common to give the density of liquids in g/cc and those of gases in g/L. Table 1-3 gives the densities of selected materials.

A property of a system closely related to density is *specific gravity*. The specific gravity is defined as the ratio of the density of a substance to the density of water. In many cases the density of water is taken at 4°C. Since specific gravity is a ratio, it is a dimensionless quantity. For routine work the density of water at 4°C is close enough to 1 g/cc, so that the specific gravity and the density of a substance are the same numerical figure.

TABLE 1-3 Densities of Selected Materials

Material	Density (g/cc)
Al	2.70
Au	19.3
B	2.50
Br_2	3.12
Ga	6.09
Hg	13.5
Na	0.971
Pb	11.3
Pt	21.4
U	18.9
CCl_4	1.60
$SiCl_4$	1.48
$SnCl_4$	2.23
Glass	2.23

1-6 ACCURACY AND PRECISION

The accuracy of a measurement may be defined as *the degree of conformity of a measure to a true value*. The *precision* of a measurement refers to its reproducibility.

The numerical value of a measurement is based in part on the reliability of the measuring instrument and in part on the judgment of the person making the measurement. When the length is measured by placing a ruler adjacent to it, the reliability depends first on the care with which the marks have been placed on the ruler. The judgment of the user enters when he decides which of the marks lies nearest to the length to be measured. If the length lies between two marks, he must *estimate* the last digit of the numerical value of the length. If a length is recorded as 28.6 mm, the final digit is an *estimate* and is uncertain. This number has three significant figures.

A measurement recorded as 28.62 implies that the uncertainty is in the last digit and that a more refined technique of measurement has been used. This number has four significant figures.

The judgment of the chemist may be involved even when more sophisticated instruments are used. Measurements made with a balance, a buret, a pH meter, or an infrared spectrophotometer, as well as many other instruments, may include estimates made by the chemist.

1-7 SIGNIFICANT FIGURES

The number of *significant figures* in a measured quantity is equal to the number of digits whose values are known plus the first digit of uncertain value. The position of the decimal point is irrelevant. A *zero* is significant unless it is used to indicate the position of the decimal point (represent the order of magnitude). A zero is never significant if there are digits only on its right, for there it is used to indicate the position of the decimal point. However, a zero is significant if it lies between two digits.

A problem arises when zeros have digits only on the left. They may or may not be significant. A mass of 2000 g may have one, two, three, or four significant figures. The zeros may represent certain values, or only indicate the position of the decimal point. It is good practice to use exponential notation in such cases.

Example 1-1

Quantity	Number of significant figures
217	3
2.17	3
0.217	3
0.000217	3
7.043	4
70.43	4
0.07043	4
1.040	4

Example 1-2

Quantity	Number of significant figures
2×10^3	1
2.0×10^3	2
2.00×10^3	3

In exponential numbers the zeros to the right of the decimal point are used to represent significant figures.

1-8 SIGNIFICANT FIGURES IN CALCULATIONS

Chemists often use measured quantities from experiments in calculations and record the final results in reports and in papers submitted to journals. The final results of such calculations must contain no more significant figures than the least precise measurement. In solving a problem the student should never include more significant figures in the answer than there are in the least precise quantity in the problem. If several measured quantities are involved, they should be rounded off, if necessary, to agree with the least precise quantity.

Rounding off a number means that figures that are not significant for the purpose at hand are dropped. There are three useful rules to follow in deciding whether to drop a digit:

1. If the digit dropped is greater than 5, add 1 to the last remaining digit.

Example 1-3

62.138 will become 62.14

2. If the digit dropped is less than 5, the last remaining digit is not changed.

Example 1-4

28.133 will become 28.13

3. If the digit dropped is 5, the last remaining digit is left unchanged if it is even; 1 is added if it is odd.

Example 1-5

1.8245 will become 1.824; 1.8235 will become 1.824

1-9 ADDITION AND SUBTRACTION

Before adding or subtracting, round off the numbers so that all have the same number of digits to the right of the decimal point.

Example 1-6

28.521	becomes	28.52
6.38		6.38
0.216		0.22
111.535		111.54

Example 1-7

18.406	becomes	18.406
2.12456		2.124
8.0745		8.074

Note that the numbers are not rounded off in succession, that is, 2.12456 became 2.124 not 2.125.

1-10 MULTIPLICATION AND DIVISION

Quantities should be rounded off to agree with the measurement containing the least number of significant figures.

Example 1-8

$$\frac{0.2536 \times 0.08205 \times 298.16}{0.974 \times 255.7} =$$

becomes

$$\frac{0.254 \times 0.0820 \times 298}{0.974 \times 256} =$$

1-11 UNITS IN CALCULATIONS

Solving problems in chemistry requires the use of numerical values of measurements, and the numerical values are expressed in either the fundamental units or units derived from them. The "answer" to a problem based on measurements must be expressed in the appropriate unit. To assure that the final unit is correct, it is wise to include the units in the steps of the calculations.

Example 1-9

Calculate the volume of a regular rectangular polyhedron that is 25.2 cm long, 15.3 cm wide, and 6.82 cm high.

The volume V of a rectangular polyhedron is given by the formula $V = A \times h$, where A is the area of the base and h is the height. The area A of a rectangle is given by $A = \ell \times w$, where ℓ is the length and w is the width. To perform this calculation the numerical values of the dimensions are substituted for the symbols in the equations:

$$V = \ell \times w \times h$$

$$V = 25.2 \times 15.3 \times 6.82$$

The units may be placed directly in the steps of the calculation along with the number:

$$V = 25.2 \text{ cm} \times 15.3 \text{ cm} \times 6.82 \text{ cm}$$

$$V = 2630 \text{ cm}^3 \text{ (cc)}$$

If the algebraic manipulation of the units leads to the unit expected in the answer, the student may be reasonably confident that the method used to find the solution is correct. This approach provides the student with a valuable self-check. It does not mean that the answer is correct, only that the procedure is reasonable.

Example 1-10

Calculate the volume occupied by 261 g of copper. The mass and volume of a substance are related by its density, and the density of copper is 8.93 g/cm^3.

$$d = \frac{m}{V}$$

$$V = \frac{m}{d}$$

$$V = \frac{261\ \not{g}}{8.93\ \not{g}/cm^3} = 29.2\ cm^3\ (cc)$$

Example 1-11

What is the mass of a cylinder of silver of radius 1.63 cm and length 8.71 cm? The density of silver is 10.5 g/cm^3.

$$d = \frac{m}{V}$$

$$m = dV$$

but

$$V = \pi r^2 h$$

$$m = d\pi r^2 h$$

$$m = 10.5\ \frac{g}{cm^3} \times 3.14 \times (1.63\ cm)^2 \times 8.71\ cm$$

$$= 10.5\ \frac{g}{\not{cm}^3} \times 3.14 \times 2.66\ \not{cm}^2 \times 8.71\ \not{cm} = 766\ g$$

1-12 CONVERSION OF UNITS

The metric system of units is used in scientific work throughout the world. The United States is the only major country in the world that still does not use the metric system in everyday life. However, it is not necessary to convert measurements from one system to another frequently.

When using conversion relationships, it is convenient to work with what are known as *unit factors* or *conversion factors*. Since 1 lb is 454 g to three significant figures, we can write this as

$$\frac{454 \text{ g}}{1 \text{ lb}} = 1$$

This is the unit factor that converts from pounds to grams. Since multiplying by 1 does not change the value of a number, all we have to do is multiply the number of pounds of a material by this factor to get grams. Similarly, for conversion of inches to centimeters the following unit factor can be used:

$$\frac{2.54 \text{ cm}}{1 \text{ in.}} = 1$$

Since the reciprocal of 1 is still 1, we can invert this expression to find the unit factor that converts from centimeters to inches:

$$\frac{1 \text{ in.}}{2.54 \text{ cm}} = 1$$

Example 1-12

What is the length in centimeters of a glass rod 7.82 in. long?

The unit factor is 2.54 cm/1 in.

$$7.82 \cancel{\text{in.}} \times \frac{2.54 \text{ cm}}{1 \cancel{\text{in.}}} = 19.8 \text{ cm}$$

Example 1-13

How many grams of bromine are there in a vial that contains 1/4 lb of this substance?

The unit factor is 454 g/1 lb

$$1/4 \cancel{\text{lb}} \times \frac{454 \text{ g}}{1 \cancel{\text{lb}}} = 114 \text{ g}$$

Example 1-14

A bottle containing a solution of ammonia that is 30% ammonia by weight contains 1 lb of solution. How many grams of ammonia are present?

$$1 \text{ lb solution} \times \frac{454 \text{ g}}{\text{lb}} = 454 \text{ g solution}$$

$$454 \text{ g solution} \times \frac{0.300 \text{ g ammonia}}{\text{g solution}} = 136 \text{ g ammonia}$$

The symbol % means "parts per 100 parts." In this instance there are 30 parts by weight of ammonia in 100 parts by weight of solution or, specifically, 30.0 g ammonia per 100 g solution. In chemistry, whenever % is used to describe the composition of a solution, it is to be assumed that percent by weight is meant unless you are told otherwise. When a % is changed to a decimal by dividing by 100, the basis is changed from 100 to 1. In this example, 30.0% becomes 0.300 g of NH_3 per 1.00 g of solution.

1-13 TEMPERATURE

It is not possible to define temperature in relation to the fundamental units of length, mass, and time; temperature is a fundamental concept by which the relative "hotness" or "coldness" of objects or environments is compared.

The temperature scale in SI is the thermodynamic or *Kelvin* temperature scale. This has an origin, or zero point, at absolute zero, and a fixed point at the triple point of water, defined as 273.16 kelvins. The triple point is that temperature at which the solid, liquid, and gaseous states all exist in equilibrium with one another at the same temperature.

The temperature scale most commonly used in science is the Celsius, or centigrade scale. The designation *Celsius* has been officially adopted, but the term *centigrade* is still frequently used. Temperatures on the Celsius (centigrade) scale are recorded as °C. The triple point of water is defined as 0.01°C on the Celsius scale. To convert from Celsius degrees to Kelvin degrees you add 273.15 to the Celsius temperature.

$$K = {}^{\circ}C + 273.15$$

Several devices, including gas thermometers, thermocouples, and the familiar mercury thermometer, are used for the measurement of temperature. The mercury-in-glass type is the most common.

The freezing point and the boiling point of water under one atmosphere of pressure are two convenient points for checking a thermometer. The freezing point of water is 0°C, while the boiling point is 100°C. The interval between these two points has 100 equal parts. *Unless otherwise noted, all temperatures given in the text are Celsius.*

On the familiar Fahrenheit scale, the freezing point of water is 32° and the boiling point of water is 212°. The interval between these two points is divided into 180 equal parts. Each part corresponds to 1° on the Fahrenheit scale. Temperatures on this scale are recorded as °F. Examples are 67°F, 81.6°F, and 1600°F.

Since the interval from the freezing point to the boiling point of water is an interval of 100 Celsius degrees or 180 Fahrenheit degrees, a *change* of 100 Celsius degrees measures the same temperature change as does a change of 180 Fahrenheit degrees. Therefore, a temperature change of 5 Celsius degrees is the same as a temperature change of 9 Fahrenheit degrees.

1-14 CONVERSION OF TEMPERATURES

It is occasionally necessary to convert a temperature from one scale to the other. Useful equations for these conversions are

$$°C = \frac{5}{9}\,(°F - 32)$$

$$°F = \frac{9}{5}\,°C + 32$$

Example 1-15

In earlier measurements of properties, such as solubility, the standard temperature was established as 18.0°C; later it was changed to 25.0°C; and now some data are measured at 30.0°C. Convert these temperatures to the Fahrenheit scale.

$$°F = \frac{9}{5}\,°C + 32$$

$$°F = \frac{9}{5} \times 18.0 + 32.0 = 64.4\ °F$$

$$°F = \frac{9}{5} \times 25.0 + 32.0 = 77.0\ °F$$

$$°F = \frac{9}{5} \times 30.0 + 32.0 = 86.0\ °F$$

Example 1-16

A temperature of 68°F has often been described as a normal, comfortable room temperature. Convert this temperature to the Celsius scale.

$$°C = \frac{5}{9}(°F - 32)$$

$$°C = \frac{5}{9} \times (68 - 32) = \frac{5}{9} \times 36 = 20°C$$

1-15 SPECIFIC HEAT

The quantity of heat required to raise the temperature of a given sample of a substance depends on the nature of the substance and the size of the particular sample. Thus it is an *extensive* property. In order to consider the quantity of heat absorbed by various substances, this quantity must be expressed as an *intensive* property, that is, one which does not depend on the particular sample. This is done by using the same *mass* of the substance. The property so measured is called the *specific heat* or the *heat capacity per gram*, the quantity of heat required to raise the temperature of 1 g of the substance 1 degree Celsius. The heat absorbed or given off by a substance is given as

$$Q = \text{mass} \times \text{specific heat} \times \Delta T$$

where ΔT is the temperature difference of the substance. It is always taken as T(final) - T(initial).

The unit of heat in the metric system is the joule (J). It is a derived unit and represents the energy associated with a force of one newton acting through a distance of one meter. It has units of $kg \times m^2/sec^2$. Other units encountered in chemistry are the calorie and the erg. They are related to the joule as follows:

1 calorie (cal) = 4.184 J

1 erg = 1×10^{-7} J

Table 1.4 gives specific heats of various substances.

Example 1-17

Calculate the quantity of heat required to raise the temperature of 50 g of water from 25°C to 50°C.

$$Q = 50 \text{ g} \times \frac{4.184 \text{ J}}{\text{g°C}} \times (50 - 25)\text{°C}$$

Q = mass × specific heat × temperature change

Q = 5230 J

$Q = 5.2 \times 10^3$ kJ

Example 1-18

How much heat is required to raise the temperature of 5.6 g of mercury from 20°C to 80°C?

$$Q = 5.6 \text{ g} \times \frac{1.38 \text{ J}}{\text{g°C}} \times (80 - 20)\text{°C}$$

Q = 460 J

TABLE 1-4 Specific Heats of Various Substances

Substance	Specific heat (J/g°C)
Alcohol	2.43
Aluminum	0.908
Cadmium	0.23
Copper	0.385
Lead	0.128
Mercury	1.38
Silver	0.233
Water	4.184
Zinc	0.387

Example 1-19

One hundred grams of water at 80°C is added to 50 g of water originally at 20°C. Assuming no heat is lost, what is the final temperature of the system?

$$Q_{gained} = 50\ g \times \frac{4.184\ J}{g°C} \times (T_f - 20)°C$$

$$Q_{lost} = 100\ g \times \frac{4.184\ J}{g°C} \times (T_f - 80)°C$$

$$Q_{gained} + Q_{lost} = 0$$

$$50 \times 4.184 \times (T_f - 20) + 100 \times 4.184 \times (T_f - 80) = 0$$

$$209.2\ T_f - 4184 + 418.4\ T_f - 33472 = 0$$

$$627.6\ T_f - 37656 = 0$$

$$T_f = \frac{37656}{627.6}$$

$$T_f = 60°C$$

PROBLEMS

1-1 State the number of significant figures in each of the following quantities.

a.	51.8	b.	0.0605 g
c.	9.89577	d.	50.0 ml
e.	5.0 ml	f.	21,000 lb
g.	16 oz	h.	6.022×10^{23} atoms
i.	3.2×10^{-23} g/atom	j.	23.40×10^{5} g

1-2 Add:

a. 94.5 g
 4.03 g

b. 79.96%
 14.2%
 5.7%

c. 2.85×10^{-3} mg
 4.19×10^{-2} mg

d. 35.5%
 33.7%
 8.63%

e. 0.0873 g
 11.61 g
 0.0048 g
 2.631 g

1-3 Subtract:

a.	6.17 ml 0.016 ml	b.	23.8 L 0.0141 L
c.	238.5 g 14.92 g	d.	0.07638 g 0.0605 g

1-4 Carry out the following manipulations, and observe the proper treatment of significant figures.

a. Add: 2.619 g + 8.96 mg + 0.0121 g + 168.6 mg
b. Subtract: 261.4 mL from 2.876 L
c. $(8.314 \times 10^7)(298.15)/(1.01 \times 10^5)$
d. $(.0820) \times (.000314)/(0.002165)$

1-5 Multiply:

a. $740 \times 250 =$
b. $0.08205 \times 298 =$
c. $0.08205 \times 298.2 =$
d. $0.08205 \times 298.15 =$
e. $6.022 \times 10^{23} \times 1.00796 =$

1-6 Divide:

a. $18.96 \div 6.939 =$
b. $48.42 \div 44.956 =$
c. $26.3 \div 207.19 =$
d. $760 \div 298.2 =$
e. $0.08205 \div 250 =$

1-7 The interatomic distance in the hydrogen molecule is given as 0.74 Å. What is this distance in nanometers, micrometers, and millimeters? (1 Å = 10^{-8} cm)

1-8 A quantity of lithium metal weighs 6.939 g. Express this weight in milligrams and kilograms.

1-9 A solution contains 1.00 mg of a substance in 1 L of the solution. How many γ of the substance are present?

1-10 Calculate, in cubic centimeters and in milliliters, the volume enclosed by a hollow cylinder whose length is 13 cm and inside diameter 22 mm.

1-11 A sample of a vapor in a 223.4-cc cell weighed 0.5763 g. Calculate the density of the vapor in grams per milliliter and grams per liter.

[A. W. Laubengayer and F. B. Schirmer, *J. Am. Chem. Soc.*, *62*, 1578 (1940)]

1-12 The density of fluoramine was measured in a pycnometer, or specific-gravity bottle, having a volume of 0.0343 mL which was calibrated with mercury. What weight of mercury was required to fill the pycnometer?

[E. A. Lawton and J. Q. Weber, *J. Am. Chem. Soc.*, *85*, 3596 (1963)]

1-13 An experiment requires that a solution be made by using 500 g of sulfuric acid. The only container available is a 400-mL beaker. Is it large enough to contain the acid? The density of the acid is 1.84 g/mL.

1-14 An experiment requires 250 g of a compound that is listed in two manufacturers' catalogs. The first manufacturer offers the compound at $12.30 per 100 g; the second offers the same compound at $60.00 per lb. From which of the two manufacturers should you purchase the compound if both supply the same quality?

1-15 Which of the following substances will occupy the largest volume?

a. 1 g of mercury
b. 2 g of lead
c. 0.5 g of aluminum

1-16 The half-life of a first-order reaction was 10 sec. What is the half-life expressed in milliseconds? What is it in microseconds?

1-17 If 108.2 g of silver oxide is required for an experiment, will 4 oz be sufficient?

1-18 In the physical chemistry laboratory, a common experiment is to determine the surface tension of a liquid by the capillary rise method. A section of a broken thermometer is often used as the capillary. A student carrying out such an experiment used a section of a thermometer 200 mm long. The thermometer section weighed 18.8480 g. When the capillary was filled with mercury, the thermometer section weighed 18.9249 g. Assuming the capillary is a uniform cylinder, what is the radius of the capillary in millimeters?

1-19 What quantity of nitric acid, in grams, is present in a bottle containing $1\frac{1}{2}$ lb of 70.0% HNO_3?

1-20 How many grams of ammonia must be present in 1.00 pint of a 30.0% aqueous solution? The specific gravity of the solution is 0.9016.

1-21 An experiment requires 383 g of toluene. A bottle containing 0.89 lb is available. Is it sufficient?

1-22 What is the volume occupied by 1.60 lb of concentrated hydrobromic acid (that is, 48% HBr) if the specific gravity of the solution is 1.50?

1-23 How many grams of hydrogen bromide are present in the solution described in Problem 1-22?

1-24 A bottle of concentrated H_2SO_4 is found to contain 96.5% H_2SO_4 by weight. The density of the acid is 1.84 g/cc.

a. How many grams of H_2SO_4 are present in 1 L of the acid solution?
b. How many grams of H_2SO_4 are present in 1 kg of the acid solution?

1-25 What is the freezing point of benzene (5.53°C) on the Fahrenheit scale?

1-26 A considerable amount of research, especially in biochemistry, is now being carried on at 37.0°C. Convert this temperature to degrees Fahrenheit. Why is this particular temperature used?

1-27 How much heat, in joules, is necessary to raise the temperature of 100 g of zinc from 25°C to 100°C?

1-28 A quantity of water weighing 5.0 g at a temperature of 50°C was mixed with an 8.0-g sample at a temperature of 10°C. What was the final temperature of the water if no heat was lost to the surroundings?

1-29 In a freshman laboratory experiment, 50.8 g of Al at 100°C was added to 50 g of water in a calorimeter. The original temperature of the water was 22°C. What is the final temperature of the mixture, assuming no heat is lost to the surroundings?

1-30 A company plans to build a power plant next to a river that has a flow rate of 1×10^{10} g/min. One-tenth of this water will be diverted through the plant to cool the condenser. If the condenser discards heat at the rate of 1×10^{10} calories per minute to the water:

a. What is the temperature change of the cooling water after it leaves the condenser?
b. What will be the temperature change of the river water when the water from the condenser is returned to the river (assume no heat loss)?
c. Give the equivalent temperature changes from part (a) and (b) in degrees Fahrenheit.

Supplementary Problems

1-31 The distance between two nitrogen atoms in a nitrogen molecule is commonly given as 1.10 Å. What is this distance in nanometers, micrometers, and millimeters? (1 Å = 10^{-8} cm)

1-32 The ionic radius of the oxide ion is listed as 1.32 Å by some crystallographers. What is the radius in millimeters, micrometers, and nanometers? (1 Å = 10^{-8} cm)

1-33 What is the weight in milligrams of 200.59 g of mercury? What is this weight in kilograms?

1-34 How many milliliters of water may be placed in a hollow cylinder 8 cm long and 12.6 mm in diameter?

1-35 What is the volume, in milliliters, of a hollow cylinder 25 cm long and 35.6 mm in diameter?

1-36 Calculate the volume occupied by 26.71 g of a substance whose density is 2.321 mg/cc.

1-37 What is the weight of 2.718 L of a gas whose density is 1.896 mg/cc?

1-38 A bottle of sulfuric acid contains 2 lb of the acid. How many pounds of HNO_3 can be stored in a bottle of the same volume? The density of sulfuric acid is 1.84 g/mL, and the density of nitric acid is 1.38 g/mL.

1-39 The 5-pint bottle has been a standard container for concentrated acids for many years. What is the approximate weight of glacial acetic acid contained in a 5-pint bottle? The density of glacial acetic acid is 1.05 g/mL.

1-40 Which is the better source of a liquid to be used in an experiment – a supplier who advertises the product at $12.00 per lb or another who offers it at $14.00 per liter? The density of the liquid is 0.782 g/mL.

1-41 Glass tubing and rod are normally sold in 4-ft lengths, and the diameter is designated in millimeters. Calculate the volume of a 4-ft length of 8-mm rod.

1-42 A reaction was carried out in a 1-ft length of tubing whose outside diameter was 30 mm. If the wall thickness was 1 mm, what was the volume, in milliliters, available for the reaction?

1-43 What is the weight of a package of 6-mm glass rod that contains 145 pieces 4 ft long?

1-44 Calculate the number of pieces of 10-mm glass tubing 4 ft long in a package weighing 34.6 lb. Assume the wall thickness of the tubing to be 1 mm.

1-45 A certain molecular species was found to exist for 3 msec. What was the life of the species in seconds and microseconds?

1-46 The density of difluoramine at -80.5°C was measured in a pycnometer with a volume of 0.0343 mL. A value of 1.587 g/mL was observed. Calculate the weight of NHF_2 contained in the pycnometer.

[E. A. Lawton and J. Q. Weber, *J. Am. Chem. Soc.*, *85*, 3596 (1963)]

1-47 What is the quantity of fluorine, in grams, in a cylinder containing 1/4 lb?

1-48 What is the weight, in grams, of a 51.5% suspension of sodium hydride in mineral oil if 1.00 lb of sodium hydride is present?

1-49 How many liters of acetone are present in a 1-gal jug? How many milliliters?

1-50 How many milliliters of pyridine are there in a 1-pint bottle? How many liters?

1-51 How many grams of sodium hypochlorite are present in 1 lb of a 5.0% solution? How many grams of water?

1-52 What is the weight of a 5.0% solution of sodium hypochlorite if it contains 63 g of water?

1-53 A bottle of concentrated hydrochloric acid has a density of 1.18 g/cc. A 25.0-ml sample of this acid is found to contain 10.50 g of HCl. What is the percent by weight of HCl in the bottle?

1-54 The freezing point of cyclohexane is 6.59°C. What is its freezing point on the Fahrenheit scale?

1-55 While looking at a thermometer outside your house, a cold-air front moves in. The temperature drops from 81° to 54°F. What is the corresponding temperature drop in degrees Celsius?

1-56 How many calories of heat must be removed from a sample of lead weighing 0.435 g to lower its temperature from 61.2 to 48.3°C?

1-57 A kilogram of alcohol at 27°C was mixed with 450 g of alcohol at another temperature. The final temperature was 25.2°C. What was the initial temperature of the 450-g sample of alcohol?

2

ELEMENTS, ATOMS, AND MOLECULES

2-1 ATOMS

All atoms of an element have the same atomic number; that is, they have the same positive electric charge on the nucleus. They also have the same number and distribution of electrons. Sodium atoms, for example, have a nuclear charge of 11, and they have 11 electrons outside the nucleus. All atoms of sodium also have the same mass, 3.82×10^{-23} g. Several other elements, such as fluorine, gold, iodine, manganese, and tantalum, also have *only one type of atom*. Most of the other elements, however, have atoms that differ in mass. All of the atoms of neon, for example, have the atomic number 10, but they do not all have the same mass. Atoms of an element with different masses are called *isotopes*. The isotopes of an element have the same atomic number but different masses. Neon, for example, has three isotopes.

2-2 ATOMIC MASS

The mass of the atoms of an element compared to the mass of the atoms of the carbon isotope designated carbon-12 (^{12}C) is the atomic mass. That is, the mass of ^{12}C has been assigned the value 12.000000 on this relative scale, and atomic masses are relative masses. A gold atom, for example, has a mass of 196.967 compared to the mass of ^{12}C, taken as 12.000000. For an element that has isotopes, the atomic mass is the *average relative mass* of the atoms.

Although atomic masses are relative masses, a unit has been adopted, the atomic mass unit (amu). One amu has been defined as 1/12th of the mass of the ^{12}C isotope. In the past the relative masses of the atoms were generally called atomic weights. The term atomic weight is still in use in some texts, but the tendency is toward the adoption of the correct designation, atomic mass.

2-3 MOLECULES

The noble gases, helium, neon, krypton, xenon, and radon, occur as individual atoms. The metallic elements, for example, sodium, magnesium, and copper, exist as atoms held together in extremely large numbers in the solid state. Several elements occur as molecules, that is, *combinations of two or more atoms held together by strong forces*. The elements hydrogen, oxygen, nitrogen, fluorine, chlorine, and iodine occur as molecules with two atoms. Phosphorus molecules have four atoms and sulfur molecules eight atoms.

2-4 MOLECULAR MASS

Molecular masses are relative masses on the same scale as the atomic masses. They are also compared to the mass of the ^{12}C isotope. A molecular mass is obtained by adding the atomic masses of the atoms in the molecule.

Example 2-1

What is the molecular mass of a normal sulfur molecule?

A normal sulfur molecule contains eight atoms. The atomic mass of sulfur from Table D-3 is 32.06. The molecular mass is: $8 \times 32.06 = 256.48$.

2-5 THE MOLE

It is possible to determine the mass of individual atoms in mass units of the SI system, but the measurements thus obtained are of no use in ordinary laboratory work. Instead, a unit for the measurement of the quantity of a substance, the mole, is used. The official definition of the mole (abbreviated mol) is: the amount of a substance that contains as many elementary entities as there are atoms in 12.0000 g of the isotope ^{12}C. This number, called Avogadro's number, is 6.022×10^{23}. A much more practical definition of the mole for everyday work with elements is: *the atomic or molecular mass expressed in grams.* The unit factor of the mole is g/mol or g mol^{-1}. A mole of calcium has a mass of 40.08 g, a mole of lead 207.19 g, and a mole of lithium 6.941 g. A mole of sulfur molecules has a mass of $8 \times 32.06 = 256.48$ g. Each of these quantities contains 6.022×10^{23} atoms or molecules, respectively.

Example 2-2

What is the mass of 2.00 mol of neon?

This calculation can be done the way unit changes were done in Chapter 1. The unit factor is g/mol. The atomic mass of neon from Table D-3 is 20.179.

$$2.00 \text{ mol} \times \frac{20.2 \text{ g}}{\text{mol}} = 40.4 \text{ g}$$

(Note that three significant figures are used for the atomic mass of neon.)

Example 2-3

Calculate the mass of 0.673 mol of the element phosphorus.

The element phosphorus occurs as molecules containing four atoms. The molecular mass is $4 \times 31.0 = 124$. The mass of a mol is 124 g/mol.

The mass of 0.673 mol is: $0.673 \text{ mol} \times 124 \text{ g/mol} = 83.4$ g.

Example 2-4

How many moles of magnesium are present in 0.478 g of the element? In this case the unit factor mol/g is necessary; it is the inverse of the factor g/mol.

$$0.478 \text{ g} \times \frac{1}{24.3 \text{ g/mol}} = 0.0197 \text{ mol, or } 1.97 \times 10^{-2} \text{ mol}$$

Since 1 mol of a substance contains 6.022×10^{23} particles, it is possible to calculate the mass of an individual atom or molecule.

Example 2-5

Calculate the average mass of an atom of boron. Boron is a mixture of two isotopes. The atomic mass of boron is 10.81.

The calculation involves the use of two unit factors, g/mol and atoms/mol: 10.81 g/mol and 6.022×10^{23} atoms/mol. The unit of the answer sought is g/atom.

10.81 g/mol must be multiplied by the inverse of the unit factor 6.022×10^{23} atoms/mol.

$$10.81 \frac{\text{g}}{\text{mol}} \times \frac{1}{6.022 \times 10^{23}} \frac{\text{mol}}{\text{atom}} = 1.7945 \times 10^{-23} \frac{\text{g}}{\text{atom}}$$

Example 2-6

Compute the mass of a molecule of fluorine, which occurs as two atoms per molecule.

$2 \times 19.0 = 38.0$ (the molecular mass of fluorine)

$$\frac{38.0 \text{ g}}{\text{mol}} \times \frac{1}{6.02 \times 10^{23}} \frac{\text{mol}}{\text{molecule}} = 6.31 \times 10^{-23} \frac{\text{g}}{\text{molecule}}$$

Example 2-7

How many sodium atoms are present in 0.152 g of the metal?

First calculate the number of moles of sodium in 0.152 g and then multiply by the unit factor atoms/mol.

$$0.152 \text{ g} \times \frac{1}{23.0} \frac{\text{mol}}{\text{g}} = 0.00661 \text{ mol, or } 6.61 \times 10^{-3} \text{ mol}$$

$$6.61 \times 10^{-3} \text{ mol} \times 6.02 \times 10^{23} \frac{\text{atoms}}{\text{mol}} = 39.8\ 10^{23} \text{ atoms, or}$$

3.98×10^{24} atoms

You may wish to combine the two steps.

$$0.152 \text{ g} \times \frac{1}{23.0} \frac{\text{mol}}{\text{g}} \times 6.02 \times 10^{23} \frac{\text{atoms}}{\text{mol}} = 3.98 \times 10^{24} \text{ atoms}$$

2-6 SYMBOLS

Symbols are convenient abbreviations used to represent elements. A uniform system of symbols has been agreed upon by the International Union of Pure and Applied Chemistry. An alphabetical list of the elements and their symbols is given in Table D-3.

When chemists discuss or write about their work, they often use symbols to represent the individual atoms of the elements.

Used in this way, symbols represent the combination, rearrangement, or separation of atoms.

In the laboratory, however, chemists use symbols to represent moles of elements: the quantities of elements used in experiments. Since a mole of an element contains 6.022×10^{23} atoms, the symbol of an element also represents Avogadro's number of atoms. The various uses of symbols are summarized below, where gold (Au) is used as an example.

Symbol	Au
Name of the element	Gold
1 atom	1 atom of gold
1 mole	197.0 g gold (1 mol)
Avogadro's number of atoms	6.022×10^{23} atoms of gold

2-7 DETERMINATION OF ATOMIC MASSES

In the past the atomic masses of the elements were determined by very precise measurements of the quantities involved in chemical reactions – a chemical method. With the development of an instrument called the mass spectrograph, it became possible to determine the relative masses of atoms with considerable accuracy – a physical method. With this instrument it is possible to determine the masses of the individual isotopes of an element as well as their relative abundances. From this information the average atomic masses of an element may be calculated.

The uncertainties in the atomic masses arise from the limitations in the accuracy of the abundance measurements, which may be observed in the following example.

Example 2-8

The natural abundances of the isotopes of silicon have been determined: ^{28}Si, 92.23%, ^{29}Si, 4.70%, ^{30}Si, 3.09%. The respective isotope masses are 27.97693, 28.97649, 29.97376 amu. Compute the atomic mass of silicon to four significant figures.

$$
\begin{aligned}
27.98 \times 0.9223 &= 25.80 \text{ amu} \\
28.98 \times 0.0467 &= 1.35 \\
29.97 \times 0.0310 &= \underline{0.93} \\
& 28.08 \text{ amu}
\end{aligned}
$$

The accepted value of the atomic mass of silicon is 28.086. (Note: if you multiply each isotope mass by the abundance in % rather than the decimal, as above, divide the sum by 100.)

The data for this and the other problems on the determination of atomic masses are from *Handbook of Chemistry and Physics*, 64th edition, 1983--1984, Chemical Rubber Co., Cleveland, Ohio.

2-8 LAW OF DEFINITE PROPORTIONS

A compound always contains the same elements in the same proportion by mass. This statement is called the law of definite proportions, or the law of constant composition. The relative proportions by mass of the elements in a compound may be expressed in a variety of ways, two of which are percent and ratio. Thus, all samples of pure silver iodide contain 45.95% silver and 54.05% iodine. This composition may also be represented as the mass ratio iodine/silver = 1.176. Such a ratio is the mass of one element equivalent to 1.000 g of another element.

Example 2-9

In three experiments, silver was converted to silver iodide. Samples of silver weighing 9.89577, 10.44012, and 9.52800 g were converted to 21.53744, 22.72226, and 20.73713 g of silver iodide, respectively. Calculate the percent composition of each sample of iodide to four significant figures.

[G. P. Baxter and O. W. Lundstedt, *J. Am. Chem. Soc.*, *62*, 1832 (1940)]

The first step is to determine the mass of iodine in each sample of silver iodide

Mass of silver iodide	21.54 g	22.72 g	20.74 g
Mass of silver	9.90 g	10.44 g	9.53 g
	11.64 g	12.28 g	11.21 g

Divide the mass of each element by the total mass of the compound, and multiply by the ratio 100. Use logarithms to obtain the proper number of significant figures.

$$\text{Percent silver} = \frac{\text{g silver}}{\text{g compd.}} \times 100$$

$$\frac{9.896\text{ g}}{21.54\text{ g}} \times 100 = 45.95\%$$

$$\frac{10.44\text{ g}}{22.72\text{ g}} \times 100 = 45.95\%$$

$$\frac{9.523\text{ g}}{20.74\text{ g}} \times 100 = 45.94\%$$

$$\text{Percent iodine} = \frac{\text{g iodine}}{\text{g compd.}} \times 100$$

$$\frac{11.64\text{ g}}{21.54\text{ g}} \times 100 = 54.05\%$$

$$\frac{12.28\text{ g}}{22.72\text{ g}} \times 100 = 54.05\%$$

$$\frac{11.21\text{ g}}{20.74\text{ g}} \times 100 = 54.05\%$$

Check by adding the percent of silver and iodine; the sum should equal 100.

Example 2-10

Use the data of Example 2-9 to calculate the iodine-to-silver ratios to four significant figures.

The ratio is obtained by dividing the mass of iodine by the mass of silver in each sample.

$$\frac{11.64\text{ g iodine}}{9.896\text{ g silver}} = 1.176\ \frac{\text{g iodine}}{1\text{ g silver}}$$

$$\frac{12.28\text{ g iodine}}{10.44\text{ g silver}} = 1.176\ \frac{\text{g iodine}}{1\text{ g silver}}$$

$$\frac{11.21\text{ g iodine}}{9.528\text{ g silver}} = 1.176\ \frac{\text{g iodine}}{1\text{ g silver}}$$

When symbols are used, the iodine/silver ratio in silver iodide is represented by I/Ag, since the formula of silver iodide is AgI.

2-9 ATOMIC MASSES FROM COMBINING RATIOS

As stated in Section 2-7, in the past atomic masses were determined by chemical means. Although they are now determined by mass spectrometry, it is wise to illustrate that the values from the two methods are the same within experimental error.

Example 2-11

Calculate the atomic mass of bromine from the mass spectral data: natural abundances, ^{79}Br, 50.69%; ^{81}Br, 49.31%; isotopic masses 78.9183 and 80.9163 amu, respectively.

$$78.92 \times 0.5069 = 40.00 \text{ amu}$$

$$80.92 \times 0.4931 = \underline{39.90}$$

$$79.90 \text{ amu}$$

Example 2-12

Calculate the atomic mass of bromine from the fact that 6.23696 g of silver combined with bromine to form 10.85722 g of silver bromide.

[G. P. Baxter, *J. Am. Chem. Soc.*, *28*, 1322 (1906)]

Two additional facts must be known to solve this problem, the formula of silver bromide (AgBr) and the atomic mass of silver (107.870).

The first step is to compute the bromine-to-silver ratio, which can then be multiplied by the atomic mass of silver in grams (the mass of a mole of silver).

$$\begin{array}{rl} 10.85722 & \text{g AgBr} \\ \underline{-6.23696} & \text{g Ag} \\ 4.62026 & \text{g Br} \end{array}$$

$$\frac{4.62026 \text{ g Br}}{6.23696 \text{ g Ag}} = 0.740787 \frac{\text{g Br}}{1 \text{ g Ag}}$$

$$107.87 \frac{\text{g}}{\text{mol}} \text{Ag} \times 0.74079 \frac{\text{g Br}}{1 \text{ g Ag}} = 79.909 \frac{\text{g}}{\text{mol}} \text{Br}$$

This answer is the mass of a mole of bromine atoms (Section 2-4). The atomic mass is 79.909 amu.

Example 2-13

Calculate the atomic mass of cadmium from the fact that 3.39038 g of the metal was obtained from 8.21123 g of cadmium bromide. The formula of cadmium bromide is $CdBr_2$.

[G. P. Baxter, M. R. Grose, and M. H. Hartmann, *J. Am. Chem. Soc.*, *38*, 857 (1916)]

First compute the mass of bromine in the sample.

$$\begin{array}{r} 8.21123 \text{ g } CdBr_2 \\ -\underline{3.39038} \text{ g Cd} \\ 4.82085 \text{ g Br} \end{array}$$

Since the formula is $CdBr_2$, the relationship required is (mol Cd) = (mol Br)/2.

$$\text{mol Br} = 4.82085 \text{ g} \times \frac{1}{79.906} \frac{\text{mol}}{\text{g}} = 0.060331 \text{ mol}$$

$$\text{mol Cd} = \frac{0.060331}{2} = 0.030165 \text{ mol}$$

$$0.030165 \text{ mol Cd} = 3.39038 \text{ g} \times 1 \frac{\text{mol}}{\text{g}}$$

$$\frac{\text{g}}{\text{mol}} \text{ Cd} = \frac{3.39038 \text{ g}}{0.030165 \text{ mol}} = 112.39 \frac{\text{g}}{\text{mol}}$$

(The accepted atomic mass of cadmium is 112.41.)

Example 2-14

Calculate the atomic mass of selenium from the silver/silver selenide ratio, 0.732081. The formula of silver selenide is Ag_2S

[*J. Am. Chem. Soc.*, *84*, 4185 (1962)]

There are two moles of silver in each mole of silver selenide.

$$\frac{2 \text{ mol Ag}}{\text{mol } Ag_2Se} = 0.732081$$

$$2 \text{ mol} \times 107.87 \frac{\text{g}}{\text{mol}} = 215.74 \text{ g}$$

$$\frac{215.74 \text{ g}}{215.74 \text{ g} + \text{g/mol Se}} = 0.732081$$

$$215.74 \text{ g} = 0.732081 \left(215.74 \text{ g} + \frac{\text{g}}{\text{mol}} \text{ Se}\right)$$

$$\text{g/mol Se} = \frac{215.74 \text{ g} - 157.94}{0.732081} = 78.953 \frac{\text{g}}{\text{mol}}$$

(Compare with the presently accepted value.)

PROBLEMS

2-1 A sample of 7.505 g cadmium chloride gave 4.602 g cadmium by electrolysis. Calculate the percent composition of cadmium chloride.

[G. P. Baxter and M. L. Hartmann, *J. Am. Chem. Soc.*, *37*, 113 (1915)]

2-2 Calculate the cadmium/chloride ratio in cadmium chloride from the data in Problem 2-1.

2-3 What is the zinc/chlorine ratio if 5.94448 g of zinc chloride gave 2.85136 g of zinc by electrolysis?

[G. P. Baxter and J. H. Hodgens, *J. Am. Chem. Soc.*, *43*, 1242 (1921)]

2-4 What is the percent zinc and chlorine in zinc chloride if the zinc/chlorine ratio is 0.92195?

[G. P. Baxter and J. H. Hodgens, *J. Am. Chem. Soc.*, *43*, 1242 (1921)]

2-5 What is the mass, in grams, of 0.178 mol of beryllium?

2-6 How many moles are present in 2.865 g of calcium?

2-7 What is the mass, in grams, of a single atom of sodium? What is the mass in milligrams?

2-8 What is the average mass of an atom of tin? What is meant by the term "average mass" in this problem?

2-9 How many atoms of zinc are present in a sample of the metal that weighs 51.8 g?

2-10 Calculate the number of moles of carbon in a diamond weighing 6.31 g.

2-11 A solution was made by dissolving 0.00612 g of sulfur in 1 ml of CS_2. How many atoms of S were present in 1 L of solution? (Assume that the volume of CS_2 did not change with the addition of S.)

2-12 How many moles of gold are present in a cube of the element measuring 2.00 in. on an edge? How many atoms of gold are in the cube?

2-13 What is the mass, in grams, of a sample of sulfur that contains one trillion atoms?

2-14 Calculate the number of sodium atoms in a rectangular block of the element 8.36×10^{-3} mm long, 2.04×10^{-4} mm wide, and 80.3 Å high.

2-15 How many atoms of uranium are there in a uranium sphere whose radius is 1.00 in.?

2-16 How many mercury atoms are present in 36.0 mL of the element at 20°C?

2-17 The atomic mass of iodine has been assigned the value 126.9044 amu based on direct determination with a mass spectrograph. Calculate a value for the atomic mass of silver from the I/Ag ratio, 1.176433.

2-18 Use the value of the atomic mass of silver from Problem 2-17 to calculate the atomic mass of chlorine from the ratio AgCl/Ag = 1.328667.

2-19 The absolute isotope abundance ratio of chlorine $^{35}Cl/^{37}Cl$ = 3.1272. Calculate the atomic mass of chlorine and compare the result with that obtained in Problem 2-18. The masses of ^{35}Cl and ^{37}Cl are 34.96885 and 36.96590 amu, respectively.

2-20 Calculate the atomic weight of krypton from the following data:

Isotopes	Abundance atom (%)	Mass (amu)
^{78}Kr	0.35	77.9204
^{80}Kr	2.27	79.9164
^{82}Kr	11.56	81.9135
^{83}Kr	11.55	82.9141
^{84}Kr	56.90	83.9115
^{86}Kr	17.37	85.9106

2-21 Calculate the atomic mass of cadmium from the data in Problem 2-1.

2-22 The absolute abundance ratio $^{14}N/^{15}N$ in atmospheric nitrogen has been determined to be 272 ± 0.3. The isotopic masses are ^{14}N = 14.003074 and ^{15}N = 15.000108 amu. Calculate the atomic mass of nitrogen.

2-23 A chemistry student heated 0.2178 g of magnesium in oxygen and obtained 0.3607 g of magnesium oxide, MgO. Using the value 16.00 as the atomic mass of oxygen, compute the atomic mass of magnesium determined by the student. Compare the result with the accepted value.

Supplementary Problems

2-24 E. W. Morley carefully determined the mass ratio of hydrogen to oxygen in water as 0.12596. Use the atomic mass of oxygen as 15.999 and calculate the atomic mass of hydrogen.
[*Z. Physik. Chem. (Leipzig)*, *20*, 417 (1896)]

2-25 In a very early experiment Dumas and Stas determined the composition of water by mass as O, 88.864%; H, 11.136%. Use the value 15.999 as the atomic mass of oxygen to calculate the atomic mass of hydrogen. Compare the answer with the result obtained 53 years later by Morley (Problem 2-24).
[*Ann. Chem.*, *8*, 189 (1843)]

2-26 What is the atomic mass of copper if the copper oxide/copper ratio is 1.125181?
[*J. Am. Chem. Soc.*, *84*, 4185 (1962)]

2-27 The silver/telluride/silver ratio was found to be 1.59145. What is the atomic mass of tellurium?
[*J. Am. Chem. Soc.*, *84*, 4189 (1962)]

2-28 What is the mass of an atom of Ta in milligrams?

2-29 What is the mass, in grams, of 1.63 mol copper?

2-30 How many moles of beryllium are contained in 32.1 g of the metal?

2-31 How many atoms of silicon are present in 1.00 g of the element? In 1.00 mg? In 1.00 ng?

2-32 How many atoms of aluminum are there in a cube of the metal 1.0×10^{-3} mm on an edge?

2-33 How many atoms of Pt are present in a piece of platinum wire 10.0 mm long and 1.00 mm in diameter?

2-34 How many moles of potassium are present in a sample containing 1.62×10^{18} atoms?

2-35 How many moles of neon are present in a sample containing 7.86×10^{4} atoms?

2-36 Calculate the number of boron atoms in a cylinder of the element 2.54×10^{-4} cm in diameter and 32.2 mm long.

2-37 How many moles of lead are there in a lead sphere 22.8 mm in diameter?

2-38 Determine the number of moles of mercury in 36.0 mL of the element at 20°C.

2-39 How many gallium atoms are present in 2.16 cc of the element at 40°C?

2-40 The mass ratio of gallium to gallium oxide was found to be 0.74396. Calculate the atomic mass of gallium.

2-41 The isotope abundances of bromine were found to be 50.537% ^{79}Br and 49.46% ^{81}Br. The isotopic masses are, respectively, 78.9183 and 80.9163 amu. Calculate the atomic mass of bromine. Compare the result with Example 2-11.

2-42 Calculate the atomic mass of zinc from the data in Problem 2-3.

2-43 Calculate the atomic mass of tellurium from the data in Problem 2-27.

2-44 The isotopic abundances of strontium are ^{84}Sr, 0.560; ^{86}Sr, 9.870; ^{87}Sr, 7.035; ^{88}Sr, 82.535%. The respective masses are 83.9134, 85.9094, 86.9089, 87.9056 amu. What is the atomic mass of strontium?

2-45 The abundances of ^{17}O and ^{18}O in atmosphere oxygen are 0.0374% and 0.2039%, respectively. The isotopic masses are ^{16}O = 15.994915, ^{17}O = 16.999134, and ^{18}O = 17.999160 amu. Calculate the atomic mass of oxygen.

3
PERCENT COMPOSITION, EMPIRICAL AND MOLECULAR FORMULAS

3-1 MOLECULAR MASSES

When two or more atoms combine, larger particles called *molecules* (see Chapter 2) are formed. The molecular mass is the average mass of the molecules of a substance compared to the mass of a ^{12}C atom. The molecular mass is the sum of the masses of the atoms in the molecule. The molecular mass of hydrogen chloride is 36.461, for example, since each molecule contains one hydrogen atom and one chlorine atom. The molecular mass of water is 18.0153, since each molecule contains two hydrogen atoms and one oxygen atom.

3-2 THE MOLE

Chemical reactions occur between atoms and molecules. In order to determine how many atoms or molecules of a substance are present, we need a conversion factor from mass to molecules (or atoms). The fundamental unit here in the SI system is the mole. The symbol for mole is mol.

As discussed in Chapter 2, the mole is defined as the amount of substance of a system containing as many elementary entities (e.g., atoms, molecules, ions) as there are atoms in 0.012 kg of ^{12}C. One mole of a substance contains 6.022×10^{23} entities. This number is known as Avogardo's number.

A mole of hydrogen chloride has a mass of 36.461 g and contains 6.022×10^{23} hydrogen chloride molecules. A mole of water contains 6.022×10^{23} molecules and has a mass of 18.0153 g. The molecular mass of ammonia is 17.036; 1 mol has a mass of 17.036 g and contains 6.022×10^{23} molecules.

Example 3-1

What is the mass, in grams, of 2.00 mol of carbon tetrachloride?
First determine the molecular mass of the compound:

$$\begin{array}{lrl} CCl_4 & 1 \times 12.0 = & 12.0 \text{ g} \\ & 4 \times 35.4 = & \underline{141.6 \text{ g}} \\ & & 153.6 \text{ g/mol} \end{array}$$

$$2 \cancel{\text{mol}} \times 154 \frac{\text{g}}{\cancel{\text{mol}}} = 308 \text{ g}$$

Example 3-2

How many moles of carbon dioxide, CO_2, are present in 6.28 g?
Determine the molecular mass:

$$\begin{array}{lrl} CO_2 & 1 \times 12.0 = & 12.0 \text{ g} \\ & 2 \times 16.0 = & \underline{32.0 \text{ g}} \\ & & 44.0 \text{ g/mol} \end{array}$$

Since 44.0 g equals 1 mol, the unit factor is 1 mol/44.0 g:

$$6.28 \cancel{\text{g}} \times \frac{1 \text{ mol}}{44.0 \cancel{\text{g}}} = 0.143 \text{ mol}$$

Example 3-3

What is the average mass of a molecule of methane, CH_4?

$$\begin{array}{lrl} CH_4 & 1 \times 12.0 = & 12.0 \text{ g} \\ & 4 \times 1.0 = & \underline{4.0 \text{ g}} \\ & & 16.0 \text{ g/mol} \end{array}$$

The unit factor comes from Avogadro's number:

6.02×10^{23} molecules = 1 mol

$$16.0 \frac{\text{g}}{\cancel{\text{mol}}} \times \frac{1 \cancel{\text{mol}}}{6.02 \times 10^{23} \text{ molecules}} = 2.66 \times 10^{-23} \frac{\text{g}}{\text{molecule}}$$

Example 3-4

How many molecules of boron trichloride, BCl_3, are present in a sample weighing 0.133 g?

First calculate the number of moles of BCl_3 present:

$$\begin{array}{lrl} BCl_3 & 1 \times 10.8 = & 10.8\ g \\ & 3 \times 35.4 = & \underline{106.2\ g} \\ & & 117.0\ g/mol \end{array}$$

$$0.133\ g \times \frac{1\ mol}{117.0\ g} = 0.00114 = 1.14 \times 10^{-3}\ mol$$

Now using Avogadro's number;

$$1.14 \times 10^{-3}\ mol \times 6.02 \times \frac{10^{23}\ molecules}{1\ mol} = 6.85 \times 10^{20}\ molecules$$

3-3 FORMULAS

Formulas are combinations of symbols that represent compounds. In speaking and writing of their work, chemists use formulas to represent molecules. The nature and number of atoms in a molecule are also apparent from the formula of a compound.

Just as a symbol may be used to represent 1 mole of an element, so a formula is often used to represent 1 mole of a compound. The senses in which formulas are used in the laboratory are illustrated in the following:

Formula	H_2SO_4
Name of compound	Sulfuric acid
One molecule	One molecule of sulfuric acid
Nature and number of atoms	Two hydrogen atoms, one sulfur atom, and four oxygen atoms combined in some manner
One mole	98.0 g sulfuric acid
Avogadro's number of molecules	6.022×10^{23} molecules of sulfuric acid

A formula indicates the number of atoms of each element in a molecule, and it is understood that the atoms are joined by one means or another. The exact arrangement of the atoms is not apparent from the formula. For example, the formula of water is usually written H_2O, but each of the hydrogen atoms is in fact joined to the oxygen atom.

The number of atoms of an element in a molecule is indicated by a *subscript*, for example, H_2, F_2, NH_3, SO_2, and N_2F_4. The subscript implies that the atoms in the molecule are joined together.

A *coefficient* is used to represent a number of atoms or molecules that are not joined, for example, 2Ne, $3H_2$, and $2NH_3$. Two neon atoms, three hydrogen molecules (each containing two atoms), and two ammonia molecules (each containing one nitrogen and three hydrogen atoms) are thus represented.

There is another class of compounds – known as ionic compounds – that do not exist as molecules. An example of this is $CaCl_2$. Its formula represents the relative number of ions of each type in the crystal. Thus, $CaCl_2$ means that in a crystal of calcium chloride there are two chloride ions for each calcium ion present. Moreover, in 1 mol of calcium chloride there are 1 mol of calcium ions and 2 mol of chloride ions.

The nature of the bonds in a compound is not apparent from the formula. The properties of the compound give evidence of the type of bonds present.

3-4 DETERMINATION OF FORMULA

A formula represents an actual substance of a definite chemical composition. This chemical composition, and hence the formula, is *determined by experiments*. If a chemist says that the formula of sulfuric acid is H_2SO_4, he is saying that someone has determined the composition of sulfuric acid by experiment and has reported the results. Of course, with even a limited knowledge of chemistry one can predict from theory what the formula of a compound is likely to be. The knowledge of the composition of a substance is based on experiment, however,

3-5 COMPOSITION FROM FORMULAS

When a known compound is prepared and analyzed, the analysis is compared with that calculated from the formula of the expected compound. If a sample of lithium nitride, Li_3N, for example, was made, the results of the analysis obtained in the laboratory would be compared with the theoretical composition calculated as follows;

$$\begin{array}{lrl} Li_3N & 3 \times 6.94 = & 20.82 \text{ g Li/mole } Li_3N \\ & 1 \times 14.01 = & \underline{14.01 \text{ g N/mole } Li_3N} \\ & & 34.83 \text{ g } Li_3N\text{/mole} \end{array}$$

$$\frac{20.82}{34.83} \times 100 = 59.77\% \text{ Li}$$

$$\frac{14.01}{34.83} \times 100 = 40.22\% \text{ N}$$

The formula Li_3N represents the mass of 1 mole. The symbols with their subscripts denote the quantity of each element in 1 mole of the compound. The ratio 20.82/34.83 is the grams of lithium per gram of compound. Multiplying this ratio by 100 changes the basis from 1 g of Li_3N to 100 g of Li_3N.

Example 3-5

W. L. Jolly, K. D. Maguire, and D. Rabinovich prepared a compound believed to be S_4N_3Cl. The analysis of the compound was reported as N, 21.02%; S, 62.09%; Cl, 17.58%. Do these results agree with the theoretical composition of S_4N_3Cl?
[*Inorg. Chem.*, 2, 1304 (1963)]

$$\begin{array}{rl} 4 \times 32.06 = & 128.24 \text{ g S/mol } S_4N_3Cl \\ 3 \times 14.01 = & 42.03 \text{ g N/mol } S_4N_3Cl \\ 1 \times 35.45 = & \underline{35.45 \text{ g Cl/mol } S_4N_3Cl} \\ & 205.72 \text{ g/mol } S_4N_3Cl \end{array}$$

$$\frac{128.2}{205.7} \times 100 = 62.31\% \text{ S in } S_4N_3Cl$$

$$\frac{\text{g S}}{\text{g } S_4N_3Cl} \times 100 = \frac{\text{g S}}{100 \text{ g } S_4N_3Cl} = \% \text{ S in } S_4N_3Cl$$

$$\frac{42.03}{205.7} \times 100 = 20.43\% \text{ N in } S_4N_3Cl$$

$$\frac{35.45}{205.7} \times 100 = 17.23\% \text{ Cl in } S_4N_3Cl$$

The compound prepared was S_4N_3Cl, because the composition found by analysis agrees well with the calculated composition. Note that, in the problem illustrated, four significant figures may be carried throughout the calculations.

3-6 FORMULAS FROM COMPOSITION

The elemental composition of a compound may be used to calculate a formula that denotes the relative number of atoms of each type in the molecule. Such formulas are called *empirical*, or *simplest*, formulas.

There are three steps necessary to determine the empirical formula of a compound from the percent composition.

1. Divide the percent composition of each element by its atomic mass. This gives the number of moles of each element in 100 g of compound.

2. Take the *smallest* number from step one and divide this into all the other results from step one. This gives the ratio of moles of the elements in 100 g of compound.

3. If in step 2 a fraction is obtained for any element, multiply all the numbers obtained in step 2 by the lowest common denominator to clear of fractions.

Example 3-6

What is the simplest formula of the compound, prepared by J. H. Junkins and his group, which contained 49.7% niobium and 50.2% fluorine?

[*J. Am. Chem. Soc.*, *74*, 3464 (1952)]

The first step is to find the number of moles of each element in 100 g of the compound.

$$\frac{49.7 \text{ g Nb/100 g compd.}}{92.9 \text{ g Nb/mol Nb}} = 0.536 \frac{\text{mol Nb}}{100 \text{ g compd.}}$$

$$\frac{50.2 \text{ g F/100 g compd.}}{19.0 \text{ g F/mol F}} = 2.64 \frac{\text{mol F}}{100 \text{ g compd.}}$$

Next find the simplest ratio of the moles of the elements in 100 g of compound.

$$\frac{0.536}{0.536} = 1.00 \qquad \frac{2.64}{0.536} = 4.93 \frac{\text{mol F}}{\text{mol Nb}}$$

Since a mole of any element contains 6.02×10^{23} atoms, the ratio can be expressed as $1 \times 6.02 \times 10^{23}$ atoms of Nb to $5 \times 6.02 \times 10^{23}$ atoms of F. Hence for each atom of niobium there are five atoms of fluorine. The simplest formula is NbF_5.

Example 3-7

Guenther described an interesting method of preparation of a compound of titanium carried out at the I.I.T. Research Institute, Chicago. Potassium chloride and titanium(IV) chloride were allowed to react in fused antimony trichloride. The analysis of the product was K, 23.1%; Ti, 14.2%; Cl, 62.7%. What is the simplest formula of the compound?

[K. F. Guenther, *Inorg. Chem.*, 3, 923–924 (1964)]

$$\text{K} \quad \frac{23.1}{39.1} = 0.592 \frac{\text{mol K}}{100 \text{ g compd.}} \qquad \frac{0.592}{0.297} = 1.99 \qquad 2$$

$$\text{Ti} \quad \frac{14.2}{47.9} = 0.297 \frac{\text{mol Ti}}{100 \text{ g compd.}} \qquad \frac{0.297}{0.297} = 1.00 \qquad 1$$

$$\text{Cl} \quad \frac{62.7}{35.4} = 1.77 \frac{\text{mol Cl}}{100 \text{ g compd.}} \qquad \frac{1.77}{0.297} = 5.96 \qquad 6$$

The simplest formula is K_2TiCl_6.

Example 3-8

A fluoride of oxygen was prepared by mixing the proper ratio of oxygen and fluorine at 60 K. This compound contains 32.1% F and 67.9% O. What is the empirical formula of the compound?

[A. G. Streng and A. V. Grosse, *J. Am. Chem. Soc.*, *88*, 169 (1966)]

$$\text{F} \quad \frac{32.1}{19.0} = 1.69 \frac{\text{mol F}}{100 \text{ g}} \qquad \frac{1.69}{1.69} = 1 \times 2 = 2$$

$$\text{O} \quad \frac{67.9}{16.0} = 4.24 \frac{\text{mol O}}{100 \text{ g}} \qquad \frac{4.24}{1.69} = 2.50 = \frac{5}{2} \times 2 = 5$$

The simplest formula is O_5F_2.

Molecular Formulas

Simplest formulas represent the *relative* number of atoms of each element in the compound in question. The molecular mass of the compound must be determined in order to learn whether the formula also indicates the total number of atoms in the molecules of the compound.

Example 3-9

A compound contains 92.3% carbon and 7.7% hydrogen. Calculation of the simplest formula gives CH.

$$\frac{92.3}{12.0} = 7.7\,\frac{\text{mol C}}{100\text{ g}} \qquad \frac{7.7}{7.7} = 1\text{ C}$$

$$\frac{7.7}{1.00} = 7.7\,\frac{\text{mol H}}{100\text{ g}} \qquad \frac{7.7}{7.7} = 1\text{ H}$$

At least two compounds have this simple mole ratio. One has a molecular mass of 26.0; the other has a molecular mass of 78.0. The mass represented by CH is 13.0. The two molecular formulas are C_2H_2 and C_6H_6.

$$\frac{26.0}{13.0} = 2 \qquad 2 \times CH = C_2H_2 \qquad \frac{78.0}{13.0} = 6 \qquad 6 \times CH = C_6H_6$$

Example 3-10

At the University of Washington, the reaction of N_2F_4 with $S_2O_6F_2$ was carried out at room temperature. A gas, boiling at -2.5°C, was collected. Its analysis gave N, 9.48%; S, 20.9%; F, 38.0%. Find the simplest formula.

[M. Lustig and G. H. Cady, *Inorg. Chem.*, *2*, 388 (1963)]

Since the oxygen content was not reported, we must obtain it by difference:

9.5%	100.0%
20.9%	68.4%
38.0%	31.6% oxygen
68.4%	

Note the application of the rule of significant figures in the addition; the number 9.48 is changed to 9.5.

$$\text{N} \quad \frac{9.48}{14.0} = 0.676\,\frac{\text{mol N}}{100\text{ g compd.}} \qquad \frac{0.676}{0.651} = 1.04$$

$$\text{S} \quad \frac{20.9}{32.1} = 0.651\,\frac{\text{mol S}}{100\text{ g compd.}} \qquad \frac{0.651}{0.651} = 1.00$$

$$\text{F} \quad \frac{38.0}{19.0} = 2.00 \frac{\text{mol F}}{100 \text{ g compd.}} \qquad \frac{2.00}{0.651} = 3.07$$

$$\text{O} \quad \frac{31.6}{16.0} = 1.98 \frac{\text{mol O}}{100 \text{ g compd.}} \qquad \frac{1.98}{0.651} = 3.04$$

The simplest formula is NSO_3F_3. This is called the simplest formula because the same analytical data would be obtained from $N_2S_2O_6F_6$, $N_3S_3O_9F_9$, etc. The molecular mass was found by experiment to be 150.8.

What is the molecular formula of the compound? The molecular mass calculated from the simplest formula is 151.1.

NSO_3F_3		
		14.0
		32.1
	$3 \times 16.0 =$	48.0
	$3 \times 19.0 =$	57.0
		151.1

Therefore, the molecular formula is NSO_3F_3.

PROBLEMS

3-1 What is the mass, in grams, of 2.21 mol of phosphorus trichloride, PCl_3?

3-2 Calculate the number of moles of water in 1 dm^3.

3-3 Calculate the number of phosphorus and oxygen atoms in 16.8 g of orthophosphoric acid, H_3PO_4.

3-4 Calculate the number of molecules of sulfur dioxide, SO_2, in 3.61 g. How many atoms of oxygen are present?

3-5 Calculate the mass, in grams, of a molecule of hydrogen bromide, HBr.

3-6 How many grams of each of the following compounds is represented by:

a. $2SO_3$ b. $4.56\ HNO_3$
c. $9.33 \times 10^{-8}\ KClO_3$ d. $1.68\ Mg_3N_2$
e. $4CuSO_4 \cdot 5H_2O$

3-7 How many calcium ions and how many chloride ions are present in 2.12 mol of $CaCl_2$?

3-8 What is the mass, in grams, of 3.01×10^{20} molecules of nitrogen trifluoride, NF_3?

3-9 What is the mass ratio potassium/chlorine in potassium chloride?

3-10 What is the mass ratio zirconium/chlorine in zirconium tetrachloride?

3-11 In a study of the use of SF_4 as a fluorinating agent at the DuPont laboratories, two compounds believed to be $KAsF_6$ and Na_2SiF_6 were prepared. The samples contained 32.9% As and 14.8% Si, respectively. Compare these results with the calculated composition.
[E. L. Meuterties, *J. Inorg. Nucl. Chem.*, *18*, 2497 (1961)]

3-12 Wold et al. prepared a series of metal sulfides by reacting H_2 and H_2S with the sulfate salts of different metals. With cobalt, they reported the preparation of Co_9S_8 based on an analysis of 67.0% Co in the sulfide. How well does this compare with the calculated composition?
[*Inorg. Chem.*, *23*, 872 (1984)]

3-13 Two methods of reduction of molybdenum pentachloride to molybdenum tetrachloride were investigated at the laboratories of the Climax Molybdenum Company of Michigan.
Method 1. *Analysis* Cl, 58.98; Mo, 40.26
Method 2. *Analysis* Cl, 58.69; Mo, 41.04
Compare the results of the two methods.
[M. L. Larson and F. W. Moore, *Inorg. Chem.*, *3*, 286 (1964)]

3-14 A unique silver, fluorine compound was prepared by Bougon, Huy, Lance, and Abazli by reacting AgF_2 with KrF_2. The reddish-brown precipitate gave, on analysis, 64.75% Ag and 34.22% F, the rest was due to impurities. What is the simplest formula for the compound?
[*Inorg. Chem.*, *23*, 3667 (1984)]

3-15 What is the simplest formula of a compound that contained 37.7% B, 47.5% N, and 13.9% H?
[G. H. Dahl and R. Schaeffer, *J. Am. Chem. Soc.*, *83*, 3033 (1961)]

3-16 The reaction of trisilylamine $(SiH_3)_3N$ with ammonia gave a compound containing 68.8% Si, 17.8% N, and 6.39% H. The rest is impurities. What is the simplest formula for the compound?
[R. Wells and R. Schaeffer, *J. Am. Chem. Soc.*, *88*, 37 (1966)]

3-17 At the I.I.T. Research Institute, the reaction of thallium(I) chloride with titanium(IV) chloride was carried out in molten antimony trichloride. The product formed was analyzed with the following results. Tl, 61.2; Ti, 7.2; Cl, 31.6. What is the simplest formula of the compound formed?
[K. F. Guenther, *Inorg. Chem.*, *3*, 923 (1964)]

3-18 Titanium metal was treated with hydrogen fluoride at 225° for 48 hr. The solid product contained 44.52% Ti and 53.18% F. What is the simplest formula?
[E. L. Mueterties, *J. Inorg. Nucl. Chem.*, *18*, 148 (1961)]

3-19 A hydrated chloride of cerium was analyzed and found to contain 37.70% cerium, 28.41% chlorine, and 33.89% water (by difference). What is the formula of the hydrate?
[L. M. Dennis and W. H. Magee, *J. Am. Chem. Soc.*, *16*, 600 (1894)]

3-20 A compound containing 24.0% cobalt, 14.4% ammonia (present as ammonium ion), and 59.9% chlorine was prepared at the University of Oklahoma. It was allowed to stand in contact with air for several days until the color was violet and the weight constant. The loss of mass from a sample of the violet hydrate on dehydration was 13.0 ± 0.2%. What was the formula of the violet hydrate?
[N. Fogel, C. C. Lin, C. Ford, and W. Grindstoff, *Inorg. Chem.*, *3*, 720–721 (1964)]

3-21 A compound containing 84.2% boron and 15.7% hydrogen was prepared in the laboratories at Indiana University. Its molecular mass was calculated from its vapor density as 76.7. What is its molecular formula?
[D. F. Gaines and R. Schaeffer, *Inorg. Chem.*, *3*, 438 (1964)]

3-22 Beers and McCarley prepared a different form of a compound of Mo by reacting $Mo(OAc)_4$ with $AlCl_3$ in chlorobenzene. Analysis of the brown precipitate gave 55.44% Mo, 41.82% Cl, with the rest due to traces of solvent. What is the simplest formula for the compound?

[*Inorg. Chem.*, *24*, 472 (1985)]

3-23 A white crystalline material melting at 122° was prepared at Duke University. An analysis gave B, 23.9; N, 31.1; F, 42.9; H, 2.2. The molecular mass was found to be 132. What is the formula of the compound?

[K. Niedenzu, *Inorg. Chem.*, *1*, 943 (1962)]

3-24 Boron trifluoride and ammonia gases were mixed and allowed to stand for several hours at 0°. The solid that formed analyzed B, 12.8; NH_3, 19.9. The molecular mass was determined by the freezing-point-depression method as 83. What is the formula of the compound?

[W. A. Jenkins, *J. Am. Chem. Soc.*, *78*, 5500 (1956)]

3-25 What is the formula of the compound with the composition 28.76% phosphorus, 17.25% fluorine, 32.40% chlorine, and the remainder oxygen? Its molecular mass was determined to be 225.

[M. M. Crutchfield, C. F. Callis, and J. R. Van Wazer, *Inorg. Chem.*, *3*, 282 (1964), footnote 8]

Supplementary Problems

3-26 What is the mass, in grams, of 0.818 mol of lithium bromide, LiBr?

3-27 How many moles of calcium oxide, CaO, are present in 4.82 g?

3-28 How many molecules of dinitrogen monoxide, N_2O, are present in a sample weighing 0.752 g?

3-29 What is the mass, in grams, of sulfur represented by each of the following:

a. S_8
b. $2\ S_2$
c. $0.25\ S_8$
d. $1.2\ S_2$
e. $8.08 \times 10^{-5}\ S_8$

3-30 What is the mass, in grams, of each of the following?

a. 3 Au
b. $2.28\ F_2$
c. $6.51 \times 10^{-3}\ P_4$
d. 7.43 Pb
e. $2.19\ N_2$

3-31 How many magnesium ions and how many bromide ions are present in 0.233 g of $MgBr_2$?

3-32 Calculate the mass ratio aluminum/oxygen in aluminum oxide.

3-33 Thallium(III) fluoride was prepared at Manchester University by the reaction of thallium(I) iodate with bromine trifluoride, and the ratio $TlF_3/TlIO_3$ was determined as 0.690. Compare this value with the theoretical ratio.
[A. A. Woolf, *J. Chem. Soc.*, *1954*, 4695]

3-34 The compound silver tetrafluoroaurate was prepared and the $AgAuF_4/Au$ ratio was determined at 1.969. Compare this value with the theoretical ratio.
[A. A. Woolf, *J. Chem. Soc.*, *1954*, 4695]

3-35 The reaction of fluorine with V_2O_5 at 475° was investigated at the University of New Hampshire. Do the analytical results agree with the composition of the expected production of the reaction, VOF_3? *Analysis* found: V, 42.2, 41.1, 40.7; F, 45.0.
[H. M. Haendler et al., *J. Am. Chem. Soc.*, *76*, 2177 (1954)]

3-36 A finely powdered mixture of ammonium fluoride and copper(II) fluoride was subjected to a pressure of 100,000 psi at 25°. The compound isolated after this treatment analyzed NH_3, 20.6%; Cu, 36.7%; F, 42.7%. Do these data support the proposed formula $(NH_4)_2CuF_4$?
[D. S. Crocket and R. A. Grossman, *Inorg. Chem.*, *3*, 645 (1964)]

3-37 What is the simplest formula of the compound, prepared at the University of Tennessee, which contained 2.97% boron and 97.66% iodine?
[W. J. McDowell and C. W. Keenan, *J. Am. Chem. Soc.*, *78*, 2069 (1956)]

3-38 Bromine vapor was passed over a mixture of zirconium dioxide and carbon at 560°. A white product containing 22.30% zirconium and 77.68% bromine was obtained. What is its simplest formula?
[R. C. Young and H. G. Fletcher, *Inorg. Syn.*, *1*, 49 (1939)]

3-39 Direct combination of tungsten and chlorine led to the production of a compound, the composition of which was determined to be W, 46.28%, Cl, 53.59%. What is the simplest formula of the compound?

[R. E. McCarley and T. M. Brown, *Inorg. Chem.*, *3*, 1233 (1964)]

3-40 McHughes, Willet, Davis, and Gard prepared a new chromium, oxygen, and fluorine compound by reacting CrO_2F_2 with XeF_2. Analysis gave 42.3% Cr and 44.6% F. What is the simplest formula for the compound?

[*Inorg. Chem.*, *25*, 426 (1986)]

3-41 A sample of a fluoride of thorium prepared by J. C. Warf contained 75.37% Th and 24.60% F. What is its simplest formula?

[*J. Am. Chem. Soc.*, *74*, 1864 (1952)]

3-42 A yellow solid melting at 147° was prepared and analyzed. Its composition was found to be Re, 74.2; F, 7.61; and the remainder oxygen. What is its formula?

[A. Engelbrecht and A. V. Grosse, *J. Am. Chem. Soc.*, *76*, 2042 (1954)]

3-43 A series of complex chlorides of rhenium of the general formula M_XReCl_6 were made at the University of North Carolina. Calculate the formula of the potassium salt from the analytical results. Found: Re, 39.25; Cl, 43.88.

[C. W. Horner, F. M. Collier, Jr., and S. Y. Tyree, *Inorg. Chem.*, *3*, 1388 (1964)]

3-44 An unstable thioiodide of phosphorus was prepared by stirring a solution of the elements at room temperature for 65 hr. *Analysis* found: I, 80.31; S, 10.01; and the remainder phosphorus. What is the simplest formula you would assign to the compound? Consult the original paper after you have reached a decision.

[A. H. Cowley and S. T. Colren, *Inorg. Chem.*, *3*, 780 (1964)]

3-45 Iodine was added to liquid chlorine cooled with a dry ice-acetone mixture. Orange crystals separated, and they were found to have the composition I, 54.5%; Cl, 45.7%. What was the compound?

[H. S. Booth and W. C. Morris, *Inorg. Syn.*, *1*, 168 (1939)]

3-46 What is the formula of the fluoride that was prepared at the University of New Hampshire from TiO_2 and found to contain 38.6% titanium?

[H. M. Haendler et al., *J. Am. Chem. Soc.*, *76*, 2177 (1954)]

3-47 What is the simplest formula of the compound that resulted from the reaction of zirconium with zirconium tetrabromide if its analysis gave Zr, 27.63; Br, 72.41?
[E. M. Larsen and J. J. Leddy, *J. Am. Chem. Soc.*, *78*, 5984 (1956)]

3-48 A compound prepared at Nagoya University in Japan had the composition I, 78.9; Te, 13.1; and the remainder potassium. What was its formula?
[D. Nakamura, K. Ito, and M. Kubo, *J. Am. Chem. Soc.*, *84*, 163 (1962)]

3-49 What is the formula of the bromide of tungsten prepared by treating a sample of WBr_2 with excess bromine at 50° for 2 weeks? *Analysis* found: W, 43.22; Br, 56.30.
[R. E. McCarley and T. M. Brown, *J. Am. Chem. Soc.*, *84*, 3216 (1962)]

3-50 What is the formula of the fluoride of rhodium prepared at the Argonne National Laboratory in Chicago? *Analysis* found: Rh, 47.8; F, 51.8.
[C. L. Chernick, H. H. Classen, and B. Weinstock, *J. Am. Chem. Soc.*, *83*, 3165 (1961)]

3-51 The reaction of boric acid, H_3BO_3, with urea, $(NH_2)_2CO$, at 600° produced a solid containing 31.6% boron, 20.5% nitrogen, and the remainder oxygen. This solid was then treated with ammonia at 1650° and another compound, containing 44.2% boron and 56.0% nitrogen, was formed. What are the simplest formulas of the two compounds?
[T. E. O'Connor, *J. Am. Chem. Soc.*, *84*, 1753 (1962)]

3-52 Calculate the simplest formula of the compound prepared by Krebs, Müller, and Zürn. *Analysis* found: P, 51.8; Cd, 47.5.
[*Z. Anorg. Allgem. Chem.*, *285*, 25–26 (1956)]

3-53 Aqueous solutions of $N_2H_4 \cdot 2HCl$ and $HgCl_2$ were mixed and a yellow precipitate was formed. *Analysis* found: Hg, 79.96; Cl, 14.2; N, 5.7. Calculate the simplest formula of the compound.
[K. Brodersen, *Z. Anorg. Allgem. Chem.*, *285*, 7 (1956)]

3-54 Potassium meta arsenate ($KAsO_3$) and potassium fluoride were mixed and heated to 800°C. The product formed contained

35.5% K, 33.7% As, 8.63% F, and the remainder oxygen. Calculate the simplest formula of the compound.

[N. K. Duff and A. K. Gupta, *Z. Anorg. Allgem. Chem.*, *285*, 93 (1956)]

3-55 What is the simplest formula of the compound formed along with $(CH_3)_2AlH$ by the reaction of $(CH_3)_3B$ with $LiAlH_4$? *Analysis* found: Li, 19.75; B, 28.9; 28.5; C, 36.6, 32.6; H, 16.68, 17.14.

[T. Wartik and H. I. Schlesinger, *J. Am. Chem. Soc.*, *75*, 838 (1953)]

3-56 Calculate the formula of the very explosive compound obtained from the reaction of silver nitrate with potassium amide in liquid ammonia. *Analysis* found: Ag, 87.4; N, 11.2

[E. C. Franklin, *J. Am. Chem. Soc.*, *27*, 834 (1905)]

3-57 A hydrate of Ag_2TeO_4 (0.7638 g) was heated to remove the water of hydration. The loss in mass was 0.0605 g. What is the formula of the hydrate?

[E. B. Hutchins, Jr., *J. Am. Chem. Soc.*, *27*, 1164 (1905)]

3-58 A mineral has been found crystallized from a small pond in Antarctica; the pond contains an extremely large amount of salt. *Analysis* gave Ca, 17.5; Cl, 32.7; H_2O, 49.2. The mineral has been found in no other location and has been named antarcticite. What is the formula of the mineral?

[T. Torii and J. Ossaka, *Science*, *149*, 975 (1965)]

3-59 A sulfide of scandium prepared at the RCA Laboratories in Princeton, New Jersey contained 48.6% scandium and 51.0% sulfur. What is its simplest formula?

[J. P. Dismukes and J. G. White, *Inorg. Chem.*, *3*, 1221 (1964)]

3-60 Calculate the simplest formula of a binary compound of sulfur that contains 42.46% chlorine.

[R. C. Brasted and J. S. Pond, *Inorg. Chem.*, *4*, 1164 (1965)]

3-61 A compound prepared at the University of Florida gave the following analytical results: F, 45.29; N, 16.86; and the remainder phosphorus. The molecular mass was found to be 250. Another compound of phosphorus had a molecular mass of 328 and contained 45.74% F and 16.83% N. What are the formulas of the two compounds?

[T. J. Mao, R. D. Dresdner, and J. A. Young, *J. Am. Chem. Soc.*, *81*, 1021 (1959)]

3-62 The reaction of fluorine with $(FCO)_2O_2$ gave a product containing 15.0% C, 40.1% O, and 45.7% F. The molecular mass obtained from the average vapor density measurements was 82. What is the formula of the product?

[R. L. Cauble and G. H. Cady, *J. Am. Chem. Soc.*, *89*, 5161 (1967)]

4
CHEMICAL EQUATIONS

4-1 EQUATIONS

Just as symbols are used to represent elements and formulas to represent compounds, so equations are used to represent chemical reactions. What has been said previously about symbols and formulas applies to equations. Thus, the equation

$$4HCl(g) + O_2(g) \rightarrow 2H_2O(g) + 2Cl_2(g)$$

is a shorthand designation for the following statement:

> Four molecules of hydrogen chloride reacted with one molecule of oxygen to form two molecules of water and two molecules of chlorine.

Chemists use equations in this manner to discuss chemical reactions.

On the other hand, we can say:

> Four moles of hydrogen chloride reacted with one mole of oxygen to give two moles of water and two moles of chlorine.

In this sense, equations are used in the laboratory to calculate the relative masses of reactants and products in chemical reactions.

These are summarized in the following table (p. 56).

The total mass of the products is equal to the total mass of the reacting substances (177.6 g), an illustration of the *law of conservation of mass*. Also, note that the number of atoms of each element is the same before and after the reaction.

$4HCl(g)$	$+ O_2(g)$	$\rightarrow 2H_2O(g)$	$+ 2Cl_2(g)$
1. Hydrogen chloride	+ oxygen	→ water	+ chlorine
2. 4 molecules	+ 1 molecule	→ 2 molecules	+ 2 molecules
3. 4 moles	+ 1 mole	→ 2 moles	+ 2 moles
4. $4 \times 6.022 \times 10^{23}$ molecules	$+ 6.022 \times 10^{23}$ molecules	$\rightarrow 2 \times 6.022 \times 10^{23}$ molecules	$+ 2 \times 6.022 \times 10^{23}$ molecules
5. 4×36.4 g	+ 32.0 g	$\rightarrow 2 \times 18.0$ g	$+ 2 \times 70.8$ g

Note that this reaction occurs in the gaseous state; the physical state of each substance is indicated by (g).

The designations for the solid (s) and liquid (l) physical states are illustrated in the equation

$$2Hg(l) + O_2(g) \rightarrow 2HgO(s)$$

The formulas of substances present in water solution are followed by (aq) for aqueous.

$$16HCl(aq) + 2KMnO_4(aq) \rightarrow 5Cl_2(g) + 2MnCl_2(aq) + 2\,KCl(aq) + 8H_2O(l)$$

4-2 BALANCING EQUATIONS

Chemists and others who require a knowledge of chemistry must remember a vast number of chemical reactions. It is not wise to attempt to remember complete equations such as we have just discussed. The *reactions* are *committed to memory* and the *equations* are *written* and balanced as needed.

For example, oxygen was first prepared by Priestley, who heated mercury(II) oxide; metallic mercury was the other product. This statement can be depicted by the equation

$$Hg + O_2 = HgO$$

Since there are two O's on the left and only one on the right, the equation is not "balanced." However, the equation

$$2Hg + O_2 = 2HgO$$

is balanced.

There are several types of equations you must learn to balance; some are easier to balance than others. It would be convenient if a simple set of rules would enable one to balance all equations to be met, but this is not the case. The method to be used depends on the type of reaction that is represented by the equation.

Chemical reactions may be classified in several ways. Two such classifications are useful for balancing equations. The first is a classification according to whether or not changes in oxidation state (transfer of electrons) occur. The second is according to the conditions under which the reactions occur. Some reactions occur in the absence of a solvent or in an indifferent solvent, i.e., one that serves only to disperse the reactants. Other reactions occur only in solution.

4-3 TYPES OF CHEMICAL EQUATIONS

A. *Simple*: No change in oxidation number occurs.

1. Anhydrous

$$TiO_2(s) + CCl_4(g) \rightarrow TiCl_4(\ell) + CO_2(g)$$

2. In aqueous solution

a. Precipitation

$$AgNO_3(aq) + KCl(aq) \rightarrow AgCl(s) + KNO_3(aq)$$

b. Neutralization

$$HCl(aq) + NaOH(aq) \rightarrow H_2O + NaCl(aq)$$

c. Gas formation

$$SiO_2(s) + 4HF(aq) \rightarrow SiF_4(g) + 2H_2O$$

d. Complex formation

$$AgCl(s) + 2NH_3(aq) \rightarrow Ag(NH_3)_2Cl(aq)$$

B. *Redox*: Changes in oxidation number occur.

1. Anhydrous

$$2KClO_3(s) \rightarrow 2KCl(s) + 3O_2(g)$$

2. In aqueous solution

$$16HCl(aq) + 2KMnO_4(aq) \rightarrow 5Cl_2(g) + 2MnCl_2(aq)$$

$$+ KCl(aq) + 8H_2O(\ell)$$

A set of rules that enables one to balance all of the simple (A) equations and most of the anhydrous redox (B.1) equations follows. Application of these rules is often called *balancing by inspection*.

4-4 RULES FOR BALANCING SIMPLE EQUATIONS

1. Balance the metals present first.
2. Balance the nonmetals except hydrogen and oxygen.
3. Balance hydrogen and oxygen.
4. Check all elements.
5. Repeat in the same order until all are balanced.

Example 4-1

$Na_2O_2(s) + H_2O(\ell) \rightarrow NaOH(s) + O_2(g)$

Rule 1 $Na_2O_2 + H_2O = 2NaOH + O_2$
Rule 2 Does not apply.
Rule 3 The hydrogen atoms balance:

$Na_2O_2 + 2H_2O = 2NaOH + O_2$

The oxygen atoms are now balanced.
Rule 4 The hydrogen atoms are no longer balanced.
Rule 5 $2Na_2O_2 + 2H_2O = 4NaOH + O_2$
The equation is balanced.

Example 4-2

$KClO_3(s) = KCl(s) + O_2(g)$

Rule 1 $KClO_3 = KCl + O_2$
Rule 2 $KClO_3 = KCl + O_2$
Rule 3 $2KClO_3 = KCl + 3O_2$
Rule 4 The K and Cl are not balanced.
Rule 5 $2KClO_3 = 2KCl + 3O_2$

Example 4-3

$ZrO_2(s) + C(s) + Br_2(\ell) = ZrBr_2(s) + CO(g)$

Rule 1 $ZrO_2 + C + Br_2 = ZrBr_4 + CO$
Rule 2 $ZrO_2 + C + 2Br_2 = ZrBr_4 + CO$
Rule 3 $ZrO_2 + C + 2Br_2 = ZrBr_4 + 2CO$

Rule 4 The C atoms are not balanced.
Rule 5 $ZrO_2 + 2C + 2Br_2 = ZrBr_4 + 2CO$

Example 4-4

$$C_2H_2(g) + O_2(g) \rightarrow CO_2(g) + H_2O(g)$$

Rule 2 $C_2H_2 + O_2 \rightarrow 2CO_2 + H_2O$

Rule 3 The H's are balanced. There is an odd number of O's on the right side and an even number on the left. It is useful in such cases to double the coefficients of the elements already balanced and proceed as before.

$$2C_2H_2 + O_2 \rightarrow 4CO_2 + H_2O$$

The H's are not balanced.

$$2C_2H_2 + O_2 \rightarrow 4CO_2 + 2H_2O$$

Only the O's remain unbalanced.

$$2C_2H_2 + 5O_2 \rightarrow 4CO_2 + 2H_2O$$

4-5 BALANCING REDOX EQUATIONS

Sometimes redox equations that occur in the absence of a solvent (B.1) may be quite difficult to balance by inspection. When several substances are involved, the key ratio required to balance the equation may not be readily apparent. In these circumstances it is wise to use the change in oxidation number method. The rules for determining the oxidation numbers of elements are given in Chapter 14 (p. 265).

4-6 OXIDATION NUMBER METHOD

1. Determine the oxidation number of each element present.

$$\overset{(0)}{Al}(s) + \overset{(0)}{S}(s) \rightarrow \overset{(+3)(-2)}{Al_2S_3}(s)$$

2. Indicate the changes in oxidation number.

oxidation
increase +3/atom

$$Al + S \rightarrow Al_2S_3$$

decrease -2/atom
reduction

3. Multiply the substances oxidized and reduced by the proper coefficients to make the increase in oxidation number equal to the decrease.

Increase $2 \times 3 = 6$

Decrease $3 \times 2 = 6$

$2Al + 3S \rightarrow Al_2S_3$

4. Continue as for simple equations.
5. Check.

Example 4-5

Balance the equation $Mg(s) + B_2O_3(s) \rightarrow MgO(s) + B(s)$

$$\text{Rule 1} \quad \overset{(0)}{Mg} + \overset{(+3)(-2)}{B_2O_3} \rightarrow \overset{(+2)(-2)}{MgO} + \overset{(0)}{B}$$

oxidation

increase +2/atom

Rule 2 $Mg + B_2O_3 \rightarrow MgO + B$

decrease -3/atom

reduction

Rule 3 $3Mg + B_2O_3 \rightarrow 3MgO + 2B$

It is not necessary to multiply B_2O_3 by 2 (the increase in oxidation number of Mg) because there are two B's in B_2O_3, and the total decrease in oxidation number is -6.

Example 4-6

Balance the equation $Ba(s) + NH_3(g) \rightarrow Ba_3N_2(s) + H_2(g)$

$$\text{Rule 1} \quad \overset{(0)}{Ba} + \overset{(-3)(+1)}{NH_3} \rightarrow \overset{(+2)(-3)}{Ba_3N_2} + \overset{(0)}{H_2}$$

oxidation

increase +2/atom

Rule 2 $Ba + NH_3 \rightarrow Ba_3N_2 + H_2$

decrease -1/atom

reduction

Rule 3 $3Ba + 2NH_3 \rightarrow Ba_3N_2 + H_2$

Example 4-7

Balance the equation $NpCl_4(s) + NH_3(g) \rightarrow NpCl_3(s) + HCl(g) + N_2(g)$

$$\text{Rule 1} \quad \overset{(+4)(-1)}{NpCl_4} + \overset{(-3)(+1)}{NH_3} \rightarrow \overset{(+3)(-1)}{NpCl_3} + \overset{(+1)(-1)}{HCl} + \overset{(0)}{N_2}$$

oxidation: increase +3/atom ($NH_3 \rightarrow N_2$)

$$\text{Rule 2} \quad NpCl_4 + NH_3 \rightarrow NpCl_3 + HCl + N_2$$

decrease -1 atom ($NpCl_4 \rightarrow NpCl_3$): reduction

Rule 3 Since there are two N's in N_2, there must be at least $2NH_3$ in the balanced equation. This makes the total increase 2(+3).

$$6NpCl_4 + 2NH_3 \rightarrow 6NpCl_3 + HCl + N_2$$

Rule 4 $6NpCl_4 + 2NH_3 \rightarrow 6NpCl_3 + 6HCl + N_2$

Example 4-8

Balance the equation $CuF_2(s) + NH_3(g) \rightarrow Cu_3N(s) + NH_4F(s) + N_2(g)$

$$\text{Rule 1} \quad \overset{(+2)(-1)}{CuF_2} + \overset{(-3)(+1)}{NH_3} \rightarrow \overset{(+1)(-3)}{Cu_3N} + \overset{(-3)(+1)(-1)}{NH_4F} + \overset{(0)}{N_2}$$

oxidation: increase +3/atom ($NH_3 \rightarrow N_2$)

$$\text{Rule 2} \quad CuF_2 + NH_3 \rightarrow Cu_3N + NH_4F + N_2$$

decrease -1/atom ($CuF_2 \rightarrow Cu_3N$): reduction

Rule 3 The total increase is 2(+3) because there are two N's in N_2.

$$6CuF_2 + NH_3 \rightarrow 2Cu_3N + NH_4F + N_2$$

Rule 4 To balance the F's:

$$6CuF_2 + NH_3 \rightarrow 2Cu_3N + 12NH_4F + N_2$$

To balance the N's and H's:

$$6CuF_2 + 16NH_3 \rightarrow 2Cu_3N + 12NH_4F + N_2$$

Balancing redox equations that occur in solution is discussed in Section 14-4.

4-7 QUANTITATIVE ASPECTS OF CHEMICAL REACTIONS

The quantities involved in a balanced equation may be described on the basis of the number of moles of each reactant and product.

Example 4-9

$$Mg_3N_2(s) + 6H_2O(l) \rightarrow 3Mg(OH)_2(s) + 2NH_3(g)$$

$$1 \text{ mol} + 6 \text{ mol} \rightarrow 3 \text{ mol} + 2 \text{ mol}$$

It is important to note that the mass of the reactants is equal to the mass of the products *in grams*. The number of atoms of each element is the same in the reactants as in the products. The number of *moles of reactants* is not necessarily equal to the number of *moles of products*.

4-8 THEORETICAL YIELD OR CALCULATED YIELD

It is possible to calculate the quantity of a product that can be obtained from a reaction by assuming that the reaction is the only one involved and that all of the product is collected. This calculated quantity is called the *theoretical yield*. Because side reactions may occur and some product may be lost in the laboratory manipulations, the *actual yield* is often less than the theoretical one. The ratio of the actual yield to the theoretical yield is represented by the *percent yield*. Examples 4-10 through 4-16 illustrate the calculation of theoretical or calculated yields.

Example 4-10

How many moles of oxygen can be obtained by the reaction of 1.18 mol of $Na_2O_2(s)$ with water?

$$\underset{2 \text{ mol}}{2Na_2O_2(s)} + 2H_2O(l) \rightarrow 4NaOH(s) + \underset{1 \text{ mol}}{O_2(g)}$$

$$1.18 \text{ mol } Na_2O_2 \times \frac{1 \text{ mol } O_2}{2 \text{ mol } Na_2O_2} = 0.590 \text{ mol } O_2$$

Example 4-11

How many grams of magnesium oxide can be prepared from 121.5 g of magnesium metal by the reaction

$$2Mg(s) + O_2(g) \rightarrow 2MgO(s)$$

Check to be sure the equation is balanced.

$$2Mg(s) + O_2(g) \rightarrow 2MgO(s)$$

$$2\ \text{mol} + 1\ \text{mol} \rightarrow 2\ \text{mol}$$

According to the equation, 2 mol of magnesium will produce 2 mol of magnesium oxide, but in this problem we have 122 g of magnesium. How many moles do we have?

$$122\ \cancel{g} \times \frac{1\ \text{mol}}{24.3\ \cancel{g}} = 5.00\ \text{mol Mg}$$

To calculate the moles of MgO produced:

$$5.00\ \cancel{\text{mol}}\ \text{Mg} \times \frac{2\ \text{mol MgO}}{2\ \cancel{\text{mol}}\ \text{Mg}} = 5.00\ \text{mol MgO}$$

To calculate the mass in grams, the moles must be multiplied by the mass of 1 mol of MgO.

MgO
24.3
16.0
40.3 g/mol

$$5.00\ \cancel{\text{mol}}\ \text{MgO} \times 40.3 \frac{g}{\cancel{\text{mol}}} = 202\ \text{g MgO}$$

This is the theoretical yield of the reaction.

The steps in the procedure may be combined.

$$122\ \cancel{g} \times \frac{1\ \cancel{\text{mol}}}{24.3\ \cancel{g}} \times \frac{2\ \cancel{\text{mol}}\ \text{MgO}}{2\ \cancel{\text{mol}}\ \text{Mg}} \times 40.3\ \frac{g}{\cancel{\text{mol}}} = 202\ \text{g MgO}$$

In a series of reactions it is not necessary to calculate the quantities of intermediate materials prepared, provided the reactions are quantitative; i.e., no side reactions occur.

Example 4-12

Sulfuric acid is manufactured by a process approximately represented by the equations

$$S(s) + O_2(g) \rightarrow SO_2(g)$$

$$2SO_2(g) + O_2(g) \rightarrow 2SO_3(g)$$

$$2SO_3(g) + H_2O(l) \quad H_2SO_4(l)$$

What mass of sulfuric acid could be obtained from 3.21 g of sulfur if all the reactions were quantitative?

According to the balanced equations, 1 mol of S produces 1 mol of SO_2; 2 mol of SO_2 give 2 mol of SO_3, which is in the ratio of 1 to 1. Therefore, 1 mol of S would lead to 1 mol of SO_3. Since 1 mol of SO_3 produces 1 mol of H_2SO_4, 1 mol of S would lead to 1 mol of H_2SO_4 through the series of reactions.

$$3.21\ \cancel{g} \times \frac{1\ \text{mol}}{32.1\ \cancel{g}} = 0.100\ \text{mol S}$$

$$0.100\ \cancel{\text{mol}}\ \text{S} \times \frac{1\ \text{mol } H_2SO_4}{1\ \cancel{\text{mol}}\ \text{S}} = 0.100\ \text{mol } H_2SO_4$$

H_2SO_4

$2 \times 1.01 = 2.02$
$1 \times 32.1 = 32.1$
$4 \times 16.0 = 64.0$
98.1 g/mol

$$0.100\ \cancel{\text{mol}}\ H_2SO_4 \times 98.1 \frac{\text{g}}{\cancel{\text{mol}}} = 9.81\ \text{g}$$

If it is known that all of an element present is transformed quantitatively from one substance to another, it is not necessary to know what the other products were in order to calculate a theoretical yield. In Example 4-12 all of the sulfur was eventually transformed to sulfuric acid, $S \rightarrow H_2SO_4$. It was not necessary to know the reaction sequence to calculate the quantity of H_2SO_4 it is theoretically possible to obtain from a given quantity of sulfur.

Example 4-13

How many grams of AgCl can be made from 0.579 g of $SeOCl_2$?

1 mol $SeOCl_2 \rightarrow$ 2 mol AgCl, assuming all of the chlorine in $SeOCl_2$ is retained in the AgCl.

$SeOCl_2$			AgCl	
	$1 \times 79.0 =$	79.0		107.9
	$1 \times 16.0 =$	16.0		35.4
	$2 \times 35.4 =$	70.8		143.3 g/mol
		165.8 g/mol		

$$0.579\ \cancel{g} \times \frac{1\ \text{mol}}{166\ \cancel{g}} = 0.00349\ \text{mol } SeOCl_2$$

$$3.49 \times 10^{-3} \text{ mol SeOCl}_2 \times \frac{2 \text{ mol AgCl}}{1 \text{ mol SeOCl}_2} = 6.98 \times 10^{-3} \text{ mol AgCl}$$

$$6.98 \times 10^{-3} \text{ mol} \times 143 \frac{\text{g}}{\text{mol}} = 0.998 \text{ g AgCl}$$

When one element in a reactant is incorporated into two or more products, all of the products must be known.

It is possible to calculate the quantity of silver nitrate that can be prepared from a quantity of silver without consideration of the reactions involved. It is not possible to make a similar calculation in the case of nitric acid. Not all of the nitrogen in the nitric acid is transformed to silver nitrate.

$$Ag + 2HNO_3 \rightarrow AgNO_3 + NO_2 + H_2O$$

Two moles of HNO_3 are required to make 1 mol of $AgNO_3$.

Example 4-14

a. Calculate the mass of silver nitrate that can be prepared by dissolving 2.75 g of silver in a solution of nitric acid.

$$2.75 \text{ g} \times \frac{1 \text{ mol}}{108 \text{ g}} = 0.0255 \text{ mol Ag}$$

$$0.0255 \text{ mol Ag} \times \frac{1 \text{ mol AgNO}_3}{1 \text{ mol Ag}} = 0.0255 \text{ mol AgNO}_3$$

$AgNO_3$
108
14
48
170 g/mol

$$0.0255 \text{ mol} \times 170 \frac{\text{g}}{\text{mol}} = 4.34 \text{ g AgNO}_3$$

b. Calculate the mass of silver nitrate that can be prepared from 3.92 g of HNO_3.

$$\underset{\phantom{2 \text{ mol}}}{Ag(s)} + \underset{2 \text{ mol}}{2HNO_3(aq)} \rightarrow \underset{1 \text{ mol}}{AgNO_3(aq)} + NO_2(g) + H_2O(l)$$

HNO_3
1.0
14.0
48.0
63.0 g/mol

$$3.92 \text{ g} \times \frac{1 \text{ mol}}{63.0 \text{ g}} = 0.0622 \text{ mol HNO}_3$$

$$0.0622\ \text{mol HNO}_3 \times \frac{1\ \text{mol AgNO}_3}{2\ \text{mol HNO}_3} = 0.0311\ \text{mol AgNO}_3$$

$$0.0311\ \text{mol} \times 170\ \frac{\text{g}}{\text{mol}} = 5.29\ \text{g AgNO}_3$$

Example 4-15

What mass of copper(II) sulfate can be obtained by the action of 2.941 g of hot concentrated sulfuric acid on an excess of copper?

$$\text{Cu(s)} + \underset{2\ \text{mol}}{2\text{H}_2\text{SO}_4\text{(aq)}} \rightarrow \underset{1\ \text{mol}}{\text{CuSO}_4\text{(aq)}} + \text{SO}_2\text{(g)} + 2\text{H}_2\text{O(l)}$$

H_2SO_4		$CuSO_4$	
2 × 1.008 =	2.016	1 × 63.55 =	63.55
1 × 32.06 =	32.06	1 × 32.06 =	32.06
4 × 16.00 =	64.00	4 × 16.00 =	64.00
	98.08 g/mol		159.55 g/mol

$$2.941\ \text{g} \times \frac{1\ \text{mol}}{98.08\ \text{g}} = 0.02998\ \text{mol H}_2\text{SO}_4$$

$$0.02998\ \text{mol H}_2\text{SO}_4 \times \frac{1\ \text{mol CuSO}_4}{2\ \text{mol H}_2\text{SO}_4} = 0.01499\ \text{mol CuSO}_4$$

$$0.01499\ \text{mol} \times 159.6\ \frac{\text{g}}{\text{mol}} = 2.392\ \text{g CuSO}_4$$

Example 4-16

A solution containing $CaCl_2$ was added to a solution of $AgNO_3$. The $AgNO_3$ solution was prepared by dissolving 1.00 g of Ag in nitric acid. What was the mass of calcium chloride in the solution if all of it reacted with the silver nitrate solution to form AgCl?

The equations are

$$\text{Ag(s)} + 2\text{HNO}_3\text{(aq)} \rightarrow \text{AgNO}_3\text{(aq)} + \text{NO}_2\text{(g)} + \text{H}_2\text{O}(\ell) \qquad (1)$$

$$2\text{AgNO}_3\text{(aq)} + \text{CaCl}_2\text{(aq)} \rightarrow 2\text{AgCl(s)} + \text{Ca(NO}_3)_2\text{(aq)} \qquad (2)$$

In the series of reactions 1 mol of $CaCl_2$ requires 2 mol of Ag.

$$1.00\ \text{g} \times \frac{1\ \text{mol}}{108\ \text{g}} = 0.00928\ \text{mol Ag}$$

$$0.00928 \text{ mol Ag} \times \frac{1 \text{ mol } CaCl_2}{2 \text{ mol Ag}} = 0.00464 \text{ mol } CaCl_2$$

$CaCl_2$
1 × 40.1 = 40.1
2 × 35.4 = 70.8
110.8 g/mol

$$0.00464 \text{ mol} \times 111 \frac{\text{g}}{\text{mol}} = 0.514 \text{ g } CaCl_2$$

4-9 ACTUAL YIELD, PERCENT YIELD

When side reactions occur or when material is lost during transfer and separation of the products, the actual quantity of product is less than that calculated. The actual yield is often expressed as a percent of the calculated yield.

Example 4-17

Calculate the percent yield of oxygen if 0.0240 g resulted from heating 0.303 g of KNO_3.

$$2KNO_3 \rightarrow 2KNO_2 + O_2$$

1. Check to be certain that the equation is balanced before starting any calculations.
2. Calculate the theoretical yield.

$$\begin{array}{lcccc} 2KNO_3 & \rightarrow & 2KNO_2 & + & O_2 \\ 2 \text{ mol} & & & & 1 \text{ mol} \end{array}$$

KNO_3
1 × 39.1 = 39.1
1 × 14.0 = 14.0
3 × 16.0 = 48.0
101.1 g/mol

O_2
2 × 16.0 = 32.0 g/mol

$$0.303 \text{ g} \times \frac{1 \text{ mol}}{101 \text{ g}} = 0.00300 \text{ mol } KNO_3$$

$$0.00300 \text{ mol } KNO_3 \times \frac{1 \text{ mol } O_2}{2 \text{ mol } KNO_3} = 0.00150 \text{ mol } O_2$$

$$0.00150 \text{ mol} \times 32.0 \frac{\text{g}}{\text{mol}} = 0.0480 \text{ g } O_2 \text{ (the theoretical yield)}$$

3. Percent yield $= \dfrac{\text{actual yield}}{\text{calculated yield}} \times 100$

$$= \frac{2.40 \times 10^{-2}}{4.80 \times 10^{-2}} = 50.0\%$$

Example 4-18

Calculate the percent yield if the reaction of 64.0 g of $NaBH_4$ with I_2 produced 15.0 g of BI_3.

[H. C. Anderson and L. H. Belz, *J. Am. Chem. Soc.*, *75*, 4828 (1953)]

1. Balanced equation

$$NaBH_4 + 4I_2 \rightarrow BI_3 + NaI + 4HI$$

2. 1 mol $NaBH_4 \rightarrow$ 1 mol BI_3

$$64.0 \not{g}\ NaBH_4 \times \frac{1\ mol}{37.8\ \not{g}} = 1.69\ mol\ NaBH_4$$

$$1.69\ \not{mol}\ NaBH_4 \times \frac{1\ mol\ BI_3}{1\ \not{mol}\ NaBH_4} = 1.69\ mol\ BI_3$$

$$1.69\ \not{mol} \times 392\ \frac{g}{\not{mol}} = 662\ g\ BI_3 \text{ (the theoretical yield)}$$

3. Percent yield: $\dfrac{15.0}{662} \times 100 = 2.27\%$

4-10 LIMITING REACTANT

In an experiment involving a chemical reaction the quantities of substances may not be in the molar ratio called for by the reaction. One or more of the reactants may be in excess; that is, there is more than is required. The amount of product or products that may be formed will be determined by the substance that is not in excess. This substance is called the *limiting reactant* or sometimes the *limiting reagent*.

Example 4-19

A sample of calcium oxide whose mass was 112 g was treated with 46.0 g of water to prepare calcium hydroxide.

a. Which compound was the limiting reactant?

$$CaO(s) + H_2O(l) \rightarrow Ca(OH)_2(s)$$
$$1\ mol \quad + 1\ mol \qquad 1\ mol$$

CaO	$Ca(OH)_2$	H_2O
40.1	40.1	2.0
16.0	32.0	16.0
56.1 g/mol	2.0	18.0 g/mol
	74.1 g/mol	

$$112\ \not{g} \times \frac{1\ mol}{56.1\ \not{g}} = 2.00\ mol\ CaO$$

$$46.0\ \not{g} \times \frac{1\ mol}{18.0\ \not{g}} = 2.56\ mol\ H_2O$$

Since the mole ratio of CaO to H_2O is 1/1, the amount of H_2O is too large: 2.00/2.56. The *limiting reactant* is therefore CaO.

b. Calculate the theoretical yield of $Ca(OH)_2$,

$$2.00\ \not{mol}\ CaO \times \frac{1\ mol\ Ca(OH)_2}{1\ \not{mol}\ CaO} = 2.00\ mol\ Ca(OH)_2$$

$$2.00\ \not{mol} \times 74.1\ \frac{g}{\not{mol}} = 148\ g\ Ca(OH)_2$$

PROBLEMS

4-1 Balance the following equations:

a. $Fe(s) + H_2O(g) \rightarrow Fe_3O_4(s) + H_2(g)$

b. $Na(s) + H_2O(\ell) \rightarrow H_2(g) + NaOH(s)$

c. $CO(g) + H_2(g) \rightarrow CH_3OH(\ell)$

d. $CaH_2(s) + H_2O(\ell) \rightarrow Ca(OH)_2(s) + H_2(g)$

e. $BaO_2(s) \rightarrow BaO(s) + O_2(g)$

f. $KNO_3(s) \rightarrow KNO_2(s) + O_2(g)$

g. $KNO_3(s) + K(s) \rightarrow K_2O(s) + N_2(g)$

h. $Ti(OH)_3(s) \rightarrow TiO_2 + H_2O + H_2(g)$

i. $Na_2CO_3(s) + FeCr_2O_4(s) + O_2(g) \rightarrow Fe_2O_3(s) + Na_2CrO_4(s) + CO_2(g)$

j. $Fe_2O_3(s) + CO(g) \rightarrow Fe_3O_4(s) + CO_2(g)$

k. $Fe_3O_4(s) + CO(g) \rightarrow FeO(s) + CO_2(g)$

l. $CS_2(\ell) + O_2(g) \rightarrow CO_2(g) + SO_2(g)$

m. $Al_4C_3(s) + F_2(g) \rightarrow AlF_3(s) + CF_4(g)$

n. $Na_2Cr_2O_7(s) + HCl(g) \rightarrow CrO_2Cl_2(g) + NaCl(s) + H_2O(\ell)$

o. $Fe_2O_3(s) + CO(g) \rightarrow Fe(\ell) + CO_2(g)$

p. $CuFeS_2(s) + O_2(g) \rightarrow Cu(s) + FeO(s) + SO_2(g)$

q. $Al(s) + Fe_2O_3(s) \rightarrow Fe(s) + Al_2O_3(s)$

r. $CS_2(g) + Cl_2(g) \rightarrow CCl_4(g) + S_2Cl_2(g)$

s. $CaC_2(s) + H_2O(\ell) \rightarrow Ca(OH)_2(s) + C_2H_2(g)$

t. $SiH_4(g) + O_2(g) \rightarrow SiO_2(s) + H_2O(\ell)$

u. $PbS(s) + O_2(g) \rightarrow PbO(s) + SO_2(g)$

v. $PbO(s) + C(s) \rightarrow Pb(s) + CO_2(g)$

w. $PbS(s) + PbO(s) \rightarrow Pb(\ell) + SO_2(g)$

x. $PbS(s) + PbSO_4(s) \rightarrow Pb(\ell) + SO_2(g)$

y. $NH_4NO_2(s) \rightarrow N_2(g) + H_2O(g)$

z. $NH_4NO_3(s) \rightarrow N_2O(g) + H_2O(g)$

4-2 The next 10 equations appeared in journal articles. Balance them.

a. $BI_3 + NH_3 \rightarrow B_2(NH)_3 + NH_4I$

[*J. Am. Chem. Soc.*, *78*, 2069 (1956)]

b. $NpF_3 + Si \rightarrow NpSi_2 + SiF_4$

c. $NpCl_4 + NH_3 \rightarrow NpCl_3 + HCl + N_2$

d. $Np_3P_4 + HCl \rightarrow NpCl_4 + PH_3$

[*J. Am. Chem. Soc.*, *75*, 1232 (1953)]

e. $CH_4 + SO_2 \rightarrow CS_2 + H_2O + CO_2$

[*Chem. Abstr.*, *53*, 18713 g (1959)]

f. $B_2Cl_4 + O_2 \rightarrow B_2O_3 + BCl_3$

g. $B_2Cl_4 + H_2O \rightarrow H_3BO_3 + HCl + H_2$

[*J. Am. Chem. Soc.*, *80*, 6153–6155 (1958)]

h. $UF_6 + HBr \rightarrow UF_5 + HF + Br_2$

i. $SOCl_2(g) + NH_3(g) \rightarrow HNSO(g) + NH_4Cl(s)$

[R. L. DeKock and M. S. Hubbard, *Inorg. Chem.*, *16*, 216 (1977)]

j. $CsIO_4(s) + BrF_5(l) \rightarrow CsIF_4O_2(s) + BrF_3O(l)$

[K. O. Christe, R. D. Wilson, and C. J. Schack, *Inorg. Chem.*, *20*, 2104 (1981)]

4-3 How many moles of sulfur dioxide and how many moles of carbon dioxide can be formed by the reaction of 6 mol of carbon disulfide with oxygen?

4-4 How many millimoles of fluorine are required to react with 2 mol of Al_4C_3 if the products are aluminum fluoride and carbon tetrafluoride?

4-5 How many grams of diborane (B_2H_6) can be produced from 0.100 mmol $LiAlH_4$, according to the reaction

$$BF_3 + LiAlH_4 \rightarrow B_2H_6 + LiF + AlF_3$$

(Balance the equation first.)

In the next five problems reactions from research papers are used. In each case, balance the equation and then calculate the theoretical yield of the designated product from the quantity of starting material given.

4-6 $P_2I_4S_2 \rightarrow P_4S_7 + PI_3 + I_2$
0.262 g ?

[A. H. Cowley and S. T. Cohen, *Inorg. Chem.*, *3*, 780 (1964)]

4-7 $Si_3Cl_8 + NH_3 \rightarrow Si_3N_4H_4 + NH_4Cl$
1.65 g ?

[M. Billy, *Compt. Rend.*, *251*, 1639 (1960)]

4-8 $ScO_3 + B \xrightarrow{1800-1850°} ScB_2 + BO$
2.72 g ?

[G. V. Samsonov, *Dok. Akad. Nauk, SSSR*, *133*, 1344 (1960)]

4-9 $UO_3 + CBr_4 \xrightarrow{110°} UOBr_3 + COBr_2 + Br_2$
0.0265 g ?

[J. Prigent, *Ann. Chim. (Paris)*, 5, 65 (1960)]

4-10 $BiS_2 + O_2 \rightarrow Bi_2O_3 + SO_2$
16.5 g ?

[M. S. Silverman, *Inorg. Chem.*, *3*, 1041 (1964)]

4-11 In the following reaction, 0.273 g of NF_2SO_3F gave 1.85 mmol of nitrite ion. Was this quantitative?

$$NF_2SO_3F + 4OH^- \rightarrow NO_2^- + SO_3F^- + 2F^- + 2H_2O$$

[M. Lustig and G. H. Cady, *Inorg. Chem.*, *2*, 388 (1963)]

4-12 As part of a research project at Iowa State University, the composition of tungsten(IV) chloride was determined by hydrogen reduction at 650°. The tungsten metal formed was weighed as such, and the chlorine was converted to hydrogen chloride by reaction with hydrogen. The hydrogen chloride was collected in a known amount of standard sodium hydroxide solution. Write equations for the reactions that occurred.

[R. E. McCarley and T. M. Brown, *Inorg. Chem.*, *3*, 1233 (1964)]

4-13 Tl_2CO_3 was fused with V_2O_5, and one product obtained was found to contain 67.12% Tl, 16.33% V, and the remainder oxygen. The other product of the reaction was CO_2. Write an equation for the reaction.

[T. Carnelley, *J. Chem. Soc.*, *26*, 334 (1873)]

4-14 The year 1986 marks the one-hundredth anniversary of the first preparation of elemental fluorine by Henri Moissan by the electrolysis of liquid hydrogen fluoride containing potassium fluoride to promote the passage of electricity. Write an equation for the reaction.

[*C. R. Hebd. Seances Acad. Sci.*, *102*, 1543 (1886)]

4-15 It had been a widely held belief that fluorine could not be generated by purely chemical means but only by the electrochemical method (Problem 4-14). At the one-hundredth anniversary of Moissan's discovery, Karl O. Christe described the first purely chemical synthesis of fluorine. The synthesis involved the reaction of K_2MnF_6 with antimony pentafluoride. The other products were $KSbF_6$ and manganese(III) fluoride.

a. Write a balanced equation for the reaction.
b. Which element was oxidized and which was reduced?

[*Chem. & Eng. News*, p. 23 (Sept. 15, 1986)]

4-16 The key compound in the chemical synthesis of fluorine (Problem 4-15) is synthesized by the reaction of potassium permanganate with hydrogen peroxide in the presence of potassium fluoride in 50% aqueous hydrofluoric acid solution. Oxygen is evolved during the reaction. Write a balanced equation for the reaction.

[K. O. Christe, *Inorg. Chem.*, *25*, 3721 (1986)]

4-17 The reaction of 6.39 g of nitrogen trifluoride with 4.94 g of fluorine and 8.54 g of antimony pentafluoride gave a 41.0% yield of NF_4SbF_6.

a. Write a balanced equation for the reaction.
b. Which was the limiting reactant?
c. How many grams of the product was obtained?

[W. E. Tolberg, R. T. Rewick, R. S. Stringham, and M. E. Hill, *Inorg. Chem.*, *6*, 1156 (1967)]

4-18 A mixture of 1.92 g of nitrogen trifluoride, 1.83 g of boron trifluoride, and 0.342 g of fluorine reacted to form 1.00 g of NF_4BF_4.

a. Write an equation for the reaction.
b. What was the limiting reactant?
c. What was the % yield of product?

[K. O. Christe, C. J. Schack, and R. D. Wilson, *Inorg. Chem.*, *15*, 1275 (1976)]

4-19 Ammonium chloride and carbon tetrachloride were heated to 475° for an hour. Hydrogen chloride gas and a yellow solid containing 36.6% C, 1.2% H, and 62.2% N were formed. Write an equation for the reaction that occurred.

[D. D. Cubbicciotti and W. M. Ratimer, *J. Am. Chem. Soc.*, *70*, 3509 (1949)]

4-20 CCl_2F_2 gas was passed over WO_2 at 525°, and a solid mixture of WOF_4 and $WOCl_4$ and CO gas were the products. The solid mixture contained W, 59.1%; F, 12.0%; Cl, 22.7%; O, 6.2%. Calculate the composition of the solid mixture. Write an equation for the reaction that occurred.

[A. D. Webb and H. A. Young, *J. Am. Chem. Soc.*, *72*, 3358 (1950)]

4-21 A mixture of mercury(II) chloride, indium, and aluminum was heated to 325°. A white solid (In, 40.51; Al, 9.69; Cl, 49.74) and mercury were the products. In a typical preparation, 3.6061 g of mercury(II) chloride, 0.761 g of indium, and 0.1792 g of aluminum were employed. Were any of the reactants in excess? What was the theoretical yield?

[R. J. Clark, E. Griswold, and J. Kleinberg, *J. Am. Chem. Soc.*, *80*, 4766 (1958)]

4-22 Protactinium metal was prepared by the reduction of 0.100 mg of protactinium(IV) fluoride with barium metal at 1400°. How many grams and how many moles of protactinium could be prepared assuming a 100% yield?

[P. A. Sellers et al., *J. Am. Chem. Soc.*, *76*, 5937 (1954)]

4-23 How many grams of neptunium nitride can be made from 1.00 mg of Np by the following reactions, assuming 100% yield?

$$2Np + 3H_2 \rightarrow 2NpH_3$$

$$NpH_3 + NH_3 \rightarrow NpN + 3H_2$$

[I. Sheft and S. Fried, *J. Am. Chem. Soc.*, *75*, 1236 (1953)]

4-24 Copper(II) chloride reacted with lithium borohydride in diethyl ether:

$$CuCl_2 + 2LiBH_4 \rightarrow Cu + 2LiCl + B_2H_6 + H_2$$

In one experiment 3.58 mmol of $CuCl_2$ and 7.15 mmol of $LiBH_4$ gave 3.17 mmol B_2H_6. What was the mole percent yield of the product?

[T. J. Klinger, *Inorg. Chem.*, 3, 1058 (1964)]

4-25 A sample of copper (0.1350 g) was dissolved in sulfuric acid, and the solution was evaporated to dryness. $CuSO_4$ (0.1340 g) was collected. What was the percent yield? The other products of the reaction were SO_2 and H_2O.
[C. Baskerville, *J. Am. Chem. Soc.*, *17*, 904 (1895)]

4-26 What was the percent yield if 1.7395 g of lead reacted with oxygen to give 1.8650 g of lead(II) oxide?
[F. J. Brisbie, *J. Chem. Soc.*, *93*, 161 (1908)]

4-27 A chocolate-colored substance (0.4512 g) prepared by the reaction of KNH_2 on mercury(II) iodide in liquid ammonia gave 0.0218 g of nitrogen when decomposed and 0.4982 g of mercury(II) sulfide by reaction with H_2S. What is the formula of the highly explosive, chocolate-colored compound?
[E. C. Franklin, *J. Am. Chem. Soc.*, *27*, 870 (1905)]

4-28 A mixture containing 5.00 g of niobium(V) chloride and 0.500 g of niobium metal was heated at 300° for 1 week. Niobium(IV) chloride was formed by the reaction that occurred. Which was the limiting reactant? What was the theoretical yield of the product?
[D. R. Taylor, J. C. Calabrese, and E. M. Larsen, *Inorg. Chem.*, *16*, 721 (1977)]

4-29 How many grams of $MoOF_4$ could be expected from the reaction of 1.10×10^{-2} mol of molybdenum(VI) fluoride with 2.40×10^{-2} mol of chromium(III) oxide if the other product was CrO_2F_2?
[P. J. Green and G. L. Gard, *Inorg. Chem.*, *16*, 1243 (1977)]

4-30 Chlorine was passed over 0.1878 g of $NaNO_3$ heated in a boat in a combustion tube. What was the percent yield of NaCl if 0.1289 g was prepared along with NO_2 and O_2?
[E. F. Smith and J. G. Hibbs, *J. Am. Chem. Soc.*, *17*, 682 (1895)]

4-31 Calculate the atomic weight of nitrogen from the fact that 12.46620 g of $AgNO_3$ gave 7.91609 g of Ag by reduction.
[O. Honigschmid, E. Zintl, and P. Thilo, *Z. Anorg. Allgem. Chem.*, *163*, 65 (1924)]

4-32

a. In the following reaction, which element was oxidized and which was reduced? $Mn_2O_3(s) + Cr_2O_3(s) \rightarrow MnCrO_3(s)$
b. Balance the equation.

4-33 A mixture of 3.11 g of lithium hydroxide and 5.08 g of aluminum oxide was heated at 500° for 4.5 days, forming 6.40 g of $LiAlO_2$ with the evolution of water.

a. Which was the limiting reactant?
b. What was the percent yield?

[A. K. Fischer, *Inorg. Chem.*, *16*, 974 (1977)]

4-34 Which element was oxidized and which was reduced in the reaction $HOCl(aq) + H_2O_2(aq) \rightarrow O_2(g) + H_2O(l) + HCl(aq)$?
[J. K. Hurst, P. A. G. Carr, F. E. Hovis, and R. J. Richardson, *Inorg. Chem.*, *20*, 2435 (1981)]

Supplementary Problems

4-35 Balance:

a. $NH_3(g) + O_2(g) \rightarrow NO(g) + H_2O(g)$

b. $NO_2(g) + H_2O \rightarrow HNO_3 + NO(g)$

c. $HNO_3(\ell) \rightarrow NO_2(g) + O_2(g) + H_2O(\ell)$

d. $N_2H_4(\ell) + H_2O_2(\ell) \rightarrow N_2(g) + H_2O(g)$

e. $Ca_3(PO_4)_2(s) + SiO_2(s) + C(s) \rightarrow CaSiO_3 + CO(g) + P_2(g)$

f. $PCl_3(s) + H_2O(\ell) \rightarrow H_3PO_3(aq) + HCl(g)$

g. $Ca_3P_2(s) + H_2O(\ell) \rightarrow PH_3(g) + Ca(OH)_2(s)$

h. $NaH_2PO_4(s) \rightarrow Na_3P_3O_9(s) + H_2O(g)$

i. $Cu_2S(s) + O_2(g) \rightarrow Cu(s) + SO_2(s)$

j. $HClO_3(\ell) \rightarrow ClO_2(g) + O_2(g) + H_2O(g)$

k. $LiH(s) + H_2O(\ell) \rightarrow LiOH(s) + H_2(g)$

l. $Ca(OH)_2(aq) + H_3PO_4(aq) \rightarrow Ca_3(PO_4)_2(s) + H_2O$

m. $Na_2O(s) + H_2O(\ell) \rightarrow NaOH(s)$

n. $CaO(s) + P_4O_{10}(s) \rightarrow Ca_3(PO_4)_2(s)$

o. $P_4O_{10}(s) + H_2O(\ell) \rightarrow H_3PO_4(\ell)$

p. $MnO_2(s) + HCl(aq) \rightarrow MnCl_2(aq) + H_2O + Cl_2(g)$

q. $KBF_4(s) + Na(\ell) \rightarrow KF(s) + NaF(s) + B(s)$

r. $Al_2O_3(s) + Mg(\ell) \rightarrow MgO(s) + Al(s)$

s. $H_3BO_3(s) \rightarrow H_2O + B_2O_3(s)$

t. $NH_3(g) + CuO(s) \rightarrow Cu(s) + H_2O(g) + N_2(g)$

u. $H_2SO_4(aq) + CaF_2(s) \rightarrow CaSO_4(s) + HF(aq)$

v. $PbS(s) + H_2O_2(aq) \rightarrow PbSO_4(s) + H_2O$

w. $Cl_2(g) + HgO(s) \rightarrow HgCl_2(s) + Cl_2O(g)$

x. $Pb(NO_3)_2(s) \rightarrow PbO(s) + NO_2(g) + O_2(g)$

y. $Zn_3As_2(s) + HCl(aq) \rightarrow ZnCl_2(aq) + AsH_3(g)$

z. $I_2 + HNO_3(aq) \rightarrow HIO_3(aq) + NO_2(g) + H_2O$

4-36 How many millimoles of CrO_2Cl_2 will result from the reaction of HCl with 2.68 g of sodium dichromate if the other products are sodium chloride and water?

4-37 How many grams of $VOCl_3$ can be made from 100 g of V_2O_5 by the reaction

$$V_2O_5 + Cl_2 \rightarrow VOCl_3 + O_2$$

4-38 At -45°, $CuCl_2$ reacted with $LiBH_4$ according to the equation

$$2CuCl_2 + 4LiBH_4 \rightarrow 2CuBH_4 + 4LiCl + B_2H_6 + H_2$$

In one experiment, 2.30 mmol $CuCl_2$ and 4.54 mmol $LiBH_4$ gave 0.98 mmol B_2H_6. Was either of the reactants in excess? What was the percent yield of B_2H_6?

[T. J. Klinger, *Inorg. Chem.*, *3*, 1058 (1964)]

4-39 Zirconium metal was heated in an atmosphere of chlorine. The product of the reaction was analyzed and found to contain 39.0% Zr and 60.6% Cl. One sample of the product weighing 12.0 g was mixed with 0.35 g of zirconium metal and heated to 460°. The product of this reaction contained 45.9% Zr and 53.7% Cl. Write equations for the reactions that occurred.

[H. L. Schlafer and H. W. Wille, *Z. Anorg. Allgem. Chem.*, *327*, 253 (1964)]

4-40 Ten grams of high-purity silicon, a 90% yield, was made by the reaction of sodium with $SiCl_4$. What other product was formed? How many grams of $SiCl_4$ were used?

[B. Kamenas and D. Ordenic, *Z. Anorg. Allgem. Chem.*, *321*, 113 (1963)]

4-41 A sample of a compound containing 56.45% tungsten and 43.10% chlorine was heated to 450−500° in a sealed, evacuated tube. Tungsten(V) chloride and a compound containing 72.11% tungsten and 27.67% chlorine were formed. What is the equation for the reaction that occurred?

[R. E. McCarley and T. M. Brown, *Inorg. Chem.*, *3*, 1233 (1964)]

4-42 Fluorine gas was passed over tin(II) sulfide at 500°, and the solid fluoride that was formed contained 61.00% tin. Assuming the other product was SF_6, write the equation for the reaction.

[H. M. Haendler et al., *J. Am. Chem. Soc.*, *76*, 2178 (1954)]

4-43 Carbon tetrachloride vapor was passed over a sample of Gd_2O_3 in a tube furnace at 650°, and phosgene and a solid product containing 40.2% chlorine were formed. Write an equation for the reaction.

[J. F. Miller, S. E. Miller, and R. C. Himes, *J. Am. Chem. Soc.*, *81*, 4450 (1959)]

4-44 The oxide of tungsten, WO_3, was reduced with hydrogen gas. A sample of tungsten weighing 2.49330 g was obtained from 3.14520 g of WO_3. What was the percent yield?

[G. S. Thomas, *J. Am. Chem. Soc.*, *21*, 373 (1899)]

4-45 In the reaction of 64.0 g of $NaBH_4$ with 1054 g of elemental iodine 15.0 g of boron triiodide was obtained along with sodium iodide and hydrogen iodide.

a. Write the equation for the reaction.
b. Which was the limiting reactant?
c. Calculate the % yield.

[H. C. Anderson and L. H. Belz, *J. Am. Chem. Soc.*, *75*, 4828 (1953)]

4-46 What was the percent yield in the reaction

$$\underset{0.6029\ g}{BaPt(CN)_4 \cdot 4H_2O} + ZnSO_4 \rightarrow ZnPt(CN)_4 + \underset{0.4261\ g}{BaSO_4} + H_2O$$

[L. A. Levy, *J. Chem. Soc.*, *101*, 1084 (1912)]

4-47 In a study of the thermal decomposition of $K_3Fe(C_2O_4)_3$, 25.0 mg of that compound produced 2.24 mg of carbon dioxide along with FeC_2O_4 and $K_2C_2O_4$. Calculate the % yield of carbon dioxide.

[J. D. Danforth and J. Dox, *Inorg. Chem.*, *10*, 1623 (1971)]

4-48 What was the percent yield in the following reaction if 1.07 g of $Co_2(CO)_8$ gave 0.30 g of $[Co_2S(CO)_5]_n$? Assume n = 1.

$$Co_2(CO)_8 + S \rightarrow CO + [Co_2S(CO)_5]_n$$

[L. Marko, G. Bor, and G. Almasy, *Chem. Ber.*, *94*, 847 (1961)]

4-49 Trifluoroacetic anhydride $(CF_3CO)_2O$ was prepared by treating 100 g of trifluoroacetic acid CF_3COOH with 100 g of P_4O_{10}. The anhydride collected weighed 65.6 g. What was the percent yield? (The P_4O_{10} removes water to form H_3PO_4.)

4-50 When stibine, SbH_3, was heated at pressures greater than 2/3 atm, it decomposed explosively into the elements. If the yield was 99.90% of the theoretical, how much hydrogen, in grams, was obtained from 0.8231 g of stibine?

[W. Jolly et al., *J. Inorg. Nucl. Chem.*, *14*, 190 (1960)]

4-51 A sample of UCl_3 (10.2 g) was obtained, along with $ZnCl_2$, from the reaction of UCl_4 with zinc at 450° for 18 hr. If the yield was 86.0% of the theoretical, how much UCl_4, in grams, was used in the reaction?

[H. S. Young, *Chem. Abstr.*, *53*, 18710 i (1959)]

4-52 A compound, believed to be CdO_2, was heated, and cadmium(II) oxide and oxygen gas were formed. The loss in mass was 11.5%. Does this loss in mass substantiate the belief that the compound was CdO_2?

[C. W. W. Hoffmann, R. C. Ropp, and R. W. Mooney, *J. Am. Chem. Soc.*, *81*, 3830 (1959)]

4-53 Which element is oxidized and which reduced in the reaction

$SOF_2(g) + 2ClF(g) \rightarrow SOF_4(g) + Cl_2(g)$

[C. J. Schack and R. D. Wilson, *Inorg. Chem.*, *9*, 311 (1970)]

4-54 How many grams of a 3.00% hydrazoic acid solution, HN_3, was required to make 14.7 g of potassium azide, KN_3, from potassium hydroxide if the yield was 91.5%?

[A. W. Browne, *Inorg. Syn.*, *1*, 80 (1939)]

4-55 In the following reaction, is either of the reactants in excess? What is the percent yield?

$B_{10}H_{14}$ +	$2(CH_3CH_2)_3NBH_3$ →	$[(CH_3CH_2)_3NH]_2B_{12}H_{12}$ + $3H_2$
1.727 g	5.688 g	4.6 g

[N. N. Greenwood and J. H. Morris, *Proc. Chem. Soc.*, *1963*, 338]

4-56 A 40-mg sample of XeF_2 reacted with water and gave 0.25 mmol Xe, 0.12 mmol O_2, and 0.55 mmol HF. Which of the quantities reported was in excess of the quantity predicted by the equation for the reaction that presumably occurred?

[L. V. Streng and A. G. Streng, *Inorg. Chem.*, *4*, 1370 (1965)]

4-57 The reaction of magnesium metal with 0.055 g of SiH_3Br in ether solution produced a 6.0% yield of SiH_4. The other products were hydrogen gas, elemental silicon, and magnesium bromide.

a. Write a balanced equation for the reaction.
b. Which elements were oxidized or reduced?
c. Calculate the actual yield of SiH_4.

[E. R. Van Arsdale and J. Gavis, *J. Am. Chem. Soc.*, *74*, 3196 (1952)]

4-58 Which element was oxidized and which reduced in the reaction

$SO_2(g) + ClF(g) \rightarrow ClSO_2F(g)$

[J. Schack and R. D. Wilson, *Inorg. Chem.*, *9*, 311 (1970)]

4-59 Calculate the theoretical yield of $B_{10}H_{13}Na$ to be expected from the reaction of 1.56 g of $B_{10}H_{14}$ with 2.784 g of a 60.62% sodium hydride solution in n-propyl ether. The reaction was

$$B_{10}H_{14} + NaH \rightarrow B_{10}H_{13}Na + H_2$$

[R. J. F. Polchak, J. H. Norman, and R. E. Williams, *J. Am. Chem. Soc.*, *83*, 3381 (1961)]

4-60 Diboron tetrachloride (1.400 mmol) and an excess of boron tribromide were allowed to stand at room temperature. The diboron tetrabromide (0.3663 g) that formed was distilled away. What was the percent yield?
[G. Urry et al., *J. Am. Chem. Soc.*, *76*, 5297 (1954)]

4-61 LaF_3 (22.83 mg) was reduced to the metal with Li. What was the percent yield if 15.29 mg of La was formed?
[J. G. Stites, M. R. Salutsky, and B. D. Stone, *J. Am. Chem. Soc.*, *77*, 240 (1955)]

4-62 $SiCl_4$ (65 g) reacted with H_2S to form HCl and 26 g of the solid, $HSSiCl_3$. What was the percent yield of the silicon compound?
[W. C. Schumb and W. J. Bernard, *J. Am. Chem. Soc.*, *77*, 362 (1955)]

4-63 A sample of gold(III) fluoride was prepared at Manchester University by the reaction of bromine trifluoride with gold. The AuF_3/Au ratio was found to be 1.287. How does this compare with the theoretical AuF_3/Au ratio?
[A. S. Woolf, *J. Chem. Soc.*, *1954*, 4695]

4-64 A sample of beryllium bromide weighing 6.04443 g was dissolved in water, and to the resulting solution an excess of silver nitrate was added. The precipitate of AgBr that formed was collected and carefully dried; it weighed 13.44565 g. By using the accepted values of the atomic masses of Ag and Br, from Table D-3, calculate a value for the atomic mass of beryllium.
[G. E. F. Lundell, *J. Am. Chem. Soc.*, *70*, 3531 (1948)]

4-65 T. W. Richards and G. S. Forbes dissolved silver in nitric acid and collected the $AgNO_3$ by evaporation. In one experiment, of a series of six, 4.60825 g Ag gave 7.25706 g $AgNO_3$. Calculate the atomic mass of silver from the equation

$$Ag + HNO_3 \rightarrow AgNO_3 + NO_2 + H_2O$$

[*J. Am. Chem. Soc.*, *29*, 808 (1907)]

4-66 Calculate the atomic mass of iron from the fact that 0.574244 g of $FeBr_3$ gave 1.00000 g of AgBr.
[*J. Am. Chem. Soc.*, *84*, 4185 (1962)]

4-67 Calculate the atomic mass of calcium if 1.03950 g of $CaCl_2$ gave 2.68467 g of AgCl.
[O. Honigschmid and K. Kempter, *Z. Anorg. Allgem. Chem.*, *195*, 1 (1931)]

4-68 What is the atomic mass of vanadium calculated from the observation that 4.42279 g of $VOCl_3$ gave 10.97187 g of AgCl?
[H. V. A. Broscoe and H. F. V. Little, *J. Chem. Soc.*, *105*, 1310 (1941)]

5
GASES

The introductory study of gases, i.e., substances in the gaseous state, involves the relationships between four measurable quantities: mass, volume, temperature, and pressure. These are the four properties sufficient to define a state. Of these four quantities, the first three were previously defined. The SI unit of pressure is the pascal, symbol Pa. It is defined as a force per unit area of 1 Newton (N) per square meter. In chemistry two other units are more commonly used. They are the atmosphere:

$$1 \text{ atmosphere (atm)} = 101\ 325 \text{ Nm}^{-2}$$

and millimeters of mercury or torrs:

$$1 \text{ atm} = 760 \text{ mm Hg}$$

The latter two units are not part of the international system of units. Another unit used in meteorology that is now appearing in chemistry is the bar. The bar is very close to 1 atm.

$$1 \text{ bar} = 1 \times 10^5 \text{ Pa} = 0.9869 \text{ atm}$$

(Because the bar is so close to 1 atm and it is an exact multiple of pascals, it has been proposed that standard pressure be changed to one bar.)

One of the outstanding characteristics of substances in the gaseous state is the sensitivity of the volume to changes in pressure and temperature. The quantitative relationships between the volume of a given mass of a gas and the pressure and temperature are described by two physical laws.

5-1 BOYLE'S LAW

This law may be stated:

> The volume of a given mass of gas, at constant temperature, varies inversely with the pressure.

The volume is an inverse function of the pressure

$$V \propto \frac{1}{P} \quad \text{or} \quad V = \frac{k}{P}$$

The law may be stated another way:

> The product of the pressure and volume of a given mass of a gas is constant at a fixed temperature: $PV = k$

It follows from this that $P_1V_1 = P_2V_2 = P_3V_3$, etc. $= k$.

Example 5-1

If a gas occupies a volume of 23.0 dm^3 at 100 kPa, what volume will it occupy at 125 kPa? (Note that 1 dm^3 is equal to 1 L.)

Method 1: We can use the pressure factor to solve this problem. Since the pressure increased, the volume should decrease. The pressure factor is thus 100 kPa/125 kPa.

$$23.0\ dm^3 \times \frac{100\ \text{kPa}}{125\ \text{kPa}} = 18.4\ dm^3$$

Method 2: By using algebra, we can solve Boyle's law to get an appropriate formula. Substitution of the correct values for the pressures and the initial volume will give the final volume.

$P_1 = 100$ kPa $\qquad P_2 = 125$ kPa

$V_1 = 23.0\ dm^3 \qquad V_2 = ?\ dm^3$

$$P_1V_1 = P_2V_2$$

$$V_2 = \frac{P_1V_1}{P_2}$$

$$V_2 = \frac{(100\ \text{kPa})(23.0\ dm^3)}{(125\ \text{kPa})} = 18.4\ dm^3$$

5-2 GAY-LUSSAC'S LAW (CHARLES' LAW)

At a constant pressure, the volume of a given mass of a gas increases 1/273 of the volume at 0°C for each 1° rise in temperature.

$$V_t = V_o + \frac{1}{273} V_o t$$

where V_t is the volume at some temperature t°C and V_o is the volume at 0°C.

$$V_t = V_o \left(1 + \frac{t}{273}\right)$$

$$V_t = V_o \left(\frac{273 + t}{273}\right)$$

$$V_t = \frac{V_o}{273} (273 + t)$$

if

$$T = 273 + t$$

then

$$V_t = \frac{V_o}{273} T$$

where T is the absolute temperature (more correctly the thermodynamic temperature) in degrees Kelvin (K).

$$V = k'T$$

The volume is directly proportional to the absolute temperature. For a given quantity of a gas at constant pressure: $V_1 = k'T_1$, $V_2 = k'T_2$, $V_3 = k'T_3$, etc.

$$\frac{V_1}{T_1} = \frac{V_2}{T_2} = \frac{V_3}{T_3}$$

Example 5-2

What was the volume of a quantity of a gas at 27°C if its volume was 400 mL at 0°C? The pressure remained constant.

Method 1: Since the temperature increased, the volume increased. Here we must use a temperature factor greater than 1, that is, 300/273.

The temperature must be in degrees Kelvin.

20°C + 273 = 300 K

0°C + 273 = 273 K

$$400 \text{ mL} \times \frac{300 \text{ K}}{273 \text{ K}} = 438 \text{ mL}$$

Method 2: Solve Charles' law to get an expression for the unknown volume. Substitute the appropriate data into the formula.

$V_1 = 400$ mL $\qquad T_1 = 0°C + 273 = 273$ K

$V_2 = ?$ mL $\qquad T_2 = 27°C + 273 = 300$ K

$$\frac{V_1}{T_1} = \frac{V_2}{T_2}$$

$$V_2 = \frac{V_1 T_2}{T_1}$$

$$V_2 = \frac{(400 \text{ mL})(300 \not{K})}{(273 \not{K})} = 438 \text{ mL}$$

5-3 EQUATION OF STATE

Boyle's law and Charles' law may be combined to give what is called an *equation of state*. Suppose a given mass of a gas has a volume V_1 and a pressure P_1. Suppose that the pressure is changed to P_2 while the temperature remains constant at T_1. The volume will assume a new value V'_1. Then,

$$P_1V_1 = P_2V'_1$$

$$V'_1 = \frac{P_1V_1}{P_2}$$

Suppose further that the pressure is now kept constant at P_2 and the temperature is changed from T_1 to T_2. The volume will change to a new value, V_2. Then,

$$\frac{V'_1}{T_1} = \frac{V_2}{T_2}$$

but

$$V'_1 = \frac{P_1V_1}{P_2}$$

$$\frac{P_1V_1}{P_2T_1} = \frac{V_2}{T_2}$$

$$\frac{P_1V_1}{T_1} = \frac{P_2V_2}{T_2}$$

By similar arguments

$$\frac{P_1V_1}{T_1} = \frac{P_2V_2}{T_2} = \frac{P_3V_3}{T_3} = \cdots$$

or, in general,

$$\frac{PV}{T} = \text{constant}$$

The above is known as the combined-gas law.

Example 5-3

What volume will be occupied by a sample of gas at 40°C and 720 mm if its volume at 20°C and 750 mm was 250 mL?

Method 1: Since the pressure decreased, the volume would increase and the original volume should be multiplied by the factor greater than 1, that is 750/720. The temperature increased; therefore, the volume would increase. The original volume should be multiplied by the factor greater than 1, or (273+40)/(273+20).

$$250 \text{ mL} \times \frac{750 \text{ mm}}{720 \text{ mm}} \times \frac{313 \text{ K}}{293 \text{ K}} = 278 \text{ mL}$$

Method 2: We must algebraically solve the combined-gas law to get an expression for the final volume.

$P_1 = 750$ mm	$P_2 = 720$ mm
$V_1 = 250$ mL	$V_2 = ?$ mL
$T = 20 + 273 = 293$ K	$T = 40 + 273 = 313$ K

$$\frac{P_1V_1}{T_1} = \frac{P_2V_2}{T_2}$$

$$V_2 = \frac{P_1V_1T_2}{P_2T_1}$$

$$V_2 = \frac{(750 \text{ mm})(250 \text{ mL})(313 \text{ K})}{(720 \text{ mm})(293 \text{ K})} = 278 \text{ mL}$$

5-4 QUANTITY OF A GAS

Previously we have considered the relationship of volume, pressure, and temperature for a given quantity of a gas. The volume of a gas is obviously related to the quantity present. We would now like to introduce quantity as a variable. In the general equation

$$\frac{PV}{T} = \text{constant}$$

the value of the constant depends on the mass of the gas present. According to Avogadro's law, equal volumes of gases under the same conditions of temperature and pressure contain the same number of molecules. We have previously stated that *1 mole* of any substance contains the same number of molecules. We can then introduce a universal constant R, the *gas constant*, which has the same value for all gases.

$$\frac{PV}{T} = R$$

or PV = RT, where V is the volume occupied by 1 mole, the *molar volume*, at the pressure P and the temperature T. For n moles of gas,

$$PV = nRT$$

$$n = \frac{m}{M}$$

where m is the mass in grams and M is the molecular mass (grams per mole).

In this derivation we have assumed that all gases exhibit the same behavior and that Boyle's and Charles' laws accurately describe this behavior for all gases. For real gases these laws can only be regarded as approximations, which can be applied at relatively low pressures and moderately high temperatures. It is useful to postulate an imaginary gas or *ideal gas*. An *ideal gas* is a gas to which Boyle's and Charles' laws are strictly applicable. The equation PV = nRT is called the ideal-gas equation.

5-5 EVALUATION OF THE GAS CONSTANT

$$R = \frac{PV}{nT}$$

$$\frac{\text{Pressure} \times \text{volume}}{\text{No. of moles} \times \text{temp. in K}}$$

At standard temperature and pressure (STP), defined as 0°C and 1 atm (760 mm or torr), 1 mol of a gas occupies 22.414 L.

For 1 mol of an ideal gas:

$$R = \frac{1 \times 22.414}{1 \times 273.15} = 0.08205 \frac{\text{L-atm}}{\text{K-mol}}$$

Example 5-4

What volume will be occupied by 0.115 g of xenon at 37°C and 0.250 atm? (Assume the ideal-gas law applies.)

$$PV = \frac{mRT}{M} \qquad \begin{aligned} P &= 0.250 \text{ atm} \\ m &= 0.155 \text{ g} \\ M &= 131.3 \text{ g/mol} \\ R &= 82.05 \text{ mL-atm/K-mol} \\ T &= 37 + 273 = 310 \text{ K} \end{aligned}$$

$$V = \frac{mRT}{PM} = \frac{0.155 \times 82.0 \times 310}{0.250 \times 131} = 120 \text{ mL}$$

Example 5-5

What pressure will be exerted by 0.623 g of hydrogen at 42°C in a volume of 58.4 mL?

$$PV = \frac{mRT}{M} \qquad \begin{aligned} V &= 58.4 \text{ mL} \\ m &= 0.623 \text{ g} \\ M &= 2 \times 1.008 = 2.016 \text{ g/mol} \\ R &= 82.05 \text{ mL-atm/K-mol} \\ T &= 42 + 273 = 315 \text{ K} \end{aligned}$$

$$P = \frac{mRT}{VM} = \frac{0.623 \times 82.0 \times 315}{58.4 \times 2.02}$$

$$= 136 \text{ atm}$$

5-6 MOLECULAR MASSES OF GASES

One of the most useful applications of the ideal-gas law is the calculation of molecular masses of gases.

Example 5-6

E. W. Morley found that 43.35 L of hydrogen weighed 3.716 g at 0.0°C and 724.4 mm. Calculate the molecular mass.

[*Am. Chem. J.*, *18*, 162 (1896)]

Method 1: First correct the volume to standard pressure. The corrected volume will be *smaller*.

$$43.35 \text{ L} \times \frac{724.4 \text{ mm}}{760.0 \text{ mm}} = 41.31 \text{ L}$$

At standard conditions 1 mol of a gas occupies 22.41 L.

$$M = 3.716 \text{ g} \times \frac{22.41 \text{ L/mol}}{41.31 \text{ L}} = 2.016 \frac{\text{g}}{\text{mol}}$$

Method 2: Solve the combined-gas law for molecular mass. Substitute the appropriate numbers.

$$PV = \frac{mRT}{M} \quad \text{so} \quad M = \frac{mRT}{PV}$$

The value of R to be used is 0.08205 since the volume is in liters. Pressure must be expressed in atmospheres and the temperature in degrees Kelvin.

$$M = \frac{3.716 \text{ g} \times 0.08205 \text{ L-atm/K-mol} \times 273.2 \text{ K}}{(724.4 \text{ mm}/760.0 \text{ mm/atm}) \times 43.35 \text{ L}}$$

$$M = 2.016 \frac{\text{g}}{\text{mol}}$$

Example 5-7

A sample of a new compound in a bulb whose capacity was 268 cc weighed 0.0516 g. The pressure was 17.9 mm, and the temperature was 69°C. Calculate the molecular mass of the compound.

[G. Urry, A. G. Garrett, and H. I. Schlesinger, *Inorg. Chem.*, *2*, 396 (1963)]

Method 1: Correct the volume to standard conditions. The volume will *decrease* because of an increase in pressure and a decrease in temperature.

$$268 \text{ cc} \times \frac{17.9 \text{ mm}}{760 \text{ mm}} \times \frac{273 \text{ K}}{342 \text{ K}} = 5.04 \text{ cc}$$

Here 0.0516 g occupies 5.04 cc (mL); 1 mol will occupy 22,414 mL.

$$M = 0.0516 \text{ g} \times \frac{22\ 414 \text{ mL/mol}}{5.04 \text{ mL}} = 230 \frac{\text{g}}{\text{mol}}$$

Method 2

$$M = \frac{mRT}{PV}$$

R has the value of 82.06 when V is in milliliters.

$$M = \frac{0.0516 \times 82.06 \text{ (mL-atm/K-mol)} \times 342 \text{ K}}{(17.9 \text{ mm/760 mm/atm})(268 \text{ mL})} = 229 \frac{\text{g}}{\text{mol}}$$

5-7 GAS DENSITY AND IDEAL-GAS EQUATION

The ideal-gas equation may be rearranged:

$$PV = \frac{m}{M} RT$$

$$M = \frac{m}{V} \frac{RT}{P}$$

Density d has been defined as m/V (Section 1-5); the equation may then be written

$$M = d \frac{RT}{P} \quad \text{or} \quad d = \frac{PM}{RT}$$

Example 5-8

Calculate the density of silane, SiH_4, at 27°C and 190 mm.

SiH_4	$1 \times 28.1 =$	28.1
	$4 \times 1.0 =$	4.0
		32.1 g/mol

$$d = \frac{PM}{RT}$$

$$d = \frac{(190\ \text{mm}/760\ \text{mm/atm})(32.1\ \text{g/mol})}{(82.0\ \text{mL-atm/K-mol})(300\ \text{K})} = 3.26 \times 10^{-4}\ \frac{\text{g}}{\text{mL}}$$

Example 5-9

The density of a gas measured at 20.3°C and 749 mm was 2.66 g/L. Calculate the molecular mass.

[Gmelin's *Handbuch der Anorganischen Chemie*, System No. 9, 211 (1953)]

$$M = \frac{dRT}{P}$$

$$M = \frac{2.66\ \text{g/L} \times 0.0820\ \text{L-atm/K-mol} \times 293\ \text{K}}{749\ \text{mm}/760\ \text{mm/atm}} = 64.8\ \frac{\text{g}}{\text{mol}}$$

5-8 DALTON'S LAW OF PARTIAL PRESSURES

In a mixture of gases, each gas exerts the pressure it would exert if it alone occupied the volume, and the total pressure is equal to the sum of the partial pressures.

$$P_{total} = P_1 + P_2 + P_3 + \cdots$$

Example 5-10

A sample of argon was collected over water in a 200-mL container. The total pressure in the container at 27°C was 763 mm. What was the mass of argon? The measured pressure was the total pressure of a mixture of argon and water vapor. The pressure due to water vapor in equilibrium with water at 27°C = 26.7 mm (Appendix D).

Total pressure	763 mm
Partial pressure H_2O	27 mm
Partial pressure Ar	736 mm

$$PV = \frac{m}{M} RT$$

$$m = \frac{PVM}{RT}$$

$$P = \frac{736}{760} \text{ atm}, \qquad V = 200 \text{ mL}, \qquad M = 39.9 \text{ g/mol}$$

$$R = 82.0 \text{ mL-atm/K-mol} \qquad \text{and} \qquad T = 27 + 273 = 300 \text{ K}$$

$$m = \frac{736/760 \times 200 \times 39.9}{82.0 \times 300} = \frac{736 \times 200 \times 39.9}{760 \times 82.0 \times 300}$$

$$= 0.314 \text{ g}$$

5-9 GAY-LUSSAC'S LAW OF COMBINING VOLUMES

In chemical reactions involving gases, the volumes are in the ratio of whole numbers. In the balanced equation of the reaction, the whole numbers are represented by the coefficients. For example:

$$4HCl(g) + O_2(g) \rightarrow 2H_2O(g) + 2Cl_2(g)$$

Four volumes of hydrogen chloride will react with one volume of oxygen to form two volumes of water vapor and two volumes of chlorine. If 100 mL of HCl reacts, 25 mL of oxygen will be consumed and 50 mL of $H_2O(g)$ and 50 mL of Cl_2 will be formed.

Example 5-11

How many milliliters of ammonia gas will be formed when 50 mL of nitrogen reacts with an excess of hydrogen?

$$N_2 + 3H_2 \rightarrow 2NH_3$$

$$1 \text{ vol} \rightarrow 2 \text{ vols}$$

$$50 \times \frac{2}{1} = 100 \text{ mL}$$

5-10 MOLAR VOLUME

In Chapter 3 we saw that a formula may be used to represent 1 mol of a substance. If the substance is a gas, the formula may be used to represent one molar volume. A molar volume is the volume occupied by 1 mole of a gas. At 0°C and 1 atm the molar volume of an ideal gas is 22.4 L.

Example 5-12

How many milliliters of oxygen (measured at STP) can be prepared by the decomposition of 2.65 g of KNO_3?

KNO_3	39.1
	14.0
	48.0
	101.1

$2KNO_3 \rightarrow 2KNO_2 + O_2$

2 mol → 1 mol or 22.4 L

1 mol → 0.5 mol or 11.2 L

$$\frac{2.65\ \text{g}}{101\ \text{g/mol}} = 0.0262\ \text{mol KNO}_3 \qquad \frac{0.0262}{2} = 0.0131\ \text{mol O}_2$$

$$1.31 \times 10^{-2}\ \text{mol O}_2 \times 22.4\ \frac{\text{L}}{\text{mol}} = 2.94 \times 10^{-1} = 0.294\ \text{L O}_2$$

5-11 GRAHAM'S LAW OF DIFFUSION (EFFUSION)

The rates of diffusion of gases are inversely proportional to the square root of their densities. The law may be expressed as

$$\frac{r_1}{r_2} = \sqrt{\frac{d_2}{d_1}}$$

where r_1 and r_2 are the rates of diffusion of two gases having densities d_1 and d_2, respectively.

From the ideal-gas equation

$$d = \frac{m}{V} = \frac{PM}{RT}$$

At constant temperature and pressure

$$\frac{d_1}{d_2} = \frac{(P/RT)\ M_1}{(P/RT)\ M_2}$$

$$\frac{r_1}{r_2} = \sqrt{\frac{M_2}{M_1}}$$

The rates of diffusion of gases are inversely proportional to the square root of their molecular masses. When the gases diffuse through a pinhole, it is called effusion. In this case Graham's law is called Graham's law of effusion.

Example 5-13

A sample of nitrogen diffused through a small hole at the rate of 2.65 mL/min. At what rate would NF_3 flow through the same orifice under the same conditions?

$$M_{NF_3} = 14 + 3 \times 19 = 71 \, \frac{g}{mol}$$

$$M_{N_2} = 2 \times 14 = 28 \, \frac{g}{mol}$$

$$\frac{r_{N_2}}{r_{NF_3}} = \sqrt{\frac{M_{NF_3}}{M_{N_2}}}$$

$$\frac{2.65}{r_{NF_3}} = \sqrt{\frac{71}{28}} = \sqrt{2.53} = 1.59$$

$$r_{NF_3} = \frac{2.65}{1.59} = 1.67 \, \frac{mL}{min}$$

Example 5-14

A sample of xenon required 1.0 min 8.3 sec to diffuse through a small orifice. What is the molecular mass of a gas that required only 57.0 sec for diffusion under identical conditions? The *time required* is inversely proportional to the *rate of diffusion* and is therefore directly proportional to the square root of the molecular mass.

$$\frac{t_{Xe}}{t} = \sqrt{\frac{M_{Xe}}{M}}$$

$$\frac{68.3}{57.0} = \sqrt{\frac{131}{M}} = 1.20$$

$$M = \frac{131}{1.20^2}$$

$$M = \frac{131}{1.44} = 91.0$$

5-12 BEHAVIOR OF REAL GASES

The PVT behavior of most gases is represented fairly accurately by the ideal-gas equation at relatively low pressures and high temperatures. The use of the ideal-gas equation is justified in many applications because departures from ideal behavior are not great. When more accurate calculations must be made, the discrepancies must be avoided.

5-13 EQUATIONS OF STATE FOR REAL GASES

Several equations that correct for the deviations from ideal behavior have been proposed. One of the most useful of these equations of state is the van der Waals equation:

$$\left(P + \frac{n^2a}{V^2}\right)(V - nb) = nRT$$

The constant a is the correction for the attractive force between molecules, and b is the correction factor for the volume occupied by the molecules of the gas. Values of the constants for several gases are given in Table 5-1.

TABLE 5-1 van der Waals Constants

Gas	a (L^2-atm/mol^2)	b (L/mol)
Helium	0.0341	0.0237
Argon	1.35	0.0322
Nitrogen	1.39	0.0391
Carbon dioxide	3.59	0.0427
Acetylene	4.39	0.0514
Carbon tetrachloride	20.39	0.1383

Example 5-15

Calculate the pressure exerted by *1 mol* of ammonia at 27°C in a flask whose volume is 0.800 L. Since n = 1

$$\left(P + \frac{a}{V^2}\right)(V - b) = RT$$

Solve for P:

$$P = \frac{RT}{V - b} - \frac{a}{V^2}$$

$$P = \frac{(0.0820\ \text{L-atm/K-mol})(300\ \text{K})}{0.800\ \text{L/mol} - 0.0371\ \text{L/mol}} - \frac{4.17\ \text{L}^2\text{-atm/mol}^2}{(0.800\ \text{L/mol})^2}$$

$$P = 32.24\ \text{atm} - 6.52\ \text{atm} = 25.7\ \text{atm}$$

PROBLEMS

5-1 Plot the following data for the volume of Ne at various pressures and constant temperature. Plot the volume at various pressures versus the reciprocal of the pressure. Calculate the product of the pressure and volume at each pressure.

P(mm)	V (L)
1	1.366
10	0.137
20	0.0685
30	0.0457
40	0.0343
50	0.0274
60	0.0228
70	0.0196
80	0.0171

Source: *Landolt-Bornstein Tabellen*, 5th ed., first supplement, p. 64.

5-2 What is the volume of a sample of carbon dioxide at 735 mm if its volume at 723 mm is 223 mL?

5-3 A sample of gas occupied 400 mL at STP. What volume will it occupy at 740 mm and 87°C?

5-4 Calculate the molecular mass of a gas if 0.638 g occupies a volume of 223 mL at 23°C and 758 mm.

5-5 A sample of a gas occupied 0.400 dm^3 at STP. What volume will it occupy at 10.0 kPa and 27°C?

5-6 What is the density of nitrogen trifluoride, at standard conditions, in:

a. grams per cubic meter
b. grams per cubic decimeter

5-7 Calculate the molecular mass for each of the following gases from the densities. Compare the results of your calculations with the molecular mass obtained from the most recent values of the atomic masses. The densities are in grams per liter at STP.

a. Ar, 1.7824
b. He, 0.1785
c. H_2, 0.08987
d. Kr, 3.708
e. Ne, 0.9002
f. N_2, 1.2506
g. O_2, 1.4290
h. Rn, 9.73
i. Xe, 5.851

[N. A. Lange, *Handbook of Chemistry*, 10th ed., McGraw-Hill, New York, 1966, pp. 100–107]

5-8 Calculate the molecular mass of stibine from its density at standard conditions, which was found to be 5.685 g/L.
[Stock and Gutman, *Chem. Ber.*, *37*, 887 (1904)]

5-9 When the space probe Viking landed on Mars, it found an atmospheric pressure of 6–7 millibars at the surface of the planet. What is this pressure in:

a. atmospheres
b. torrs

5-10 A mixture of B_2H_6 (3.28 mM) and PF_3 (19.21 mM) was collected in a vessel whose capacity was 62.5 mL. What was the total pressure at 25°C?
[B. W. Parry and T. C. Bissot, *J. Am. Chem. Soc.*, *78*, 1526 (1956)]

5-11 A fluoride of platinum was prepared at the Argonne National Laboratory. A sample weighing 0.1418 g was placed in a thin-walled nickel bulb at a pressure of 68.81 mm and a temperature of 26°C. The volume of the bulb was 126.63 cc. What is the molecular mass of the compound?

[B. Weinstock, J. G. Mahn, and E. C. Weaver, *J. Am. Chem. Soc.*, *83*, 4313 (1961)]

5-12 Acetylhypofluorite, $H_3CC(O)OF$, is an explosive compound and decomposes easily. When a 25.9-mg sample was allowed to decompose at STP, 14.74 cc of gas was produced.

a. How many moles of reactant were originally present?
b. How many moles of product were formed?
c. If the products formed an equimolar mixture, what products were probably formed?

[E. Appleman, M. Mendelsohm, and H. Kim, *J. Am. Chem. Soc.*, *107*, 6515 (1985)]

5-13 A compound was found to contain 64.70% carbon, 10.97% hydrogen, and 24.33% oxygen. A sample weighing 0.1396 g occupied a volume of 41.56 mL at 147°C and 658.8 mm. What is the formula of the compound?

[E. Frankland and B. F. Duppa, *J. Chem. Soc.*, *20*, 108 (1867)]

5-14 A pale-yellow solid containing 23.0% boron and 75% chlorine was prepared from boron trichloride. A 0.0516-g sample was vaporized at 69°, and the vapor occupied a volume of 268 cc at 22.2 mm. What is the formula of the compound?

[G. Urry, W. G. Garrett, and H. I. Schlesinger, *Inorg. Chem.*, *2*, 396–400 (1963)]

5-15 Diboron tetrachloride was treated with sodium hydroxide, and the following reaction occurred:

$$B_2Cl_4 + 10NaOH \rightarrow 2NaBO_3 + 4NaCl + 4H_2O + H_2$$

If 13.26 cc of hydrogen formed, how much B_2Cl_4, in grams, was consumed?

[G. Urry et al., *J. Am. Chem. Soc.*, *76*, 5296 (1954)]

5-16 In a preparation of ^{14}C-labeled methane, 2.00 mmol of barium carbonate containing 4% radioactive carbon atoms was treated with sulfuric acid. The carbon dioxide formed was freed of water vapor and reduced with hydrogen at 330° over a nickel-thorium dioxide catalyst. The yield of methane, based on barium carbonate, was 95.0%. What volume of methane, measured at 20°C and 750 mm, was obtained?

[W. H. Beamer, *J. Am. Chem. Soc.*, *70*, 3900 (1948)]

5-17 $LiBH_4$ (500 mg) reacted with 2.4247 g of iodine to produce boron triiodide, lithium iodide, hydrogen iodide, and 133 cc of hydrogen. What was the percent yield of hydrogen, based on $LiBH_4$, for the equation: $2LiBH_4 + 5I_2 \rightarrow 2BI_3 + 2LiI + 2HI + 3H_2$?

[F. Klanberg and H. W. Kohischutter, *Chem. Ber.*, *94*, 786 (1961)]

5-18 A mixture of the gases N_2 and CF_3COOF was contained in a metal trap at room temperature (23°C) and a pressure of 43.7 mm. The CF_3COOF was decomposed by a spark to CF_4 and CO_2. The pressure was then 58.4 mm. What was the initial partial pressure of CF_3COOF?

[G. L. Gard and G. H. Cady, *Inorg. Chem.*, *4*, 595 (1965)]

5-19 The density of acetic acid in the vapor state at its normal boiling point of 118.5°C, is 3.15 g/L. What is the molecular mass of the vapor at this temperature?

5-20 The vapor density of the compound containing 26.8% vanadium and 73.02% chlorine was determined in 1870. The weight of the bulb filled with air at 15° and 751 mm was 19.8945 g. Filled with the vapor at 205° and 758.1 mm, it weighed 20.3015 g. Its capacity was 112 cc. The density of air at 0°C and 1 atm is 1.293 g/L. What is the formula of the compound?

[H. E. Roscoe, *J. Chem. Soc.*, *23*, 345–347 (1870)]

5-21 A small quantity of a material believed to be ammonium ozonide was prepared at Armour Research Foundation. After standing for 90 min at 78°C, 74.1% had decomposed according to the equation

$$4NH_4O_3 \rightarrow 2NH_4NO_3 + O_2 + 4H_2O$$

How many milliliters of oxygen would be formed if 10.0 g of NH_4O_3 had been prepared and the pressure was 740 mm?

[I. J. Solomon et al., *J. Am. Chem. Soc.*, *84*, 34 (1962)]

5-22 In a study of the thermal decomposition of potassium chlorate it was found that, at 551°C, 97.5% of the theoretical quantity of oxygen was released after 60 min. How many milliliters of oxygen would result from heating 1.00 g of potassium chlorate? Assume that the gas volume was measured at 25°C and 743.1 mm.
[A. Glasner and L. Weidenfeld, *J. Am. Chem. Soc.*, *74*, 2467 (1952)]

5-23 The molecular mass of a highly explosive vapor was calculated from the density measurements at Cornell University. A sample weighing 0.0223 g had a volume of 13.8 cc at 747.6 mm and 19°C. What was the molecular mass of the compound?
[L. M. Dennis and H. Isliam, *J. Am. Chem. Soc.*, *29*, 222 (1907)]

5-24 Calculate the molecular mass of the gas whose normal density is 1.978 g/L.

5-25 A sample of an oxide of nitrogen weighing 0.64304 g was passed over nickel and catalytically decomposed, giving 0.34390 g of oxygen and 0.29920 g of nitrogen. What was the volume of the original sample of gas measured at 27°C and 740 mm?
[R. W. Gray, *J. Chem. Soc.*, *87*, 1616 (1905)]

5-26 What was the percent yield if 71.4 mL of hydrogen (at STP) was obtained from 0.8588 g of $SiHBr_3$ by the reaction

$$SiHBr_3 + 5NaOH \rightarrow 3NaBr + Na_2SiO_3 + H_2 + 2H_2O$$

[W. C. Schumb, *Inorg. Syn.*, *1(1939)*, 41]

5-27 A compound melting at -92.6° was prepared by the action of an electric discharge on boron trichloride. A sample weighing 0.0995 g was found to contain 0.0878 g of chlorine and 0.0131 g of boron. Another sample, collected in a 191.6-cc bulb, weighed 0.0810 g at 48.9 mm and 26.5°C. What is the formula of the compound?
[G. Urry et al., *J. Am. Chem. Soc.*, *76*, 5296 (1954)]

5-28 Gallium metal and chlorine combined to form a compound that melted at 77° and boiled at 200°. *Analysis* found: Ga, 39.55; Cl, 60.35. A sample weighing 0.5763 g occupied a volume of 223.4

cc at a pressure of 206.7 mm and a temperature of 177.5°C. What is the formula of the compound?

[A. W. Laubengayer and F. B. Schirmer, *J. Am. Chem. Soc.*, *62*, 1578 (1940)]

5-29 The accepted value of the molar volume of an ideal gas is 22.4136 L. Calculate the volume occupied by 1.0000 mol of each of the following gases at STP from the densities in g/L.

Ar, 1.7824
Cl_2, 3.214
F_2, 1.695
He, 0.1785
H_2, 0.08987
Kr, 3.708
Ne, 0.9002
N_2, 1.2506
O_2, 1.4290
O_3, 2.144
Rn, 9.73
Xe, 5.851

5-30 Calculate the rate at which arsine (AsH_3) gas will effuse through an orifice if ammonia effuses at the rate of 0.481 L/min under identical conditions.

5-31 Calculate the ratio of rates of effusion through a small orifice of silane, SiH_4 and diborane B_2H_6.

5-32 Using the van der Waals equation, calculate the pressure exerted by 0.399 g of argon at -23°C in a volume of 300 cc.

5-33 Use the van der Waals equation to calculate the pressure of 1.00 mol of CCl_4 in a 250-ml flask at 17°C.

Supplementary Problems

5-34 Plot the following data for the volume of Ne at constant pressure versus temperature. Determine the slope of the curve from the graph and compare it with the slope calculated from Gay-Lussac's law (Charles' law).

t (°C)	V (L)
400	2.47
300	2.11
200	1.74
100	1.37
0	1.01
-50	0.82
-100	0.64
-150	0.45
-183	0.33

Source: *Landolt-Bornstein Tabellen*, 5th ed., first supplement, pp. 64–65.

5-35 What volume will be occupied by a sample of hydrogen at 153 kPa, if the volume at 0.974 atm is 325 cc?

5-36 What volume will be occupied by a sample of hydrogen at 1.51 atm if the volume at 740 mm is 325 mL?

5-37 A sample of hydrogen sulfide was collected in a 250-mL flask at a pressure of 740 mm and 37°C. What volume would it occupy at 3.65 atm and 40°C?

5-38 At STP, an ideal gas occupies a volume of 0.022414 m^3. Determine the value of the gas constant, R, when the pressure is in pascals.

5-39 Calculate the density of diborane, B_2H_6, at 40° and 1.81 atm.

5-40 Calculate the molecular masses of the following gases and compare them with the values obtained from the latest values of the atomic masses of the elements. Account for any discrepancies. The densities are in grams per liter.

a. Ammonia, 0.7708
b. Stibine, 5.685
c. Argon, 1.782
d. Arsine, 3.50
e. Boron trifluoride, 3.00
f. Hydrogen bromide, 3.644
g. Hydrogen chloride, 1.639
h. Fluorine, 1.71

[*Landolt-Bornstein Tabellen*, 5th ed., first supplement]

5-41 Oxygen was collected by the displacement of water at 25°C. The vapor pressure of water at this temperature is 3173 Pa. If the volume of the container was 5.00 dm^3 and the total pressure in the system was 101200 Pa, how many moles of oxygen were collected?

5-42 A compound was prepared at Cornell University by the reaction of ammonium chloride with boron trichloride. *Analysis* found: Cl, 57.62, 57.80, 57.92; B, 17.88, 17.36, 17.62; N, 22.83, 22.84. A sample weighing 0.2182 g in a 668.5-mL bulb had a pressure of 45.0 mm at 141.1°C. What was the compound formed?

[C. A. Brown and A. W. Laubengayer, *J. Am. Chem. Soc.*, *77*, 3699 (1955)]

5-43 The molecular mass of chromyl fluoride was calculated from vapor-density measurements as 121.4. The temperature was 25° and the pressure 500 mm. What was the density of the vapor?

[C. A. Engelbrecht and A. V. Grosse, *J. Am. Chem. Soc.*, *74*, 5262 (1952)]

5-44 In a study of the hydrolysis of pentaborane, B_5H_9, a sample (6.56 cc at STP) was shaken with 20% water in dioxane. The volume of hydrogen collected was 78.4 cc. If the other product was orthoboric acid, write a balanced equation consistent with the results of the experiment.

[I. Shapiro and H. G. Weiss, *J. Am. Chem. Soc.*, *76*, 6020 (1954)]

5-45 Samples of zirconium trichloride, tribromide, and triiodide were hydrolyzed, and the hydrogen that formed was collected.

$$2ZrX_3 + 2H_2O \rightarrow 2Zr(OH)X_3 + H_2$$

Compound	Mass (g)	Volume H_2 (corr.) (mL)
$ZrCl_3$	0.9183	50.3
$ZrBr_3$	1.2591	39.2
ZrI_3	1.3121	29.7

What was the percent yield of hydrogen in each case?

[E. M. Larsen and J. J. Leddy, *J. Am. Chem. Soc.*, *78*, 5985 (1956)]

5-46 A mixture containing 1.3 mmol of HNF_2 gas and equal quantity of chlorine was led into a flask containing 5.0 g of potassium fluoride and allowed to stand for 18 hr at room temperature. The gas $ClNF_2$ (66% yield) and the solid KF·HCl were formed. What was the volume percent of $ClNF_2$ in the gaseous mixture present after the reaction? What was the weight percent of $ClNF_2$ in the gas mixture after the reaction?

[W. C. Firth, Jr., *Inorg. Chem.*, *4*, 254 (1965)]

5-47 When pure sulfuric acid was heated to 450°, its molecular mass, calculated from a vapor-density measurement, was found to be 50.03. What conclusion can be drawn from this?

5-48 A sample of the compound in Problem 5-49 weighing 0.0947 g was subjected to a series of reactions with the eventual preparation of 0.3140 g of silver chloride. If 1 mol of the unknown produced 1 mol of silver chloride, are the results of this experiment consistent with the molecular mass found in Problem 5-49?

5-49 A bulb whose capacity was 267.43 cc at 0°C was filled with a gas at a pressure of 760 mm. The difference in weight between the filled bulb and a tare weight of similar displacement was 0.35851 g. When the bulb was filled with oxygen under the same conditions, the difference in weight between the bulb and tare was 0.38228 g. Calculate the molecular mass of the gas.

[R. W. Gray, *J. Chem. Soc.*, *87*, 1608 (1905)]

5-50 A bulb filled with a gaseous element weighed 0.32280 g at 10°C and 760 mm; when filled with oxygen under the same conditions, it weighed 0.36884 g. What is the molecular mass of the gaseous element?

[R. W. Gray, *J. Chem. Soc.*, *87*, 1616 (1905)]

5-51 What is the molecular mass of a gas if 503.5 mL of it weighs 1.10 g at 545.5 mm and 0°C?

[D. R. Martin and R. E. Dial, *J. Am. Chem. Soc.*, *72*, 853 (1950)]

5-52 The volume of 0.0808 g of a boron hydride was 128.8 mL at 17.3°C and 405.7 mm. The same sample of gas was then decomposed and 63.6 mg of boron and 197.3 mL of hydrogen (at STP) were obtained. What is the molecular mass of the boron hydride? What is its formula?

[A. Stock and E. Kuss, *Chem. Ber.*, *56*, 801 (1923)]

5-53 Difluoramine (24.6 cc) was maintained in a high-pressure ampoule for 68 hr at 0° in the presence of a catalyst. Some HNF_2 (10.0 cc) was recovered, but the remainder had been converted to N_2F_4 (5.5 cc). Calculate the percent yield of N_2F_4 assuming its formation by the reaction $5NHF_2 = 2N_2F_4 + NH_4F + HF$.
[E. A. Lawton and J. Q. Weber, *J. Am. Chem. Soc.*, *85*, 3597 (1963)]

5-54 The compound described in Problem 5-28 was mixed with gallium metal and heated to 175° in a vacuum. *Analysis* found: Ga, 49.42; Cl, 50.41. Write an equation for the reaction that occurred.

5-55 Calculate the mass of each component present in a mixture of fluorine and xenon in a 2-L glass flask. The partial pressure of Xe was 350 mm and the total pressure was 724 mm at 25°.
[L. V. Streng and A. G. Streng, *Inorg. Chem.*, *4*, 1370 (1965)]

5-56 Calculate the percent yield of product if the components of the mixture in Problem 5-55 reacted to form 0.50 g XeF_2.

5-57 A mixture of the gases PH_3 and BF_3 weighing 1.03 g exerted a pressure of 538.0 mm in a volume of 503.51 mL at 0°. What was the mole ratio BF_3/PH_3 of the mixture?
[D. R. Martin and R. E. Dial, *J. Am. Chem. Soc.*, *72*, 853 (1950)]

5-58 What was the weight of $SF_5ONF_2(g)$ contained in a 200-mL glass vessel if the pressure was 56.5 mm when the vessel was immersed in a water bath maintained at 85°C?
[W. H. Hale, Jr. and S. M. Williamson, *Inorg. Chem.*, *4*, 1341 (1965)]

5-59 A sample of $SF_5OF(g)$ was contained in a glass vessel at 25° and a pressure of 80 mm. A quantity of N_2F_4 that was added brought the total pressure to 160 mm. The reaction that occurred produced a variety of products, NF_3, NO, SiF_4 (by reaction with glass), SF_6, SO_2F_2, SOF_4, SF_5ONF_2, and NO_2. The yield of SF_5ONF_2 was 40 mol% with respect to the reactant, SF_5OF. All of the SF_5OF and N_2F_4 were consumed in the reaction. What was

the mass of SF_5ONF_2 produced if the volume of the vessel was 1 L? (The other product of the reaction in question was NF_3.)
[W. H. Hale, Jr., and S. M. Williamson, *Inorg. Chem.*, *4*, 1342 (1965)]

5-60 What time will be required for a sample of ethane, C_2H_6, to effuse through an orifice if a similar sample of butane, C_4H_{10}, required 2 min 23 sec?

5-61 Calculate the rate of flow of krypton through the walls of a porous tube if helium flows at the rate of 2.73×10^{-4} mL/sec under identical conditions.

5-62 What will be the ratio of the rate of flow through a porous wall of UF_6 containing ^{234}U to that of UF_6 containing ^{238}U?

5-63 According to calculations made by using the van der Waals equation, what will be the pressure of 2.00 mol of nitrogen at 27°C in a 500-mL flask?

6
CHEMICAL BONDS

The atomic theory proposes that the atoms of elements are combined in compounds. Chemists use the term *chemical bond* to identify the forces between atoms in compounds, and the nature of these forces is inferred from the properties of the compounds. Substances may be roughly divided into five types according to their properties. A detailed examination of their structural properties indicates that substances of a given type also have the same type of bonds.

6-1 CLASSIFICATION OF COMPOUNDS

1. Ionic, for example, NaCl, KNO_3, MgO
2. Covalent, for example, CCl_4, SiH_4, NF_3
3. Polar covalent, for example, HCl, HNO_3, $SnCl_4$
4. Metallic, for example, TiN, VC, Fe_2N
5. Adamantine, for example, SiC, BN

6-2 THE IONIC BOND

Ionic compounds are crystalline aggregates of ions, i.e., atoms, or groups of atoms, with an electric charge. The cations (+ charge) and anions (-) are arranged in a three-dimensional array approximately represented as spheres in a closely packed arrangement. Cations are formed by the loss of electrons by atoms; anions are formed by the gain of electrons by atoms. The electrostatic attraction between these ions of opposite charge leads to the formation of ionic crystals.

6-3 THE COVALENT BOND

Many compounds exist as gases, liquids, or low-melting solids. These compounds contain molecules; the attractive forces between them are relatively weak. The atoms within a molecule are joined by electron-pair or *covalent bonds*. In a covalent bond, atoms share pairs of electrons, attain a stable electronic configuration, and form molecules.

6-4 LEWIS STRUCTURES

Molecules of covalent compounds are represented by structural formulas called Lewis structures or electron-dot formulas. The usual symbols designate atoms, and dots are used for electrons.

6-5 RULES FOR WRITING LEWIS STRUCTURES

1. Distinguish between ionic and covalent bonds.

Example 6-1

Lithium fluoride, LiF, *ionic*

$Li + F \rightarrow Li^{+} + F^{-}$

Electrons are transferred, *not shared.*

Example 6-2

Carbon tetrafluoride, CF_4, *covalent*

```
    F
    ••
F : C : F
    ••
    F
```

Example 6-3

Sodium hydroxide, NaOH, *ionic*

Na^{+} $O{:}H^{-}$

Note that the bond within $O\text{-}H^{-}$ is covalent. Some compounds have both ionic and covalent bonds.

Example 6-4

$NaBH_4$

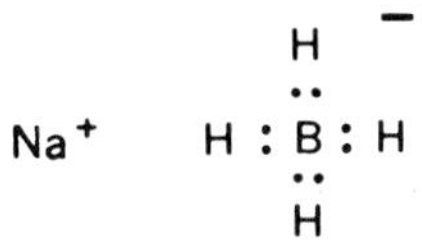

2. The proper arrangement of the atoms must be known.

   a. In some cases only one arrangement is possible.

Example 6-5

HCl

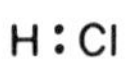

Example 6-6

NH_3

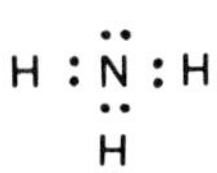

b. In other cases chemical and physical evidence must be available to support a structure.

Example 6-7

H_2O_2

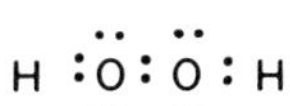

correct

Example 6-8

H_2O_2

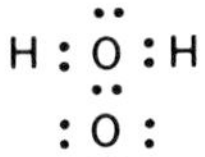

incorrect

Chemists may predict the Lewis structure of many compounds with confidence because of experience. Students must be given more information about compounds in order to know the proper arrangement. In many cases, more than one structure is possible, and even experienced chemists must have evidence to support the arrangement proposed.

The compound nitrosylchloride has the formula NOCl and is usually written that way. A determination of its structure has shown that the Cl is attached to N, and not O. The N—Cl distance is 1.95 Å, and the Cl—N—O angle is 116°. The appropriate Lewis structure is

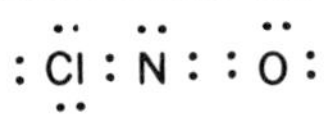

3. The correct number of electrons must be included. The correct number of electrons for each atom is given by the periodic group number.

Example 6-9

PCl_3

$5 + (3 \times 7) = 26$

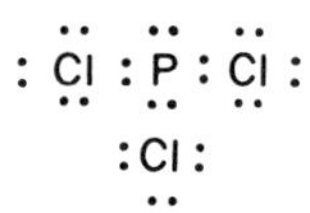

4. Arrange the electrons to provide full octets of electrons, if possible.

Example 6-10

Methane CH_4

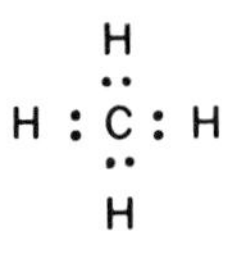

Example 6-11

NF_3

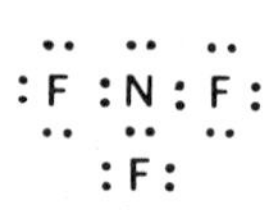

A covalent bond is usually represented by a dash (—). Students should use dots until they are thoroughly familiar with writing of structures. The use of dashes often leads to errors in the application of Rule 3.

6-6 MULTIPLE BONDS

Multiple bonds involve two or three pairs of electrons and are used when necessary to complete octets.

Example 6-12

Ethylene, C_2H_4

```
H : C :: C : H     A double bond (::)
    ..   ..
    H    H
```

Example 6-13

Nitrous acid, HNO_2

```
     ..
H :  O : N :
     ..  ::
         O :
         ..
```

Example 6-14

Acetylene, C_2H_2

```
H : C ::: C : H    A triple bond (:::)
```

Example 6-15

Nitrogen, N_2

```
: N ::: N :
```

6-7 COORDINATE COVALENT BONDS

When one atom supplies both electrons of the electron pair in the formation of a covalent bond, the bond is called a *coordinate covalent bond*. Other names for this type of bond are *coordinate bond* and *dative bond*. Once formed, the coordinate covalent bond cannot be distinguished from a normal covalent bond.

Example 6-16

NH_3BF_3

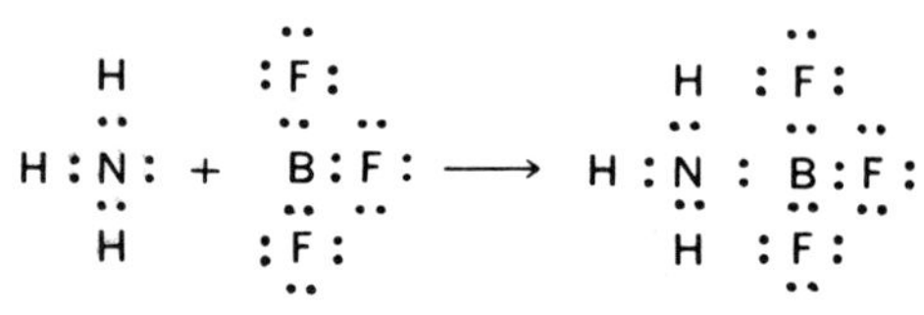

Example 6-17

Chloric acid, $HClO_3$

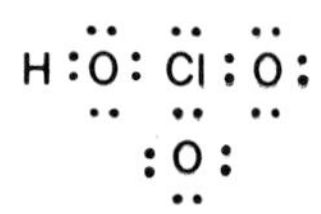

Two coordinate covalent bonds are present between Cl and the two O's not bonded to H.

6-8 DEPARTURES FROM THE OCTET RULE

1. Less than an octet

    a. Two electrons

Example 6-18

Hydrogen

Example 6-19

Hydrogen chloride

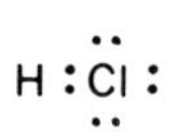

b. Quartet

Example 6-20

Beryllium chloride vapor

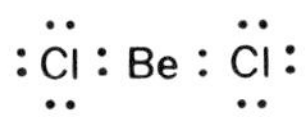

c. Sextet

Example 6-21

Boron trichloride

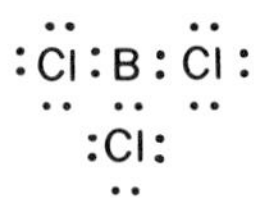

2. More than an octet

Example 6-22

Phosphorus pentachloride

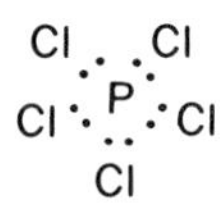

Example 6-23

Sulfur hexafluoride

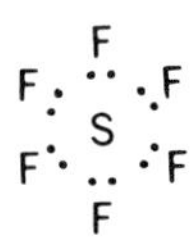

Example 6-24

Sulfur tetrafluoride

6-9 RESONANCE

In the case of many molecules, more than one Lewis structure that satisfies the rules may be drawn. Since we are attempting to represent a complex system of atoms and electrons with letters of the alphabet and dots, the limitations of our symbols cause this difficulty. An accurate representation of the bonds is not possible; more than one Lewis structure is possible. These are called *canonical structures*.

Example 6-25

Nitric acid, HNO_3

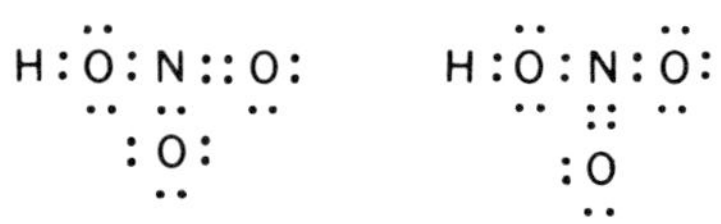

or

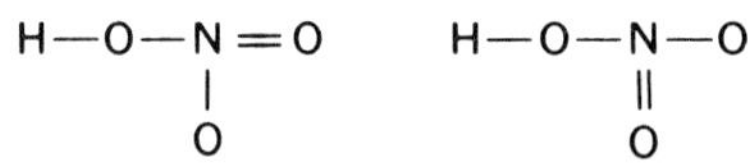

Note the positions of the double and single bonds in the two structures.

Another description of the molecules describes the bonds as being neither single nor double, but bonds that lie somewhere between. They may be represented as

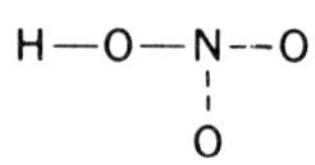

or more commonly

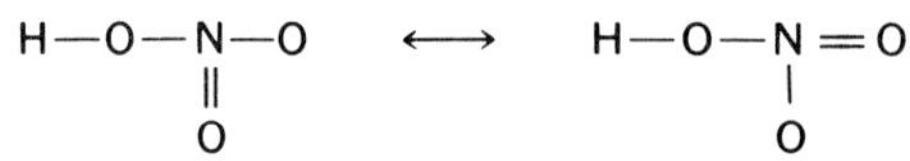

The two structures are called canonical forms, and the molecule is described as a *resonance hybrid* of the forms. The term *resonance* refers to this view of the bonding in molecules of this type. Resonance does not imply that the bonds are alternately double or single bonds; it is a description of the bond as lying somewhere between a single bond and a double bond.

6-10 FORMAL CHARGES

Formal charges are based on the assumption that the electrons in covalent bonds are shared equally. *Formal charges* are calculated by assigning to each atom one-half of the electrons shared in covalent bonds plus all of its unshared electrons. The *formal charge* on an atom is the charge it would have if it were an isolated ion with the same number of valence electrons. This does not mean that the charges are fully present, but that there is a nonuniform distribution of charge. The formal charge on each atom may be determined by

$$FC = VE - \frac{1}{2}BE - NE$$

FC = formal charge
VE = valence electrons
BE = bonding electrons
NE = nonbonding electrons

(The number of valence electrons is equal to the group number in the older periodic table. In the newer version it is equal to the group number for groups 1 and 2, but the group number minus 10 for groups 13 through 18.)

Example 6-26

Indicate the formal charges on each of the atoms in

a. CO b. NH_4^+ c. OH^- d. CCl_4 e. $HClO_2$

a. :C:::O: C 4-3-2 = -1 O 6-3-2 = +1 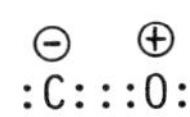

b. 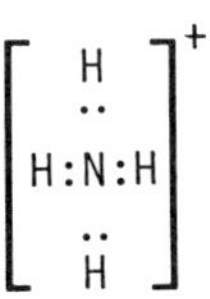 N 5-4-0 = +1 H 1-1-0 = 0 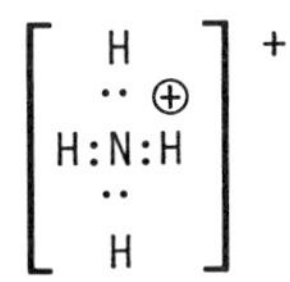

c. 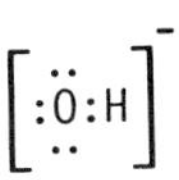 O 6-1-6 = -1 H 1-1-0 = 0 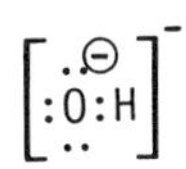

d. 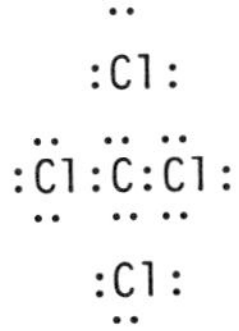 C 4-4-0 = 0 Cl 7-1-6 = 0

e. H:O:Cl:O: H 1-1-0 = 0 O 6-2-4 = 0 H:O:Cl:O: (⊕ on Cl, ⊖ on O)

O 6-1-6 = -1 Cl 7-2-4 = +1

6-11 THE SHAPE OF MOLECULES

The structure of molecules, i.e., the spatial arrangement of atoms in a molecule, is of genuine importance in chemistry. A sound understanding of the properties of compounds requires a knowledge of the structure of molecules. The spatial arrangement of atoms in the molecules of most compounds of the nontransitional elements has been determined by a variety of methods including X-ray and electron diffraction as well as infrared and nuclear magnetic resonance spectrometry. The distances between atoms and the angles between them can be determined with considerable precision.

A theoretical basis for the observed molecular structures and electron distributions, which requires an extensive knowledge of mathematics, is still under development. It is possible, however, to predict with reasonable accuracy the shapes of the molecules of compounds of the nontransitional elements with the aid of a few relatively simple principles. These principles were first proposed by Sidgwick and Powell, but they were applied and advanced by Gillespie and Nyholm. Students are urged to study the discussion of these principles.

[*Q. Rev. Chem. Soc.*, (*London*), *11*, 339 (1957); R. J. Gillespie, *J. Chem. Ed.*, *47*, 18 (1970)]

6-12 ELECTRON-PAIR REPULSION THEORY

In a molecule, the *arrangement* of the pairs of electrons about an atom, whether shared (bonding) or unshared (nonbonding), depends only on the number of pairs.

The arrangements of electron pairs given in Table 6-1 are those predicted if it is assumed that the electron pairs repel each other so that they arrange themselves as far apart as possible.

The arrangements of electron pairs listed in Table 6-1 lead to *regular shapes* of molecules when all pairs are bonded to other atoms or groups that are alike. When the atoms or groups are not alike or when some of the positions are occupied by *unshared pairs* or lone pairs, deviations from regular shapes result. The *regular shapes* to be expected are given in Table 6-2.

Table 6-3 shows the shapes of molecules with unshared pairs of electrons.

TABLE 6-1 Arrangement of Electron Pairs in Regular Structures

No. of electron pairs	Arrangement[a]
2	Linear
3	Triangular plane
4	Tetrahedron
5	Trigonal bipyramid
6	Octahedron
7	Pentagonal bipyramid

[a]The word "arrangement" is used to designate the positions of electron pairs. The word "shape" is used to designate the positions of atomic nuclei.

6-13 DEVIATIONS FROM REGULAR SHAPES

Deviations from the regular shapes can be attributed to decreasing order of repulsion among electron pairs; unshared pair-unshared pair > unshared pair-bond pair > bond pair-bond pair. This can be abbreviated u-u > u-b > b-b.

Applying the principle that the shape of a molecule depends on the number of pairs of electrons to methane leads to the tetrahedral structure. (See Fig. 6-1.) Here the H-C-H bond angle is 109°28'. In a similar fashion the ammonia molecule is expected to be a trigonal pyramid. The H-N-H bond angle observed is only 107°, however, not 109°28'. The greater u-b repulsions decrease the H-N-H angle below the regular 109°28'. The b-b repulsions are less than the u-b repulsions.

The H-O-H bond angle in water is 105° in the V-shape molecule, a lower value than that in NH_3 because there are two unshared pairs of electrons, leading to greater repulsion and a smaller angle.

When considering more complicated molecules, repulsions of electron pairs at angles greater than 90° may be ignored. Applying the principles to SF_4 leads to the prediction that the

TABLE 6-2 Regular Shapes of Molecules

No. of electron pairs	Arrangement of electron pairs	Shape of molecule	Example
2	Linear	Cl—Be—Cl	$BeCl_2$ (vapor)
3	Triangle	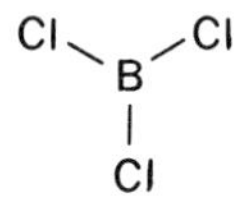	BCl_3
4	Tetrahedron	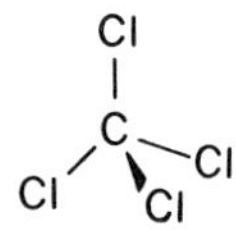	CCl_4
5	Trigonal bipyramid	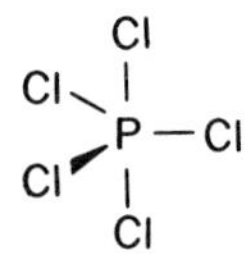	PCl_5
6	Octahedron	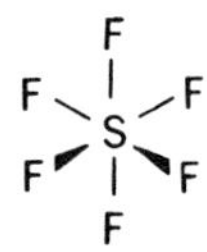	SF_6
7	Pentagonal bipyramid	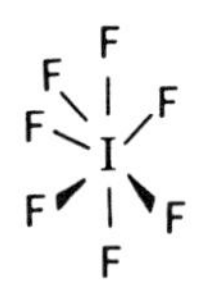	IF_7

five pairs of electrons will occupy the apices of a trigonal bipyramid. The position of the unshared pair will be either axial [Fig. 6-2(a)] or equatorial [Fig. 6-2(b)]. In Fig. 6-2(a) there are three u-b repulsions at 90°, and in Fig. 6-2(b) there are only two. The second structure is predicted.

TABLE 6-3 Shapes of Molecules with Unshared Pairs of Electrons

No. of electron pairs	No. of unshared pairs	Shape	Example
4	1	Trigonal pyramid	NH_3
	2	V shape	H_2O
5	1	Irregular tetrahedron	SF_4
	2	T shape	BrF_3
	3	Linear	ICl_2^-
6	1	Square-based pyramid	IF_5
	2	Square plane	ICl_4^-
7	1	Irregular octahedron	$(SbBr_6)^{3-}$

Three possibilities may be considered for the shape of ClF_3, as in Fig. 6-3(a), (b), and (c). The first has six u-b repulsions at 90°, the second has one u-u and three u-b repulsions at 90°, and the third has four u-b repulsions at 90°. The third is the preferred structure.

Example 6-27

Predict the shapes of the following molecules and ions. In each case write the Lewis structure before attempting to decide on a shape.

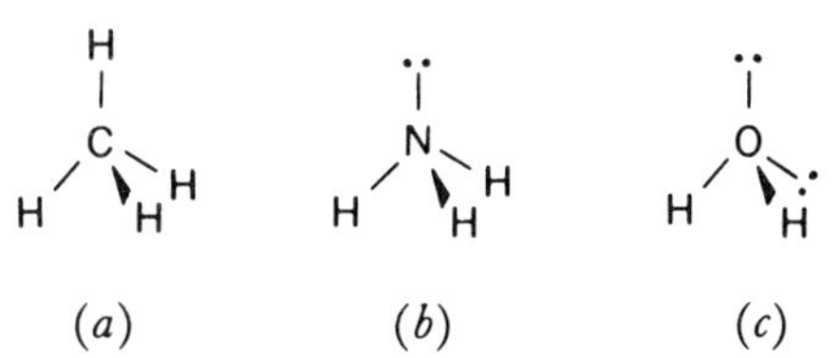

FIGURE 6-1 (a) Methane, (b) ammonia, and (c) water molecules.

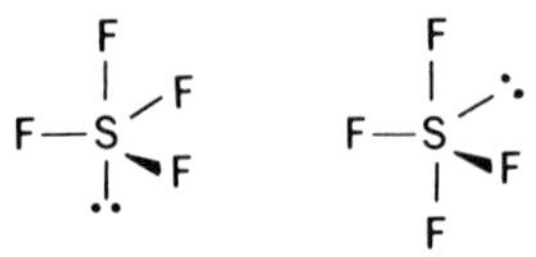

(a) (b)

FIGURE 6-2 (a) Axial and (b) equatorial positions of unshared pairs of complicated molecules.

a. BeH_2

H : Be : H

Two pairs of electrons—linear arrangement of electrons; therefore, the molecule is linear: H-Be-H.

b. BBr_3

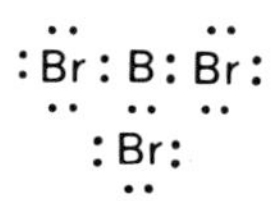

Three pairs of electrons—trigonal plane arrangement; therefore, the molecule is a triangular plane.

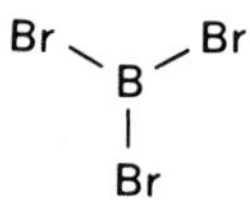

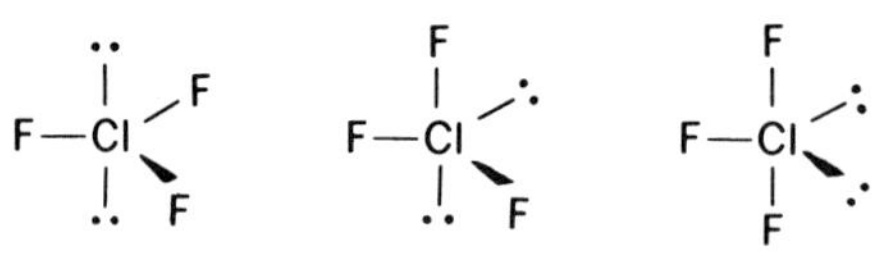

(a) (b) (c)

FIGURE 6-3 Possible shapes of ClF_3. (a) Six u-b repulsions at 90°, (b) one u-u and three u-b repulsions at 90°, and (c) four u-b repulsions at 90°.

c. H_2Te

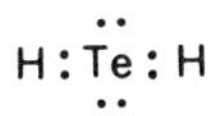

Four pairs of electrons—tetrahedral arrangement; therefore, the molecule is V-shaped.

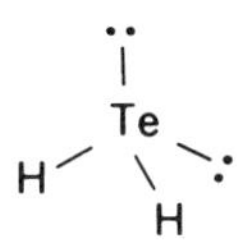

d. SiH_4

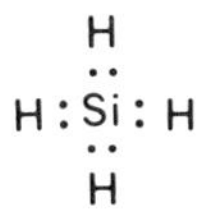

Four electron pairs—tetrahedral arrangement; therefore, the molecule is tetrahedral in shape.

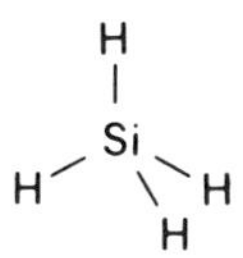

e. PH_3

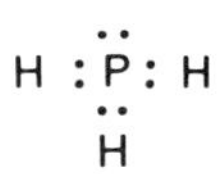

Four electron pairs—tetrahedral arrangement; therefore, the molecule is trigonal pyramidal in shape.

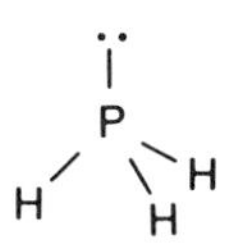

f. $TeCl_4$

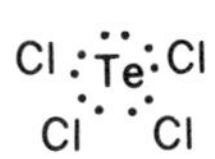

Five electron pairs—trigonal bipyramidal arrangement; therefore, the molecule is a distorted tetrahedron.

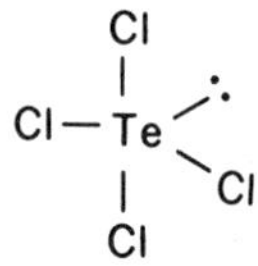

g. $CsIF_4$ $Cs^+IF_4^-$

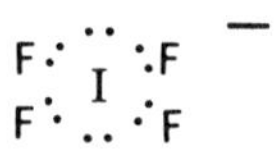

Six electron pairs—octahedral arrangement; therefore, a square-planar shape.

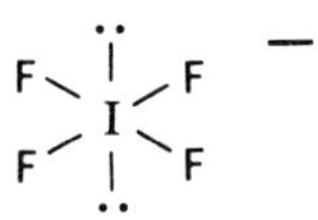

6-14 MOLECULES CONTAINING MULTIPLE BONDS

The shapes of molecules containing multiple bonds can also be predicted. The electron-pair double or triple bonds are assumed to occupy only one of the positions in the arrangements listed in Table 6-1.

Example 6-28

a. Carbon dioxide, CO_2

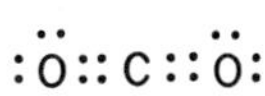 (linear)

There are two bonds and no unshared pairs.

b. Carbonyl fluoride, COF_2

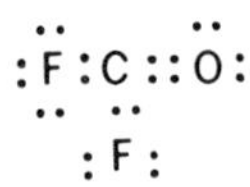 (triangular plane)

There are three bonds and no unshared pairs.

c. Ozone, O_3

```
 ..  ..  ..
:O : O::O :    (V shape)
 ..
```

There are two bonds and one unshared pair.

6-15 AN ALTERNATE REPRESENTATION OF TYPES OF STRUCTURES

The different types of molecules are often represented by the generalized symbols listed below, in which A = the central atom, X = attached ligands, and E = unshared electron pairs. A ligand is an atom or group of atoms connected to another atom by a covalent bond, e.g., Cl, CN, OH, etc.

Type of molecule	Example	Type of molecule	Example
AX_2	BeH_2, CO_2	AX_2E	SO_2
AX_3	BF_3, CO_3^{2-}	AX_3E	NF_3
AX_4	NH_4^+, SiH_4	AX_2E_2	H_2O
AX_5	PF_5	AX_4E	SF_4
AX_6	SF_6	AX_3E_2	BrF_3
		AX_5E	IF_5
		AX_4E_2	XeF_4
		AX_2E_4	XeF_2

6-16 BONDING ORBITAL TYPES

Type of molecule	Bonding or hybrid orbital	
AX_2	sp	Linear
AX_3	sp^2	Triangular
AX_4	sp^3	Tetrahedral
AX_5	sp^3d	Trigonal bipyramidal
AX_6	sp^3d^2	Octahedral

PROBLEMS

6-1 Write Lewis structures for:

a. Carbon tetrabromide
b. Silane, SiH_4
c. Oxygen difluoride
d. Hydrogen selenide, H_2Se
e. Ammonium chloride
f. Hypochlorous acid, HOCl
g. Sulfuric acid
h. Perchloric acid
i. Hypobromous acid
j. Sodium sulfite
k. Carbon dioxide
l. Dichlorine monoxide
m. Nitrogen trifluoride
n. Phosphine, PH_3
o. Calcium nitrite

6-2 Write Lewis structures for:

a. Mercury(II) chloride
b. Boron tribromide
c. Tellurium tetrachloride
d. Antimony pentachloride
e. Selenium hexafluoride

6-3 Write Lewis structures for the following compounds. Include canonical forms (resonance forms), if they are appropriate.

a. Ozone
b. Carbon monoxide
c. Formaldehyde, H_2CO
d. Lithium carbonate
e. Dinitrogen monoxide, NNO
f. Sodium nitrate
g. Sulfur dioxide, OSO
h. Hydrazoic acid, HNNN
i. Potassium nitrite
j. Sulfur trioxide

6-4 Write Lewis structures for the compounds listed below.

a. Diboron tetrachloride

B–B = 1.75 Å   B–Cl = 1.72 Å

Cl–B–Cl angle = 122°   Cl–B–B angle = 119°

[M. J. Linevsky, E. R. Shull, D. E. Mann, and T. Wartik, *J. Am. Chem. Soc.*, *75*, 3287 (1953)]

b. Cyanogen $(CN)_2$. The molecules are symmetrical and linear.

C–C = 1.37 Å   C–N = 1.16 Å

[L. Pauling, H. D. Springall, and K. L. Palmer, *J. Am. Chem. Soc.*, *61*, 933 (1939)]

c. Nitryl fluoride, NO_2F. The molecules are planar.

F—N = 1.35 Å N—O = 1.23 Å O—N—O angle = 125°

[D. F. Smith and D. W. Magnuson, *Phys. Rev.*, *87*, 226 (1952)]

d. $POCl_3$

P—O = 1.45 Å P—Cl = 1.99 Å Cl—P—Cl angle = 103.5°

[G. R. Badgley and R. L. Livingston, *J. Am. Chem. Soc.*, *76*, 261 (1954)]

e. $Na_2S_2O_4$

S—S = 2.389 Å

[J. D. Dunitz, *J. Am. Chem. Soc.*, *78*, 878 (1956)]

f. Formyl fluoride, HCOF

C—F = 1.351 Å C—O = 1.192 Å O—C—F angle = 121.9°

[M. E. Jones, K. Hedberg, and C. Shomaker, *J. Am. Chem. Soc.*, *77*, 5278 (1955)]

6-5 A compound prepared by the reaction of fluorine with sulfur trioxide contained 27.0% S, 32.0% F, and the remainder oxygen. The molecular weight calculated from gas-density measurements was 118.6. One of the fluorine atoms is bonded to oxygen. Write a Lewis structure representing the molecules of the compound. How is the structure related to that of sulfuric acid?

6-6 Draw diagrams representing the expected shapes of the molecules of each of the following compounds. Name the geometrical figures in each case.

a. Hydrogen selenide
b. Oxygen difluoride
c. Ammonium ion
d. Dichlorine monoxide
e. Mercury(II) chloride
f. Boron triiodide
g. Nitrogen trifluoride
h. Selenium tetrachloride
i. Selenium hexafluoride
j. Arsenic pentafluoride

6-7 Predict the shapes of the molecules of the following compounds.

a. Hypobromous acid
b. Selenium dichloride
c. Silicon tetrafluoride
d. Tetrahydroborate ion, BH_4^-
e. Antimony trichloride
f. Tin tetrachloride
g. Gallium triiodide [*Inorg. Chem.*, *4*, 1775 (1965)]
h. Phosphorus pentafluoride
i. Bromine trifluoride
j. Tellurium hexafluoride

6-8 Draw diagrams representing the shapes of the molecules of the compounds listed.

a. Hydrogen cyanide, HCN
b. Phosgene (carbonyl chloride), $COCl_2$
c. Sulfur dioxide
d. Phosphorus oxychloride, $POCl_3$
e. Nitrate ion
f. Thionyl fluoride, SOF_2
g. Nitrosyl chloride, NOCl
h. Nitryl fluoride, NO_2F

6-9 A crystal structure analysis of potassium selenocyanate was made by D. D. Swank and R. D. Willett. The C–N distance was recorded as 1.117 Å, the Se–C distance as 1.829 Å, and the Se–C–N angle as 178.8°. Write a Lewis structure for the compound. Show that the structure is consistent with the shape you might predict from electron-pair repulsions.
[*Inorg. Chem.*, *4*, 501 (1965)]

6-10 A 14:1 mixture of F_2 and ClF_3 was heated to 350°C for 1 hr at 250 atm. The infrared spectrum of the product was similar to that of BrF_5. Write a Lewis structure for the expected product of the reaction. Draw a diagram representing the shape of the molecules.
[D. F. Smith, *Science*, *141*, 1039 (1964)]

6-11 From an electron-diffraction investigation, Harvey and Bauer determined the S–S distance to be 2.21 Å in the compound disulfur decafluoride. Write a Lewis structure for the molecule and draw a diagram representing the shape.
[*J. Am. Chem. Soc.*, *75*, 2840 (1953)]

6-12 To a solution of 13.0 g of BCl_3 in liquid SO_2 as a solvent at -30°, 26 g of anhydrous sodium thiocyanate, NaSCN, was added. Sodium chloride and 7.5 g of a colorless liquid were obtained. *Analysis* found: C, 19.55; N, 22.92.
[D. B. Sowerby, *J. Am. Chem. Soc.*, *84*, 1831 (1962)]

a. What was the liquid product of the reaction?
b. Which of the starting materials was present in excess?
c. Write a Lewis structure for the liquid product.

6-13 Thionyl fluoride, SOF_2, was treated with silver(II) fluoride and an oxygen-containing product was formed. *Analysis* found: S, 25.6; F, 62.0. The molecular mass from the mass spectrum was 124.

[F. B. Dudley, G. H. Cady, and D. F. Eggers, Jr., *J. Am. Chem. Soc.*, *78*, 1554 (1956)]

a. Write a Lewis structure representing the molecules of the compound.
b. Draw a diagram representing the shape of the molecules.

6-14 The fluorination of a gaseous mixture of SNF and SNF_2 gave as a product a colorless gas. *Analysis* found: F, 54.4; S, 30.4. Molecular mass; 1.3352 g occupied 471.3 mL at 499.3 mm and 290.5 K.

[O. Glemser and H. Schroder, *Z. Anorg. Allgem. Chem.*, *284*, 97 (1956)]

The infrared spectrum of the compound was similar to that of OPF_3.

[H. Richert and O. Glemser, *Z. Anorg. Allgem. Chem.*, *307*, 328 (1961)]

Predict the shape of the molecules of the compound. Compare your prediction with the results of a microwave investigation of W. H. Kirchoff and E. B. Wilson of Harvard.

[*J. Am. Chem. Soc.*, *84*, 334 (1962)]

6-15 Draw diagrams representing the shapes of the molecules of the noble gas compounds:

a. Xenon tetrafluoride
b. Xenon difluoride
c. Xenon trioxide
d. Krypton difluoride
e. Sodium perxenate

6-16 Write Lewis structures for the following compounds that have peroxy linkages.

a. Hydrogen peroxide
b. Dioxygen difloride
c. CF_3O_2F
d. CF_3O_2H
e. $(CF_3)_2O_2$

[C. J. Marsden, D. D. Des Marteau, and L. S. Bartell, *Inorg. Chem.*, *16*, 2 (1977)]

6-17 Draw a diagram representing the structure of CF_3O_2H.

6-18 Arsenic pentafluoride was shown to react with anhydrous hydrogen fluoride forming the anion AsF_6^-. If a small amount of water was present, it was possible to separate solid H_3OAsF_6.

a. Write an appropriate Lewis structure for the compound.
b. Draw an appropriate diagram representing its structure.

[K. O. Christe, C. J. Schack, and R. D. Wilson, *Inorg. Chem.*, *14*, 2224 (1975)]

6-19 Write Lewis structures and draw diagrams representing the structures of PF_2I and PF_2CN.
[A. W. Cox, Jr., A. J. ZoZulin, J. D. Odum, and J. R. Dung, *Inorg. Chem.*, *16*, 2048 (1977)]

6-20 Ammonia gas reacts with boron trichloride forming the white solid adduct NH_3BCl_3. Write a Lewis structure and draw a diagram representing its structure.

6-21 Ammonia gas reacts with silicon tetrafluoride gas by donating a pair of electrons to form a white solid adduct. Write a Lewis structure and draw a diagram representing the structure of the adduct.
[B. S. Ault, *Inorg. Chem.*, *20*, 2817 (1981)]

6-22 Boron trifluoride reacts with anhydrous hydrogen fluoride forming fluoroboric acid, which may be neutralized by ammonia gas yielding ammonium tetrafluoroborate. Write a Lewis structure and draw a diagram representing the structure of ammonium tetrafluoroborate.

6-23 Many experiments require a solvent that will dissolve salts and yield solutions that conduct electricity. A commonly used solvent for this purpose is the organic compound acetonitrile, CH_3CN. Write the Lewis structure and draw a diagram representing its structure.

6-24 Assign the appropriate molecular type (Section 6-14) to each of the compounds in Problem 6-6.

6-25 What type of hybrid orbital may be assigned to the central atom of each of the compounds in Problem 6-7?

7
SOLUTIONS

7-1 EXPRESSING CONCENTRATIONS OF SOLUTIONS

The method of expressing the concentration of a solution depends on the intended use of the solution. For many uses it is sufficient to express the concentration as a *weight percent*: the weight of solute per 100 parts by weight of solution.

Example 7-1

An experiment required 250 g of a 23.2% aqueous solution of glucose. What weight of glucose was required? What weight of water?

A 23.2% solution means that 1 g of solution contains 0.232 g of glucose. The unit factor is thus:

$$\frac{0.232\ \text{g glucose}}{\text{g solution}}$$

$$250\ \text{g solution} \times \frac{0.232\ \text{g glucose}}{\text{g solution}} = 58.0\ \text{g glucose}$$

The weight of water is found by subtracting the weight of glucose from the total weight:

$$250\ \text{g solution} - 58\ \text{g glucose} = 192\ \text{g water}$$

Other relationships used to express very dilute concentrations are parts per million (ppm) and parts per billion (ppb). Both are usually derived from weight fractions. Often the symbol (w/w) is used to show this.

If the fraction by weight of solute in solution is multiplied by a million (10^6) the concentration is expressed as ppm, while if the multiplier is a billion (10^9) the concentration is expressed as ppb.

Example 7-2

A 0.500-g sample of fish was dissolved in 50 mL of concentrated nitric acid; 1 mL of this solution was taken, neutralized, and diluted to 25 mL. This final solution was found to contain 0.04 ng of mercury per milliliter of solution. Express the mercury in the fish in ppm and ppb.

$$0.04 \times 10^{-9} \frac{\text{g Hg}}{\text{mL}} \times 25 \text{ mL} = 1 \times 10^{-9} \text{ g Hg}$$

This came from 1 mL of the 50-mL solution, so

$$50 \text{ mL} \times 1 \times 10^{-9} \frac{\text{g Hg}}{\text{mL}} = 50 \times 10^{-9} \text{ g Hg}$$

There were 50×10^{-9} g of mercury in 0.500 g of fish.

$$\frac{50 \times 10^{-9}}{0.500} \times 10^{9} = 100 \text{ ppb}$$

$$= 0.10 \text{ ppm}$$

In much of the recent work done in water pollution studies, a weight/volume unit of concentration is used. This is also expressed as parts per million or parts per billion, based on the number of milliliters of solution. Here the symbol (w/v) is used to distinguish this from the weight fraction unit previously discussed.

Example 7-3

The concentration of copper in salt water has been measured by atomic absorption spectroscopy. A typical result in the Atlantic Ocean, in μmol/liter, is 0.030. Express this in ppm and ppb.
[B. P. Faricand, R. R. Sawyer, S. G. Ungar, and S. Adler, *Geochim. Cosmochim. Acta, 26*, 1023 (1962)]

$$1 \text{ mol Cu} = 63.55 \text{ g}$$

$$1\ \mu\text{g} = 10^{-6} \text{ g}$$

Using the unit factors we can convert μmol/L to g solute per mL of solution.

$$1 \frac{\mu\text{mol}}{\text{Liter}} \times \frac{1 \times 10^{-6}\ \text{g}}{\mu\text{g}} \times \frac{63.55\ \text{g Cu}}{1\ \text{mol}} \times \frac{1\ \text{liter}}{1000\ \text{mL}} = \frac{63.55 \times 10^{-9}\ \text{g Cu}}{\text{mL}}$$

$$1 \frac{\text{mol}}{\text{Liter}} = 63.55 \times 10^{-9}\ \text{g} \frac{\text{Cu}}{\text{mL}}$$

Using this unit factor:

$$0.030 \frac{\text{mol}}{\text{L}} \times \frac{63.55 \times 10^{-9}\ \text{g Cu/mL}}{1\ \text{mol/L}} = 1.09 \times 10^{-9}\ \text{g} \frac{\text{Cu}}{\text{mL}}$$

To convert to ppb multiply the fraction by 10^9.

$$1.09 \times 10^{-9}\ \text{g} \frac{\text{Cu}}{\text{mL}} \times 10^9 = 1.09\ \text{ppb}$$

To convert to ppm multiply the fraction by 10^6.

$$1.09 \times 10^{-9}\ \text{g} \frac{\text{Cu}}{\text{mL}} \times 10^6 = 0.00109\ \text{ppm}$$

For chemical experiments and calculations it is usually more convenient to use solutions based on the natural unit of chemistry, the mole.

7-2 MOLAR SOLUTIONS

A molar solution (M) contains 1 mol of solute per liter of solution. To prepare a molar solution, the correct quantity of solute is dissolved in less than sufficient solvent to prepare the solution. When the solute has dissolved, additional solvent is added to make the proper volume of solution. The usual vessel for accurate preparation of solutions is the volumetric flask. (See Fig. 7-1.)

The symbol 3 M HNO_3 refers to a three-molar solution of nitric acid, a solution that contains 3 mol of nitric acid in 1 L of solution. A 0.10 M H_2S solution contains 0.10 mol of H_2S per liter of solution.

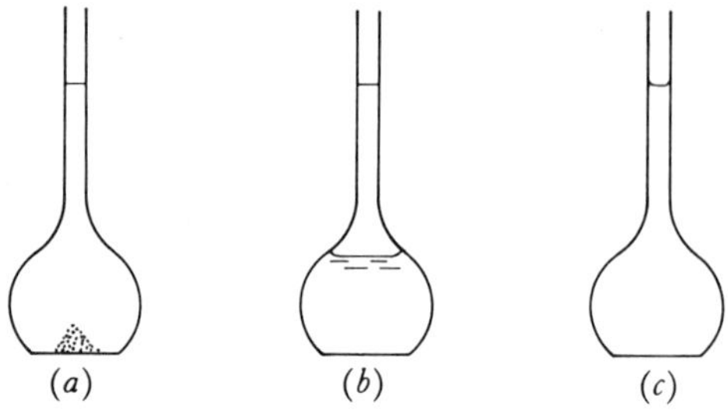

FIGURE 7-1 Preparation of solutions: (a) accurately weighed solute, (b) solvent to dissolve, (c) fill to mark.

Example 7-4

Calculate the mass of solute in grams required to make 1 L of 1.72×10^{-2} M $HgCl_2$.

$$\begin{array}{lrr} HgCl_2 & 1 \times 200.6 = & 200.6 \\ & 2 \times \ \ 35.4 = & \underline{70.8} \\ & & 271.4\ g/mol \end{array}$$

$$1.72 \times 10^{-2}\ \frac{mol}{L} \times 271.4\ \frac{g}{mol} = 4.66\ \frac{g}{L}$$

The concentration of a solution in moles per liter is referred to as the *molarity* of the solution.

Example 7-5

What is the molarity of a solution that contains 171 g of sucrose (sugar) in 2.00 L of solution?

Since the molarity is the number of moles of solute in 1 L of solution, the first step is to calculate the number of moles of sucrose in 171 g.

Find the molecular mass of the sugar:

$$\begin{array}{llr} C_{12}H_{22}O_{11} & C\ \ 12 \times 12 = & 144 \\ & H\ \ 22 \times \ \ 1 = & 22 \\ & O\ \ 11 \times 16 = & \underline{176} \\ & & 342\ g/mol \end{array}$$

$$171\ \not{g} \times 1\ mol/342\ \not{g} = 0.500\ mol\ of\ sugar$$

Since this amount was in 2 L of solution, the molarity is

$$\frac{0.500 \text{ mol sugar}}{2.00 \text{ L}} = 0.250 \text{ M}$$

It is often much more convenient to use the smaller unit of volume, the milliliter. Since 1 mL = 1/1000 L, the molarity of a solution may be expressed in terms of milliliters and *millimoles* (mmol).

$$\text{Molarity} = \frac{\text{millimoles solute}}{\text{milliliters solution}}$$

The solution in Example 7-5 contains 0.250 mmol/mL. A 0.100 M H_2SO_4 solution will contain 0.100 mmol of H_2SO_4 per milliliter of solution

$$\frac{\text{mole} \times 1000 \text{ mmol/mol}}{\text{liter} \times 1000 \text{ mL/L}} = \frac{\text{mmol}}{\text{mL}}$$

7-3 CALCULATION OF QUANTITY OF SOLUTE

Frequently it is necessary to calculate the quantity of solute in a given volume of solution from a knowledge of its concentration.

Example 7-6

What mass of hydrogen chloride is present in 1.88 L of 0.150 M HCl solution?

Each liter contains 0.150 mol HCl, and the volume is 1.88 L.

$$0.150 \frac{\text{mol HCl}}{\text{L}} \times 1.88 \text{ L} = 0.282 \text{ mol HCl}$$

The molecular mass of HCl is 36.4 g/mol

$$0.282 \text{ mol HCl} \times 36.4 \text{ g/mol HCl} = 10.3 \text{ g HCl}$$

in general:

Number of moles of solute = molarity × volume in liters

and

Mass of solute in grams = molarity × volume × molecular mass

Example 7-7

Calculate the quantity of potassium nitrate present in 200 mL of 0.125 M solution.

Mass solute = molarity × volume × mole mass

KNO_3	$1 \times 39.1 =$	39.1
	$1 \times 14.0 =$	14.0
	$3 \times 16.0 =$	48.0
		101.1 g/mol

200 mL = 0.200 L

$$\text{Mass solute} = 0.125\,\frac{\text{mol}}{\text{L}} \times 0.200\,\text{L} \times 101\,\frac{\text{g}}{\text{mol}} = 2.52\text{ g}$$

If the volume is left as milliliters, then the product of molarity and volume will give the number of *millimoles* of solute. The mass of 1 mmol rather than 1 mol will be required.

$$1\text{ mmol } KNO_3 = \frac{101}{1000} = 0.101\text{ g}$$

$$\text{Mass solute} = 0.125\,\frac{\text{mmol}}{\text{mL}} \times 200\,\text{mL} \times 0.101\,\frac{\text{g}}{\text{mmol}} = 2.52\text{ g}$$

Example 7-8

Calculate the mass of solute present in 15.2 mL of 0.212 M HBr.

HBr	1.01
	79.91
	80.92 g/mol

$$1\text{ mmol HBr} = \frac{80.9}{1000} = 0.0809\text{ g}$$

$$\text{Mass solute} = 0.212\,\frac{\text{mmol}}{\text{mL}} \times 15.2\,\text{mL} \times 0.0809\,\frac{\text{g}}{\text{mmol}} = 0.260\text{ g}$$

Example 7-9

What is the mass of lithium chloride, in grams, in 61.8 mL of 1.03 M LiCl?

LiCl 6.9
  35.4
  42.3 g/mol

$$1 \text{ mmol LiCl} = \frac{42.3}{1000} = 0.0423 \text{ g}$$

$$\text{Mass LiCl} = 1.03 \frac{\text{mmol}}{\text{mL}} \times 61.8 \text{ mL} \times 0.0423 \frac{\text{g}}{\text{mmol}} = 2.69 \text{ g}$$

Example 7-10

What is the molarity of concentrated sulfuric acid that contains 95.2% H_2SO_4? The density of the solution is 1.53 g/mL.

$$\text{Mass 1 L} = 1.53 \frac{\text{g}}{\text{mL}} \times 1000 \frac{\text{mL}}{\text{L}} = 1530 \frac{\text{g}}{\text{L}}$$

$$\frac{\text{Mass } H_2SO_4}{\text{L}} = \frac{\text{mass}}{\text{L}} \times \frac{\% \; H_2SO_4}{100}$$

$$= 1530 \frac{\text{g}}{\text{L}} \times 0.952 = 1450 \text{ g}$$

H_2SO_4 2 × 1.0 = 2.0
  1 × 32.1 = 32.1
  4 × 16.0 = 64.0

molecular mass = 98.1 g/mol

$$\text{Molarity} = \frac{1450 \text{ g/L}}{98.1 \text{ g/mol}} = 14.8 \frac{\text{mol}}{\text{L}} = 14.8 \text{ M}$$

Example 7-11

Calculate the molarity of a solution of hydrochloric acid from the data that appeared on the label of the bottle in which it was received from the manufacturer: Assay 36.5%, sp. gr. 1.1854. Since the molarity is based on 1 L of solution, base your calculations on 1 L.

$$\text{Mass 1 L} = 1000\ \frac{\text{mL}}{\text{L}} \times 1.18\ \frac{\text{g}}{\text{mL}} = 1180\ \frac{\text{g}}{\text{L}}$$

$$\text{Mass HCl/L} = 1180 \times 0.365 = 430\ \frac{\text{g HCl}}{\text{L}}$$

$$\begin{array}{ll} \text{HCl} & 1.0 \\ & \underline{35.4} \\ & 36.4\ \text{g/mol} \end{array}$$

$$\frac{430}{36.4} = 11.8\ \frac{\text{mol HCl}}{\text{L}} = 11.8\ \text{M}$$

7-4 DILUTION

The dilution of more concentrated solutions is an everyday experience in the laboratory. The calculation of the volume of concentrated solution required to make a given volume of dilute solution is based on the fact that the quantity of solute in the two solutions is the same. The concentration (moles of solute per liter of solution) varies, but the number of moles remains the same:

No. moles solute in dilute soln. = no. moles solute in conc. soln.

$$M_d \times V_d = M_c \times V_c$$

where M_d is the molarity of the dilute solution, V_d is the volume of the dilute solution, M_c is the molarity of the concentrated solution, and V_c is the volume of the concentrated solution.

Example 7-12

Calculate the volume of 12.0 M HCl required to make 250 mL of 1.50 M HCl solution by dilution with water.

$$M_d \times V_d = M_c \times V_c$$

millimoles HCl = millimoles HCl

$$1.50 \times 250 = 12.0 \times V_c$$

$$V_c = \frac{1.50\ \cancel{M} \times 250\ mL}{12.0\ \cancel{M}} = 31.2\ mL\ of\ 12.0\ M\ HCl$$

Example 7-13

What is the concentration in moles per liter of a solution made by adding 20.0 mL of H_2O to 15.0 mL of 2.50 M HNO_3?

$$M_d \times V_d = M_c \times V_c$$

$$M_d \times (20.0 + 15.0) = 2.50 \times 15.0$$

$$M_d = \frac{2.50 \times 15.0}{35.0} = 1.07\ \frac{mmol}{mL}$$

$$1.07\ \frac{mmol}{mL} = 1.07\ \frac{mol}{L}$$

7-5 MOLAL SOLUTIONS

The concentration of a solution based on the volume varies with temperature. For this and other reasons, solutions whose concentrations are based on the mass of solute and solvent are sometimes preferred. A *molal* solution contains 1 mol of solute per 1000 g of *solvent*. Note carefully that the basis is 1000 g of *solvent*, not *solution*.

$$\text{Molality} = \frac{\text{moles solute}}{\text{1000 g solvent}}$$

Example 7-14

Calculate the molality of a 16.2% solution of alcohol in water.

A 16.2% solution contains 16.2 g of solute (alcohol) and 100.0 - 16.2 = 83.8 g of solvent (water).

Since the basis of the molal solution is 1000 g of solvent, change the mass of solute to that basis. Multiply 16.2 by the fraction 1000/83.8.

$$16.2 \times \frac{1000}{83.8} = 194 \text{ g}$$

This is the mass of solute that would be present in the solution if the mass of solvent were 1000 g. Next calculate the number of moles of solute in 1000 g of solvent.

$$\begin{array}{lll} C_2H_6O & 2 \times 12.0 & = 24.0 \\ & 6 \times 1.01 & = 6.06 \\ & 1 \times 16.0 & = \underline{16.0} \\ & & 46.1 \text{ g/mol} \end{array}$$

$$\frac{194}{46.1} = 4.21 \text{ mol/1000 g, or } 4.21 \text{ m(molal)}$$

Example 7-15

Calculate the molality of a concentrated HCl solution that contains 36.5% HCl.

The percent HCl is the number of grams of HCl in 100 g of solution. Therefore, in 100 g of solution, the mass of water is given by 100 less the percent HCl.

$$100.0 - 36.5 = 63.5 \text{ g } H_2O \text{ in } 100 \text{ g of solution}$$

Thus, there are 36.5 g of HCl in 63.5 g of H_2O. To find the mass of HCl in 1000 g of water we multiply by the factor

$$\frac{36.5 \text{ g HCl}}{63.5 \text{ g } H_2O}$$

$$1000 \text{ g } \cancel{H_2O} \times \frac{36.5 \text{ g HCl}}{63.5 \text{ g } \cancel{H_2O}} = 575 \text{ g HCl}$$

The molecular mass of HCl is 36.4, so the number of moles of HCl is

$$\frac{575 \cancel{\text{g}}}{36.4 \cancel{\text{g}}/\text{mol}} = 15.8 \text{ mol} = 15.8 \text{ m}$$

This is the molality, since the 15.8 mol represents the number of moles of HCl in 1000 g of H_2O. Compare this result with that of Example 7-11.

7-6 MOLE FRACTION

The mole fraction X of a solution may be expressed as the mole fraction of solute or the mole fraction of solvent.

$$X \text{ solute} = \frac{n}{n + n^0}$$

$$X \text{ solvent} = \frac{n^0}{n^0 + n}$$

where n is the number of moles of solute and n^0 is the number of moles of solvent in the solution.

Example 7-16

Calculate the mole fraction of solute in a 2.50 m aqueous solution. This solution contains 2.50 mol of solute in 1000 g of H_2O. The number of moles of H_2O is given by 1000/18.0, since 1 mol of H_2O weighs 18.0 g.

$$X = \frac{2.50}{2.50 + (1000/18.0)} = \frac{2.50}{2.50 + 55.6}$$

$$X = \frac{2.50 \text{ mol}}{58.1 \text{ mol}} = 0.0430$$

Example 7-17

Calculate the mole fraction of solvent in a solution containing 89.9 g of mannitol, $C_6H_{14}O_6$, in 1000 g of H_2O.

[J. C. W. Frazer, B. F. Lovelace, and T. H. Rogers, *J. Am. Chem. Soc.*, *42*, 1801 (1920)]

$$\begin{array}{lrl} C_6H_{14}O_6 & 6 \times 12.0 = & 72.0 \\ & 14 \times 1.0 = & 14.0 \\ & 6 \times 16.0 = & \underline{96.0} \\ & & 182.0 \text{ g/mol} \end{array}$$

$$H_2O \quad 2 \times 1.0 = 2.0$$
$$1 \times 16.0 = \underline{16.0}$$
$$18.0 \text{ g/mol}$$

$$\frac{89.9}{182.0} = 0.494 \text{ mol } C_6H_{14}O_6$$

$$\frac{1000}{18.0} = 55.6 \text{ mol } H_2O$$

$$X \text{ solvent} = \frac{55.6}{55.6 + 0.494} = \frac{55.6}{56.1} = 0.990$$

PROBLEMS

7-1 In order to determine the concentration of phosphorus present in a sample using uv spectroscopy, a solution was prepared containing 2.200 g of KH_2PO_4 per liter of solution. A stock solution was made by diluting 20.00 mL of the first solution to 1 L with distilled water. Express the concentration of P in the stock solution in ppm.

[T. Hurford and D. Boltz, *Anal. Chem.*, *40*, 379 (1968)]

7-2 In analyzing for trace amounts of chromium, a standard stock solution was prepared by dissolving 1.414 g of dry primary standard $K_2Cr_2O_7$ in 500 mL of distilled water. How many ppm of Cr are present in the standard solution?

[J. A. Hurlbert and C. D. Chriswell, *Anal. Chem.*, *43*, 465 (1971)]

7-3 A boron standard (w/v) was prepared by dissolving 5.7200 g of boric acid to 500 mL of solution. What is the concentration of boron in this standard in ppm?

[Castillo, J. R., Mir, J. M., Bendicho, C., and Martinez, C., *At. Spectrosc.*, *6*, 152 (1985)]

7-4 Calculate the mass of solute required to make 1 L of each of the following solutions.

a. 0.475 M sodium carbonate
b. 0.9251 M $NaNO_3$
c. 0.047 M copper(II) sulfate
d. 0.0531 M zinc chloride
e. 0.896 M cesium nitrate

7-5 Calculate the mass of solute required to make 500 mL of each of the following solutions.

a. 2.12 M sodium hydroxide
b. 8.5 M KCNS
c. 2.38 mmol lead nitrate
d. 1.03×10^{-4} M cadmium nitrate
e. 0.0571 M aluminum sulfate

7-6 How many grams of each of the following must be weighed to make the respective solution?

a. Potassium hydroxide, 750 mL of 0.200 M KOH
b. $CuSO_4 \cdot 5H_2O$, 200 mL of 2.5 M $CuSO_4$
c. $Pb(NO_3)_2$, 500 mL of 1000 ppm of Pb^{2+}
d. $Na_2S_2O_3 \cdot 2H_2O$, 3.00 L of 0.100 M $Na_2S_2O_3$
e. $NaClO_4 \cdot H_2O$, 50.0 mL of 1.25×10^{-4} M $NaClO_4$

7-7 Calculate the molarity of each of the following solutions:

a. 18.0% $CaCl_2$, sp. gr. 1.1578
b. 32.0% $(NH_4)_2SO_4$, sp. gr. 1.1836
c. 14.0% KBr, sp. gr. 1.1070

7-8 Calculate the volume of the concentrated reagent required to make the corresponding dilute solution in each instance.

a. 15.0 M NH_3 to make 200 mL of 1.84 M NH_3
b. 12.0 M HCl to make 300 mL of 4.0 M HCl
c. 16.0 M HNO_3 to make 50.0 mL of 0.250 M HNO_3
d. 12.3 M HCl to make 100 mL of 1.02×10^{-3} M HCl
e. 15.2 M HNO_3 to make 250 mL of 2.00 mmol HNO_3

7-9 What is the final concentration of each of the following solutions, when the dilutions, in water, as indicated are made?

a. 3.00 mL of concentrated HCl (12.0 M) diluted to 250 mL
b. 2.50 mL of concentrated HNO_3 (16.0 M) diluted to 50 mL
c. 1 mL of 1000 ppm Ca^{2+} diluted to 250 mL
d. 24.5 mL of concentrated NH_3 (15.7 M) diluted to 200 mL

7-10 How many milliliters of concentrated H_2SO_4 would be required to make 20.0 L of 6.00 M acid? The concentrated sulfuric acid contains 95.0% H_2SO_4, and its specific gravity is 1.8407.

7-11 Calculate the molarity of concentrated hydrochloric acid that contains 38.0% HCl. The specific gravity is 1.1885.

7-12 What is the molarity of concentrated nitric acid if its specific gravity is 1.4251 and it contains 71.15% HNO_3?

7-13 The assay of a bottle of aqueous ammonia reads: "Min. 23.0%, Max. 30.0%." The specific gravity is given as: "Min. 0.8957, Max. 0.9016." What are the limits of the molarity of the solution?

7-14 Calculate the molality of 32.31% nitric acid.

7-15 Calculate the molality of 11.8 M HCl. The density of the solution is 1.1854 g/mL.

7-16 The solubility of potassium iodate in water was found to be 0.0919 mol/L. The density of the solution was 1.0160 g/mL. Calculate the molality of the saturated solution.

[W. D. Larson and J. J. Renier, *J. Am. Chem. Soc.*, *74*, 3184 (1952)]

7-17 A 27.76% solution of calcium dichromate, $CaCr_2O_7$, was found to have a density of 1.2568 g/mL. Calculate the molality and molarity of the solution.

[W. H. Hartford, K. A. Lane, and W. A. Meyer, Jr., *J. Am. Chem. Soc.*, *72*, 3655 (1950)]

7-18 Tin(II) chloride dihydrate (112.8 g) was mixed with 22.5 g of 49.6% HF and heated on a steam bath. When the mixture was allowed to cool, 21.8 g of product was collected. *Analysis* found: Sn, 68.7, 68.3, 68.5; F, 10.7, 10.7, 10.8; and the remainder chlorine. The solubility of the product in water was 55.0%.

a. Calculate the molality of the solution.

b. Which of the reagents used to prepare the crystalline product was in excess?

c. How many grams of tin(II) chloride dihydrate would be required to make 100 g of a 1 m solution of the product? Assume that the reaction and recovery of the product are quantitative.

[W. H. Nebergall, G. Baseggio, and J. C. Muhler, *J. Am. Chem. Soc.*, *76*, 5353 (1954)]

7-19 Calculate the volume of oxygen, at STP, liberated by the decomposition of hydrogen peroxide in 25.0 mL of 0.01121 M H_2O_2. The reaction is

$$2H_2O_2(aq) \rightarrow 2H_2O + O_2(g)$$

7-20 Calculate the mole fraction of solute and the mole fraction of solvent in a 31.2% solution of sugar, $C_{12}H_{22}O_{11}$.

7-21 What is the mole fraction of mannitol, $C_6H_{14}O_6$, in 1000 g of H_2O if the solution contains 162 g of solute?
[J. C. W. Frazer, B. F. Lovelace, and T. H. Rogers, *J. Am. Chem. Soc.*, *42*, 1801 (1920)]

Supplementary Problems

7-22 Silver can be determined in the 0.01–10-ppm range using atomic fluorescence spectroscopy. In such an experiment West and Williams prepared a stock solution of silver by dissolving 0.9128 g of $AgNO_3$ and diluting to 100 mL with distilled water. What was the concentration, in ppm, of silver in the stock solution?
[*Anal. Chem.*, *40*, 335 (1968)]

7-23 Calculate the quantity of solute, in grams per liter, of each of the following solutions, which were used in research projects described in the literature.

a. 0.9646 M $LiNO_3$
b. 6.67×10^{-3} M $Co(NO_3)_2$
c. 0.1511 M NaOH
d. 2.31×10^{-3} M $Cu(ClO_4)_2$
e. 1.793×10^{-2} M $CuCl_2$
f. 0.45 M chromium(III) perchlorate
g. 0.475 M sodium bicarbonate
h. 9.93 mmol $TlNO_3$
i. 2.36×10^{-2} mmol cadmium nitrate
j. 2.94×10^{-3} M zinc chloride

7-24 In an experiment to determine the amount of calcium in fruit juice, a standard stock solution of 500 ppm of calcium is needed. How many grams of dry calcium carbonate should be taken to make 1 L of this solution?
[A. Strohl, *J. Chem. Ed.*, *62*(#4), 343 (1985)]

7-25 To prepare the working curve for the experiment in Problem 7-24, 50 mL of the standard calcium solution in Problem 7-24 is diluted to 250 mL to make a secondary standard (or working standard). From this secondary standard a series of working solutions are prepared by taking —— mL of this solution and diluting to 100 mL.

a. What is the concentration of the secondary standard solution in ppm?

b. What is the concentration of each solution in the working curve, in ppm, when the following volumes of secondary solution are taken?

i. 5 mL
ii. 10 mL
iii. 15 mL

7-26 Calculate the mass of solute, in grams, in each of the following solutions.

a. 200 mL of 0.1 M perchloric acid
b. 0.125 L of 0.16 M $NaNO_3$
c. 1.67 L of 7.10×10^{-3} M manganese(II) nitrate
d. 50.0 mL of 8.96×10^{-3} M $CuCl_2$
e. 723 mL of 9.44×10^{-4} M copper(II) chloride

7-27 Calculate the mass of iodine monochloride present in 5.00 mL of 6.82×10^{-3} M solution in acetonitrile CH_3CN.

7-28 Calculate the molarity of each of the following solutions.

a. 200 mL of solution containing 8.31 g of $LiNO_3$
b. 75.0 mL of solution containing 0.00641 g of $AgNO_3$
c. 2.50 L of solution containing 1.74 g of alcohol, C_2H_6O

7-29 Calculate the solubilities in moles per liter of the following compounds from the solubility in grams per 100 mL as reported by Wheeler, Perros, and Naeser.
[*J. Am. Chem. Soc.*, *77*, 3489 (1955)]

a. Na_2PtF_6, 20.49 g/100 mL
b. $(NH_4)_2PtF_6$, 7.32 g/100 mL
c. K_2PtF_6, 0.750 g/100 mL
d. $RbPtF_6$, 0.278 g/100 mL
e. $CsPtF_6$, 0.484 g/100 mL

7-30 Glacial acetic acid is 99.7% acetic acid, and its specific gravity is 1.049. What volume of glacial acetic acid would be required to make 5.65 L of 4.57 M CH_3COOH?

7-31 Calculate the volume of concentrated phosphoric acid required to make 10.0 L of 6.00 M H_3PO_4. The concentrated acid contains 85.0% H_3PO_4, and its specific gravity is 1.689.

7-32 What is the molarity of a solution of hydrobromic acid whose assay is 47.0% and specific gravity 1.50?

7-33 In each of the following cases, how many mL of the concentrated acid must be taken to prepare the desired solution?

a. 100 mL of 3 M HCl from 12 M HCl
b. 250 mL of 0.1 M HNO_3 from 15 M HNO_3
c. 125 mL of 0.5 M H_2SO_4 from 18 M H_2SO_4

7-34 Calculate the molarity of 88.0% formic acid, HCOOH. The specific gravity is 1.2012.

7-35 Calculate the molarity of 70.0% perchloric acid. Its specific gravity is 1.674.

7-36 What is the molality of an 18.9% NaCl solution?

7-37 The solubility of KPF_6 in water at 25°C was found to be 8.35% by Sarmousakis and Low. Calculate the molality of the solution.
[*J. Am. Chem. Soc.*, *77*, 6518 (1955)]

7-38 A saturated solution of manganese(II) sulfate was prepared by Richards and Frapie. A sample of the solution weighing 4.4318 g contained 1.7470 g of the salt. What was the molality of the solution?
[*J. Am. Chem. Soc.*, *26*, 77 (1901)]

7-39 Crystals were collected after the reaction of stannous oxide with hydrofluoric acid. *Analysis* found: Sn, 75.8; F, 24.3. The solubility of the crystals in water was 29.6% at 18°. Calculate the molality of the solution.
[W. H. Nebergall, J. C. Muhler, and H. G. Day, *J. Am. Chem. Soc.*, *74*, 1604 (1952)]

7-40 Approximately 45.0 L of 4.360 m hydrochloric acid was required for several experiments. How many liters of concentrated acid (37.5% HCl) was required? The specific gravity of the concentrated acid was 1.1875.

7-41 The mole fraction of alcohol, C_2H_6O, in an aqueous solution was 0.250. What mass of alcohol was present in a solution containing 100 g of H_2O?

8
COLLIGATIVE PROPERTIES

8-1 COLLIGATIVE PROPERTIES

Four properties of solutions that depend only on the number of solute particles are called *colligative properties*. They are vapor-pressure lowering, boiling-point elevation, freezing-point depression, and osmotic pressure.

8-2 VAPOR-PRESSURE LOWERING

Vapor pressure is defined as the pressure of a vapor in equilibrium with its liquid. A vapor is the gaseous state of a substance that is normally a liquid at ordinary temperatures, e.g., water vapor.

The vapor pressure of a solution is lower than the vapor pressure of pure solvent. The extent of the depression of the vapor pressure is related to the composition of the solution by Raoult's law when a nonvolatile solute is present. Raoult's law may be represented by

$$P = P^0X$$

where P is the vapor pressure of the solution, P^0 is the vapor pressure of the pure solvent, and X is the mole fraction of solvent in the solution:

$$X = \frac{n^0}{n + n^0}$$

where n^0 is the number of moles of solvent and n is the number of moles of solute.

Example 8-1

Calculate the vapor pressure of water at 25.0° over a solution containing 171 g of sugar ($C_{12}H_{22}O_{11}$) in 900 g of H_2O.

$$P = P^0X$$

$$X = \frac{n^0}{n + n^0}$$

$$n^0 = \frac{900}{18} = 50.0 \text{ mol } H_2O$$

$$n = \frac{171}{342} = 0.500 \text{ mol } C_{12}H_{22}O_{11}$$

$$P = 23.8 \times \frac{50.0}{50.5} = 23.6 \text{ mm}$$

$$p^0 = 23.8 \text{ torr (from Table D-1)}$$

$$n^0 = 900 \text{ g} \times \frac{1 \text{ mol}}{18.0 \text{ g}} = 50.0 \text{ mol } H_2O$$

$$n = 171 \text{ g} \times \frac{1 \text{ mol}}{342 \text{ g}} = 0.500 \text{ mol } C_{12}H_{22}O_{11}$$

$$p = 23.8 \text{ torr} \times \frac{50.0 \text{ mol}}{0.500 \text{ mol}} = 23.6 \text{ torr}$$

Example 8-2

Calculate the vapor pressure at 30.0° of a 10.0% aqueous solution of glucose, $C_6H_{12}O_6$.

$$P = P^0X$$

$$X = \frac{n^0}{n + n^0}$$

$$n^0 = \frac{90.0}{18.0} = 5.00 \text{ mol } H_2O$$

$$n = \frac{10.0}{180} = 0.556 \text{ mol } C_6H_{12}O_6$$

$$P = 31.8 \times \frac{5.00}{5.06} = 31.4 \text{ mm}$$

$$p^0 = 31.8 \text{ torr (from Table D-1}$$

$$n^0 = 90.0\ g \times \frac{1\ mol}{18.0\ g} = 5.00\ mol\ H_2O$$

$$n = 10.0\ g \times \frac{1\ mol}{180\ g} = 0.0556\ mol\ C_6H_{12}O_6$$

$$p = 31.8\ torr \times \frac{5.00\ mol}{0.556\ mol} = 31.4\ torr$$

8-3 BOILING-POINT ELEVATION

The normal boiling point of a liquid is the temperature at which its vapor pressure equals 1 atm. It follows from Raoult's law that the boiling point of a solution will be higher than the boiling point of pure solvent. The extent of the boiling-point elevation is given by

$$\Delta t_b = K_b m$$

where Δt_b is the amount of the temperature increase, K_b is the molal boiling-point constant that is characteristic for each solvent, and m is the *molal* concentration of the solution. The value of K_b for water is 0.512. See Table 8-1 for the molal boiling point elevation constants of some common solvents.

[*Handbook of Chemistry and Physics*, 64th ed., CRC Press, Boca Raton, FL (1983-84), p. D-188]

TABLE 8-1 Molal Boiling-Point Elevation Constants

Solvent	K_b	Normal boiling point (°C)
Acetic acid	3.07	118.5
Benzene	2.53	80.1
Carbon disulfide	2.34	46.3
Chloroform	3.63	61.2
Dichloromethane	2.52	40.5
Water	0.512	100.0

Example 8-3

Calculate the expected boiling-point elevation of a solution containing 0.504 g of anthracene, $C_{14}H_{10}$, in 42.0 g of benzene and compare it with the value 0.175° observed by E. Beckmann.
[*Z. Physik. Chem.*, 6, 439 (1890)]

The boiling-point elevation of a solution is given by

$$\Delta t_b = K_b m$$

$$m = \frac{\text{moles solute}}{1000 \text{ g solvent}}$$

Calculate the mass of $C_{14}H_{10}$ in 1000 g of solvent.

$$0.504 \text{ g} \times \frac{1000 \text{ g benzene}}{42.0 \text{ g benzene}} = 12.0 \text{ g } C_{14}H_{10}$$

$$\frac{12.0 \text{ g } C_{14}H_{10}}{1000 \text{ g benzene}} \times \frac{1 \text{ mol}}{178 \text{ g}} = \frac{0.0674 \text{ mol } C_{14}H_{10}}{1000 \text{ g benzene}}$$

$$\Delta t_b = 2.53 \frac{°C}{\text{mol}/1000 \text{ g benzene}} \times 0.0674 \frac{\text{mol}}{1000 \text{ g benzene}}$$

$$\Delta t_b = 2.53°C \times \frac{1000 \text{ g benzene}}{\text{mol}} \times 0.0674 \frac{\text{mol}}{1000 \text{ g benzene}}$$

$$\Delta t_b = 0.171°C$$

The steps may be combined giving a more direct calculation.

$$\Delta t_b = k_b \times \frac{m_{\text{solute}} \times 1000_{\text{solvent}}}{M_{\text{solute}} \times m_{\text{solvent}}}$$

$$\Delta t_b = 2.53 \frac{°C \times \text{mol}_{\text{solute}}}{1000 \text{ g}_{\text{solvent}}} \times \frac{0.504 \text{ g} \times 1000 \text{ g}}{178 \text{ g/mol} \times 42.0 \text{ g}} = 0.170°C$$

8-4 FREEZING-POINT DEPRESSION

Solutions have lower freezing points than pure solvents. The extent of the freezing-point depression depends on the *molal* concentration of the solute: it is given by the equation

$$\Delta t_f = K_f m$$

TABLE 8-2 Molal Freezing-Point Depression Constants

Solvent	K_f	Normal freezing point (°C)
Acetic acid	3.90	16.604
Benzene	4.90	5.5
Naphthalene	6.85	80.22
Phenol	7.40	43
Water	1.86	0.00

where Δt_f is the freezing-point lowering, K_f is the molal freezing-point constant of the solvent, and m is the molal concentration. K_f is also called the *cryoscopic constant.* See Table 8-2 for the molal freezing-point depression constants of some common solvents.

[*Handbook of Chemistry and Physics*, 64th ed., CRC Press, Boca Raton, FL (1983-84), p. D-188]

Example 8-4

Calculate the freezing point of a solution containing 50.0 g of ethylene glycol, $C_2H_6O_2$, in 700 g of water. The freezing-point depression of the solution is given by

$$\Delta t_f = K_f m$$

Calculate the mass of $C_2H_6O_2$ in 1000 g of water.

$$50.0 \text{ g} \times \frac{1000 \text{ g}}{700 \text{ g}} = 71.4 \text{ g}$$

$$\frac{71.4 \text{ g}}{1000 \text{ g}} \times \frac{1 \text{ mol}}{62.0 \text{ g}} = 1.15 \frac{\text{mol}}{1000 \text{ g}} C_2H_6O_2$$

$$\Delta t_f = 1.86 \frac{°\text{C}}{\text{mol}/1000 \text{ g}_{\text{solvent}}} \times 1.15 \frac{\text{mol}}{1000 \text{ g}_{\text{solvent}}} = 2.14°\text{C}$$

The steps may be combined giving a more direct calculation.

$$\Delta t_f = k_f \times \frac{m_{solute} \times 1000_{solvent}}{M_{solute} \times m_{solvent}}$$

$$\Delta t_f = 1.86 \frac{°C \times mol_{solute}}{1000\ g_{solvent}} \times \frac{50.0\ g \times 1000\ g}{62.0\ g/mol \times 700\ g} = 2.14°C$$

The predicted freezing point of the solution is -2.14°C, since the freezing point of water is 0°C.

8-5 OSMOTIC PRESSURE

When a solution is separated from pure solvent by a semipermeable membrane, a pressure develops in the solution. An illustration of the existence of this pressure is given by the rise of solution in the stem of an inverted thistle tube, containing a sugar solution held there by a cellophane membrane, when it is immersed in water. A semipermeable membrane is one through which solvent molecules, but not solute molecules, can pass. The pressure thus developed is called *osmotic pressure*.

The osmotic pressure π of a *dilute* solution may be expressed by the equation

$$\pi = mRT$$

where m = molality

$$R = 0.0820 \frac{L\ atm}{K\ mol}$$

T = temperature, K

If the solution is dilute enough, the difference between the molarity and molality of a solution may be insignificant. The osmotic pressure may be expressed as

$$\pi = MRT$$

in fact, many freshman books use this form of the formula instead of using molality.

Example 8-5

Calculate the osmotic pressure of a 0.100 m aqueous solution at 20°C.

$$\pi = mRT$$

$$0.100\ \frac{mol}{L} \times 0.0820\ \frac{L\ atm}{K\ mol} \times 293\ K = 2.40\ atm$$

8-6 DETERMINATION OF MOLECULAR MASSES

All four of the colligative properties of solutions may be used for the experimental determination of molecular masses of solutes. Cryoscopic (freezing-point depression) measurements are the most common for the usual range of molecular masses because of fewer experimental difficulties (see Fig. 8-1). Osmotic-pressure measurements are especially useful for substances with very high molecular masses such as protein and other polymeric materials.

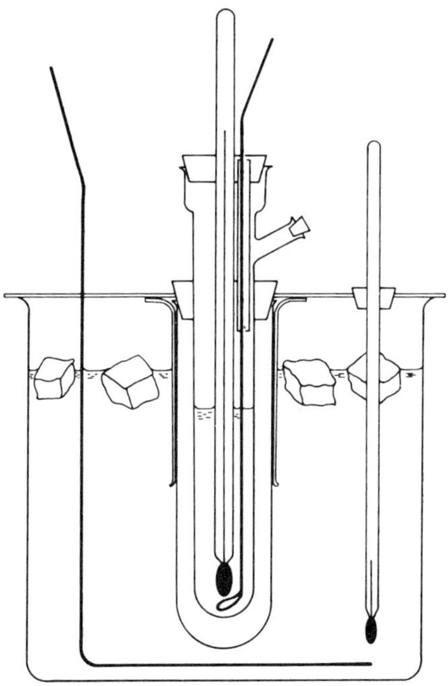

FIGURE 8-1 Apparatus for determining the freezing-point depression from which molecular mass is calculated.

Example 8-6

What is the molecular mass of a compound if the vapor pressure of a solution containing 0.8696 g in 29.740 g of benzene had a vapor pressure of 631.9 mm? The vapor pressure of pure benzene at the same temperature was 639.7 mm.

[M. A. Rosanoff and R. A. Dunphy, *J. Am. Chem. Soc.*, *36*, 1417 (1914)]

The mole fraction of solvent in the solution can be calculated from $P = P^0X$

$$n^0 = 29.74\ g \times \frac{1\ mol}{78.05\ g} = 0.3810\ mol$$

$$n = 0.8696\ g \times \frac{1\ mol}{Mg} = \frac{0.8696}{M}\ mol$$

The molecular mass of benzene C_6H_6 is 78.0. Let M equal the molecular mass of solute

$$X = \frac{631.9\ torr}{639.7\ torr} = \frac{0.3810\ mol}{(0.3810 + 0.8696/M)} = 0.9878$$

$$\frac{0.3810}{0.9878} = 0.3810 + \frac{0.8696}{M}$$

$$0.3857 - 0.3810 = \frac{0.8696}{M}$$

$$M = \frac{0.8696}{0.0047} = 185\ \frac{g}{mol}$$

Example 8-7

Calculate the molecular mass of a compound, 1.632 g of which raised the boiling point of 36.74 g of benzene 0.728°. The molality of the solution can be calculated from $\Delta t_b = K_b m$.

[E. Beckmann, *Z. Physik. Chem. (Leipzig)*, *6*, 439 (1890)]

$$\Delta t_b = 2.53\ \frac{°C}{1\ mol\ solute/1000\ g\ solvent}\ m = 0.728°C$$

$$m = \frac{0.728}{2.53} = 0.288\ \frac{mol\ solute}{1000\ g\ solvent}$$

$$0.288\ \frac{mol\ solute}{1000\ g\ benzene} = \frac{1.63\ g\ solute}{M\ g/mol} \times \frac{1000\ g\ benzene}{36.7\ g\ benzene}$$

$$M = \frac{44.4}{0.288} = 154\ \frac{g}{mol}$$

Example 8-8

Calculate the molecular mass of a compond if a solution containing 1.6491 g in 25.9256 g benzene had a freezing point of 5.12°C. The molality of the solution m is $\Delta t_f/K_f$ from $\Delta t_f = K_f m$.

[J. K. Ruff, *Inorg. Chem.*, *4*, 1448 (1965)] (See also Fig. 8-1.)

$$\Delta t_f = 5.48 - 5.12 = 0.36°$$

$$\Delta t_f = 4.90 \frac{°C}{1 \text{ mol solute}/1000 \text{ g solvent}} \text{ m} = 0.36°C$$

$$m = \frac{0.36}{4.90} = 0.0735 \frac{\text{mol solute}}{1000 \text{ g solvent}}$$

$$0.0735 \frac{\text{mol solute}}{1000 \text{ g benzene}} = \frac{1.65 \text{ g solute}}{M \text{ g/mol}} \times \frac{1000 \text{ g benzene}}{25.9 \text{ g benzene}}$$

$$M = \frac{63.7}{0.0735} = 867 \frac{\text{g}}{\text{mol}}$$

Example 8-9

A solution containing 0.0150 g of polymer material per gram of water had an osmotic pressure of 1.52×10^{-3} atm at 25°C. Estimate the molecular mass of the polymeric material.

$$m = \frac{\pi}{RT} = \frac{1.52 \times 10^{-3}}{0.0820 \times 298} = 6.22 \times 10^{-5}$$

$$m = \frac{6.22 \times 10^{-5} \text{ mol}}{1000 \text{ g } H_2O}$$

The solution contained 0.0150 g/g H_2O.

$$0.0150 \times 1000 = \frac{15.0 \text{ g}}{1000 \text{ g } H_2O}$$

Since 6.22×10^{-5} mol weighed 15.0 g, to find the mass of 1 mol, multiply

$$15.0 \times \frac{1}{6.22 \times 10^{-5}} = 2.41 \times 10^{5} \frac{\text{g}}{\text{mol}}$$

8-7 IDEAL SOLUTION

Raoult's law and the equations used to express the colligative properties are approximate, just as the gas laws are approximate. A solution to which Raoult's law applies accurately is called an *ideal solution*. Departures from ideal behavior are pronounced when an interaction between solute and solvent molecules occurs, when the solute dissociates, or when a chemical reaction occurs. When no dissociation, reaction, or appreciable interaction occurs, solutions approach ideal behavior when they are dilute.

PROBLEMS

8-1 Calculate the vapor pressure at 20°C of an aqueous solution containing 108.2 g mannitol ($C_6H_{14}O_6$) in 1000 g of water. The experimental value obtained by Frazer, Lovelace, and Rogers was 17.439 mm.

[*J. Am. Chem. Soc.*, *42*, 1801 (1920)]

8-2 The measured pressure of a solution containing 0.5676 g of anthracene, $C_{14}H_{10}$, in 29.740 g of benzene was 634.6 mm. How does this value compare with the vapor pressure predicted by Raoult's law? The vapor pressure of pure benzene at the same temperature was 639.7 mm.

[M. A. Rosanoff and R. A. Dunphy, *J. Am. Chem. Soc.*, *36*, 1416 (1914)]

8-3 The vapor pressure of ethyl alcohol was found to be 36.6 mm. At the same temperature the vapor pressure of a solution of 3.9781 g of acetamide, C_2H_5ON, in 102 g of alcohol was 35.6 mm. Compare the observed vapor pressure of the solution with that calculated from Raoult's law.

[J. Wright, *J. Chem. Soc.*, *115*, 1165 (1919)]

8-4 Make a plot of the observed vapor pressures of the solutions of oleic acid, $C_{18}H_{34}O_2$, in diethyl ether, $C_4H_{10}O$, versus the mole fraction of solvent from the data of F. H. Campbell at 30.0°C. Calculate the vapor pressure of each solution predicted from Raoult's law and plot the results on the same graph. Discuss the results.

% $(C_2H_5)_2O$	Vapor pressure (mm)
4.96	96.7
7.17	119.9
14.36	217.9
23.55	303.7
37.40	414.8
61.93	535.6
100.00	642.1

Source: *Trans. Faraday Soc.*, *11*, 94 (1915).

8-5 Calculate the boiling-point elevation of a solution containing 0.5151 g of naphthalene, $C_{10}H_8$, in 60.81 g of chloroform. Compare your answer with the observation of E. Beckmann, who found a rise of 0.237°C.

[*Z. Physik. Chem. (Leipzig)*, *6*, 446 (1890)]

8-6 Calculate the expected boiling point of a solution that contained 0.6072 g of a compound of molecular mass 600 in 26.0520 g of CH_2Cl_2.

[J. K. Ruff, *Inorg. Chem.*, *4*, 1448 (1965)]

8-7 Calculate the expected freezing point of a solution containing 1.00 gallon of ethylene glycol, $C_2H_6O_2$, in 18.0 quarts of water. The density of ethylene glycol is 1.12 g/mL, and 1 quart contains 946.34 cc.

8-8 The molecular mass of a solid compound was determined by the freezing-point depression of an aqueous solution to be 81.0. If the solution contained 15.77 g of the compound in 1000 g water, what was the freezing-point depression of the solution?

[A. W. Laubengayer and G. F. Condike, *J. Am. Chem. Soc.*, *70*, 2224 (1946)]

8-9 Calculate the approximate osmotic pressure of a solution containing 17.1 g of sucrose, $C_{12}H_{22}O_{11}$, in 500 g of water at 20°C.

8-10 Calculate the molecular mass of a substance from the vapor-pressure lowering of water at 20.0°, which was 0.2162 mm. The mass of solute in 1000 g of water was 126.283 g.

[J. C. W. Frazer, B. F. Lovelace, and T. H. Rogers, *J. Am. Chem. Soc.*, *42*, 1801 (1920)]

8-11 Calculate the molecular mass of the solute that lowered the vapor pressure of alcohol 1.084 mm when 5.698 g was dissolved in 78.0 g of the solvent. The vapor pressure of alcohol, C_2H_6O, at the temperature of the experiment was 42.60 mm.

[J. Wright, *J. Chem. Soc.*, *115*, 1165 (1919)]

8-12 The molecular mass of a compound believed to be $Te_2(CO)_{10}$ was determined by measuring the vapor pressure of a solution containing 12.5 mg of solute in 300.9 mg of diethyl ether. The vapor-pressure depression was 3.08 mm at 22° (P^0 = 464 mm). Compare the observed molecular mass with that calculated from the formula.

[J. C. Hileman, D. K. Huggins, and H. D. Kaez, *J. Am. Chem. Soc.*, *83*, 2953 (1961)]

8-13 Calculate the molecular mass of the compound that raised the boiling point of chloroform 0.475° when 1.0171 g was dissolved in 60.81 g of solvent.

[E. Beckmann, *Z. Physik. Chem. (Leipzig)*, *6*, 446 (1890)]

8-14 A sample of sulfur weighing 0.2096 g was dissolved in 17.79 g of carbon disulfide. The observed boiling-point elevation was 0.107°. Calculate the molecular mass of the solute. What is the formula of the solute?

[A. Helff, *Z. Physik. Chem. (Leipzig)*, *12*, 200 (1893)]

8-15 What is the formula of phosphorus in carbon disulfide solution if 0.1430 g raised the boiling point of 16.87 g of carbon disulfide 0.163°?

[A. Helff, *Z. Physik. Chem. (Leipzig)*, *38*, 76 (1901)]

8-16 Hexamethylbenzene ($C_{12}H_{18}$) was used to determine the molal freezing-point constant of benzene. A solution containing 6.752 g of the solute in 1.000 kg of benzene had a freezing point depression of 0.2116°C. Calculate the value of the constant from this experiment.

[B. C. Barton, D. A. Rothrock, Jr., and C. A. Kraus, *J. Am. Chem. Soc.*, *74*, 786 (1952)]

8-17 A solution containing 6.93 g of a phosphorus compound in 879 g of benzene had a boiling point elevation of 0.014°C. What is the molar mass of the compound?

[P. J. Kirk and T. D. Smith, *J. Inorg. Nucl. Chem.*, *30*, 892 (1968)]

8-18 The freezing point of a 4.9% aqueous solution was -3.4°. What was the molecular mass of the solute?

[O. Maas and O. W. Herzberg, *J. Am. Chem. Soc.*, *42*, 2570 (1920)]

8-19 Calculate the freezing point of a 9.96% solution containing the same solute as in Problem 8-18.

8-20 Calculate the molecular mass of a compound if a solution in anhydrous acetic acid containing 74.1 g in 5904 g of solvent has a freezing point of 15.60°C.

[S. English and W. E. S. Turner, *J. Chem. Soc.*, *107*, 775 (1915)]

8-21 Estimate the molecular mass of a polymeric material if a solution containing 2.50 g in 100 g of water had an osmotic pressure of 3.50×10^{-4} atm at 25°C.

8-22 The compound S_4N_4 was fluorinated with AgF_2, and a product that contained 49.4% S, 21.5% N, and 29.3% F was obtained. The boiling point of a solution of 0.9990 g of the compound in 42.8765 g benzene had a rise of 0.235°. What is the formula of the compound?

[O. Glemser, H. Schröder, and H. Haesler, *Z. Anorg. Allgem. Chem.*, *279*, 28 (1954)]

Supplementary Problems

8-23 Calculate the vapor pressure of a solution of 1.8411 g of naphthalene, $C_{10}H_8$, in 26.53 g of benzene, C_6H_6, if the vapor pressure of pure benzene at the same temperature was 639.85 mm. The observed vapor pressure of the solution was 614.5 mm.

[M. A. Rosanoff and R. A. Dunphy, *J. Am. Chem. Soc.*, *36*, 1416 (1914)]

8-24 Frazer and Lovelace observed a vapor-pressure depression of 0.122 mm at 20.0°C when 0.500 mol of mannitol was dissolved in 1000 g water. Compare this with the calculated value from Raoult's law.

[*J. Am. Chem. Soc.*, *36*, 2448 (1914)]

8-25 Calculate the vapor pressure of ethyl alcohol at 15°C from the observation that the vapor pressure of an alcoholic solution containing 0.02823 g of benzil, $C_{14}H_{10}O_2$, per gram of solvent was 31.98 mm.

[O. F. Tower and A. F. O. Germann, *J. Am. Chem. Soc.*, *36*, 2449 (1914)]

8-26 Make a plot of the observed vapor pressures of the solutions of oleic acid, $C_{18}H_{34}O_2$, in carbon disulfide versus the mole fraction of solvent from the data of F. H. Campbell:

% CS_2	Vapor pressure (mm)
3.12	40.9
15.75	164.0
25.85	233.7
34.77	290.5
40.51	316.3
56.14	366.6
100.00	431.9

Source: *Trans. Faraday Soc.*, *11*, 94 (1915).

8-27 The molecular mass of a solute was calculated to be 118 when the vapor pressure of a solution containing 0.784 g in 48.0 g of CCl_4 was measured. The vapor pressure of CCl_4 at 15.5°C was 73.3. What was the vapor pressure of the solution?
[J. Wright, *J. Chem. Soc.*, *115*, 1165 (1919)]

8-28 The boiling-point elevation of a solution of 0.4344 g of anthracene in 44.16 g of anhydrous acetic acid was 0.140°. Calculate the predicted boiling-point elevation and compare the calculated and observed values. The formula of anthracene is $C_{14}H_{10}$.
[E. Beckmann, *Z. Physik. Chem. (Leipzig)*, *6*, 449 (1890)]

8-29 Calculate the expected boiling point of a solution containing 0.246 g of P_2I_4 in 4.859 g of carbon disulfide.
[M. Baudler and G. Fricke, *Z. Anorg. Allgem. Chem.*, *320*, 11 (1963)]

8-30 Calculate the expected freezing point of a solution containing 1.00 gallon of methanol, CH_4O, in 18.0 quarts of water. The density of methanol is 0.793 g/mL and 1 gallon contains 3.7853 L.

8-31 What is the freezing point of a 0.021 m aqueous solution of a nonelectrolyte?
[E. H. Loomis, *Z. Physik. Chem. (Leipzig)*, *37*, 411 (1901)]

8-32 Calculate the freezing point of a 0.0800 M solution if the specific gravity of the solution was 1.0054. Assume molecular mass of solute is 341 g/mol.
[E. H. Loomis, *Z. Physik. Chem. (Leipzig)*, *37*, 411 (1901)]

8-33 Calculate the freezing point of a 3.819 mol percent solution of formamide, $HCONH_2$, in water. English and Turner reported a freezing point of -2.7° for such a solution.
[S. English and W. E. S. Turner, *J. Chem. Soc.*, *107*, 775 (1915)]

8-34 The cryoscopic constant of $SC(SH)_2$ was assigned the value 6.75. To substantiate this value, the molecular mass of sulfur in a solution of $SC(SH)_2$ was determined. A molecular mass of 255 was calculated from the freezing-point depression (1.00°) of a solution containing 13.7 g of solvent. Calculate the mass of sulfur in the solution.
[G. Gattow and B. Krebs, *Z. Anorg. Allgem. Chem.*, *321*, 149 (1963)]

8-35 A solution contained 5.00 g of polymer (molecular mass 250.000) in 200 g of water at 20°C. Calculate the osmotic pressure of the solution.

8-36 A sample of an organic compound weighing 2.3446 g was dissolved in 26.53 g of benzene, C_6H_6. The vapor pressure of the solution was 607.4 mm, and the vapor pressure of pure benzene at the same temperature was 639.85 mm. What is the molecular mass of the solute?

[M. A. Rosanoff and R. A. Dunphy, *J. Am. Chem. Soc.*, *36*, 1416 (1914)]

8-37 The vapor pressure of carbon tetrachloride when pure was observed to be 87.1 mm. The vapor pressure of a solution containing 1.072 g of a solute in 49.0 g of CCl_4 was 85.4 mm. What was the molecular mass of the solute?

[J. Wright, *J. Chem. Soc.*, *115*, 1165 (1919)]

8-38 Calculate the molecular mass of the compound that lowered the vapor pressure of methanol, CH_4O, at 15°C 0.332 mm when 3.662 g was dissolved in 100 g of solvent. The vapor pressure of methanol at 15° was observed to be 73.6 mm.

[O. F. Tower and A. F. O. Germann, *J. Am. Chem. Soc.*, *36*, 2449 (1914)]

8-39 Calculate the molecular mass of the compound that depressed the vapor pressure of isopentane, C_5H_{12}, at 20.3° by 17.40 mm when 0.0697 g was dissolved in 0.891 g of solvent. The vapor pressure of isopentane at 20.3° was 580.0 mm.

[T. Wartik and H. I. Schlesinger, *J. Am. Chem. Soc.*, *75*, 339 (1953)]

8-40 Two experiments were conducted to determine the molecular mass of a phosphorus-containing compound. In the first a 0.07136 m solution gave 154.4 as the result; in the second a 0.1046 solution gave 154.2. What was the freezing-point depression of glacial acetic acid in each experiment?

[R. R. Holmes and J. A. Forstner, *Inorganic Chem.*, *2*, 381 (1963)]

8-41 A sample of an iodide containing 10.86% phosphorus and weighing 0.246 g was dissolved in 6.428 g of carbon disulfide.

The boiling-point elevation was 0.161°. What is the formula of the compound?

[M. Baudler and G. Fricke, *Z. Anorg. Allgem. Chem.*, *320*, 11 (1963)]

8-42 What is the formula of phosphorus in carbon disulfide solution if 0.2170 g of solute raised the boiling point of 19.18 g of CS_2 0.104°?

[A. Helff, *Z. Physik. Chem. (Leipzig)*, *12*, 200 (1893)]

8-43 A sample of phosphorus weighing 0.3635 g was dissolved in 16.87 g of carbon disulfide. The boiling-point elevation of the solution was 0.363°. Calculate the molecular mass of the phosphorus. What is the structure of the phosphorus molecules?

[A. Helff, *Z. Physik. Chem. (Leipzig)*, *38*, 76 (1901)]

8-44 An aqueous solution of a sugar containing 68.44 g/L had a freezing point of -0.39°C. The density of the solution was 1.0256 g/mL. Calculate the molecular mass of the sugar.

[E. H. Loomis, *Z. Physik. Chem. (Leipzig)*, *37*, 411 (1901)]

8-45 Calculate the molecular mass of the compound if a solution containing 18.21 g/L had a freezing point of -0.188°. The density of the solution was 1.0047 g/mL.

[E. H. Loomis, *Z. Physik. Chem. (Leipzig)*, *37*, 411 (1901)]

8-46 Calculate the molecular mass of sulfur in a solution containing 0.100 g in 9.80 g of $SC(SH)_2$ if the freezing-point depression of the solution was 0.268°C. The cryoscopic constant of $SC(SH)_2$ was determined as 6.75.

[G. Gatton and B. Krebs, *Z. Anorg. Allgem. Chem.*, *321*, 149 (1963)]

8-47 An aqueous solution of a polymeric material had an osmotic pressure of 0.40 mm when its concentration was 5.00 mg/g at 25°C. Estimate the molecular mass of the polymer.

9
THERMOCHEMISTRY

9-1 HEAT CAPACITY

The concept of specific heat was introduced in Chapter 1 as an example of an intensive property. When heat is withdrawn from the surroundings and added to a system, the temperature of the system rises. The quantity of heat required to raise the temperature of a system 1°C is called the *heat capacity* of the system: the *specific heat* of the substance is the heat capacity per gram. When the system is 1 mol of a pure substance, the quantity of heat required to raise the temperature 1° is called the *molar heat capacity C*. The symbols C_V and C_P are used for heat capacity at constant volume and constant pressure, respectively. The basic unit of energy in the International System of Units is the *joule* (J). The joule is the energy associated with a force of one newton acting through a distance of one meter (N-m or kg-m^2-s^{-2}). Other units commonly used are the erg (cgs) and the calorie (cal). The relationships between these units are:

1 joule (J) = 10^7 ergs

1 calorie (cal) = 4.184 J

The latter two units do not belong to the International System of Units and may someday be abandoned.

Table 9-1 lists specific heats of some common substances.

Example 9-1

The specific heat of ammonia is 4.381 J/g at -60°C. Calculate the heat required to raise the temperature of 2.1 mol from -60.0° to -47.0°. The mass of 1 mol of NH_3 is 17.01 g.

TABLE 9-1 Specific Heats of Some Common Substances

Substance	Temp. range (°C)	Specific heat cal/g (°C)	Specific heat J/g (°C)
Water	0–100	1.000	4.184
Ice	-20–0	0.505	2.113
Steam	100–200	0.475	1.987
Mercury	20–100	0.0330	0.138
Aluminum	17–100	0.217	0.908
Lead	20–100	0.0305	0.128
Alcohol	25	0.581	2.431
Cadmium	20	0.0549	0.230
Selenium (amorphous)	18–38	0.095	0.397
Copper	20	0.0912	0.382
Gallium	12–23	0.079	0.33
Sulfur (rhombic)	0–54	0.1728	0.7230
Zinc	20	0.0924	0.387
Tellurium	15–100	0.0483	0.202
Ammonia	-60	1.047	4.381
Bismuth	18	0.0303	0.127

See Table 1-4 for additional specific heats.

$$C = 17.01 \frac{\not{g}}{\text{mol}} \times 4.381 \frac{\text{J}}{\not{g}\ {}^\circ\text{C}} = 74.52 \frac{\text{J}}{\text{mol}\ {}^\circ\text{C}}$$

$$74.5 \frac{\text{J}}{\not{\text{mol}}\ \not{{}^\circ\text{C}}} \times 2.10\ \not{\text{mol}} \times (-47.0 + 60.0)\ \not{{}^\circ\text{C}} = 2034\ \text{J}$$

9-2 HEAT CONTENT, OR ENTHALPY

Thermochemistry is the study of the heat effects that accompany physical and chemical changes. In *exothermic* processes, the system evolves or gives off heat. In *endothermic* processes, the system absorbs heat. The symbol for the measured heat in such changes is q—the heat absorbed by a system is +q, and the heat lost by a system to the surroundings is -q. The heat change of a process measured at constant pressure or volume is q_p or q_v, respectively.

The heat absorbed at constant volume is equal to the change in energy of a system, ΔE. The heat absorbed at constant pressure is equal to the change in *heat content* or *enthalpy* of the system, ΔH.

The change in energy is related to the change in enthalpy by:

$$\Delta H = \Delta E + \Delta(PV)$$

where P and V are the pressure and volume, respectively.

The difference in the change in energy versus the change in enthalpy of a system as it undergoes a reaction is given by $\Delta(PV)$. This term is the work that has to be done on a system as it expands or contracts against atmospheric pressure. Since $\Delta(PV)$ for solids and liquids is very small, the difference in ΔH and ΔE for systems involving only solids or liquids is very small. For reactions involving gases $\Delta(PV)$ becomes significant. Examples of this can be found in Section 9-9.

The relationship between work, heat, and energy is given by the first law of thermodynamics.

9-3 FIRST LAW OF THERMODYNAMICS

Work and heat are two different forms of energy. Work is any other type of energy than heat. According to the first law of thermodynamics, when there is a change in energy of a system, ΔE, the heat absorbed by the system, q, is equal to the work done on the system.

$$\Delta E = q + w$$

All three of these functions, ΔE, q, and w, are algebraic functions. They have signs (directions) as well as magnitude associated with them. When heat is added to a system, q is +.

TABLE 9-2 Heat Changes Accompanying Changes in State

Process	Name	Symbol	Example
Melting	Heat of fusion	ΔH_{fus}	$H_2O(s) \rightarrow H_2O(\ell)$
Boiling	Heat of vaporization	ΔH_{vap}	$H_2O(\ell) \rightarrow H_2O(g)$
Sublimation	Heat of sublimation	ΔH_{subl}	$I_2(s) \rightarrow I_2(g)$
Crystal structure change	Heat of transition	ΔH_{trans}	$S(r) \rightarrow S(m)$

Whenever heat is given off by a system q is -. In the same way, when work is done on a system, w is +, and whenever work is done by a system, w is -.

For a system at constant pressure, the heat lost or gained by the system is equal to the enthalpy change of the system; i.e.,

$$\Delta H = q_p$$

Since much of the work done in chemistry is carried out in beakers or flasks, at constant atmospheric pressure, any heat changes measured give us ΔH. For this reason enthalpy changes are used more often than energy changes in chemistry.

9-4 HEAT CHANGES ACCOMPANYING CHANGES IN STATE

Changes in the heat content of a system accompany changes in state. Four such changes, with the names of the heat content changes, are listed in Table 9-2.

Example 9-2

Calculate the heat required to change 1.12 mol of ice at 0°C to water at 0°C.

$$H_2O(s) \rightarrow H_2O(\ell)$$

The heat of fusion is 333.5 J/g; 1 mol of H_2O has a mass of 18.0 g.

$$1.12\ \text{mol} \times 18.0\ \frac{\text{g}}{\text{mol}} \times 333.5\ \frac{\text{J}}{\text{g}} = 6720\ \text{J}$$

Example 9-3

Calculate the heat required to change 2.2 mol of $H_2O(\ell)$ at 80.0°C to $H_2O(g)$ at 100.0°C. The average specific heat of water over this range is 4.184 J/g. The heat of vaporization of water is 2258 J/g.

Heat to raise temperature of $H_2O(\ell)$

$$2.20\ \text{mol} \times 18.0\ \frac{\text{g}}{\text{mol}} \times 20°\text{C} \times 4.184\ \frac{\text{J}}{\text{g}\ °\text{C}} = 3,330\ \text{J}$$

Heat to change $H_2O(\ell) \rightarrow H_2O(g)$

$$2.20\ \text{mol} \times 18.0\ \frac{\text{g}}{\text{mol}} \times 2258\ \frac{\text{J}}{\text{g}} = 89,400\ \text{J}$$

Total heat required 92,750 J

or with the appropriate number of significant figures, 92,700 J.

Example 9-4

The heat accompanying the change of amorphous boron to crystalline boron (ΔH°_{trans}) is -0.4 kcal/mol.

[Natl. Bur. Std. (U.S.) Circ. 500]

Calculate the difference in heat content between 50 g of B(c) and 50 g of B(amorph) at 25°C.

$$B(c) \rightarrow B(amorph)$$

$$\Delta H^\circ = +0.4\ \text{kcal/mol}$$

The sign is reversed from that given in the statement of the problem because the direction of the change is reversed.

$$\frac{50}{11} = 4.5 \text{ mol B}$$

$$4.5 \times 0.4 = 1.8 \text{ kcal}$$

This means that 1.8 kcal of heat must be added to change 50 g of crystalline boron at 25°C to 50 g of amorphous boron at 25°C.

9-5 HEAT OF REACTION

The difference in heat content between the products of a chemical reaction and the reactants is called the *heat of reaction* or *enthalpy of reaction*.

$$\Delta H = \Delta H(\text{products}) - \Delta H(\text{reactants})$$

In the reaction $2H_2(g) + O_2(g) \rightarrow 2H_2O(\ell)$, $\Delta H = -136.64$ kcal
In the reaction $3UO_2(s) + O_2(g) \rightarrow U_3O_8(s)$, $\Delta H = -76.01$ kcal

These are typical enthalpies of reaction.

Absolute values of the heat content H of substances cannot be determined. The differences in heat content ΔH can be obtained, however, from measured heats of reaction.

Consistent with the convention that heat absorbed by a system is *positive*, ΔH_{react} is *negative* for an exothermic reaction.

9-6 HEAT OF COMBUSTION

Of the many types of reactions for which heats of reaction have been determined, those involving reactions of oxygen are among the most important. These heats of reaction, called *heats of combustion*, are determined by oxygen bomb calorimetry.

In a combustion reaction, there is always one molecule of the substance undergoing the reaction, written as a reactant, and sufficient oxygen present as a reactant to complete the reaction. With compounds that contain only carbon and hydrogen, the products are always water (ℓ) and carbon dioxide (g). Examples are

	$\Delta H°_{298}$ kcal/mol	kJ/mol
Methane:		
$CH_4(g) + 2O_2(g) \rightarrow CO_2(g) + 2H_2O(\ell)$	-212.798	-890.36

	$\Delta H°_{298}$ kcal/mol	kJ/mol
Ethane:		
$C_2H_6(g) + \frac{7}{2}O_2(g) \rightarrow 2CO_2(g) + 3H_2O(\ell)$	-372.820	-1559.88
Benzene:		
$C_6H_6(\ell) + \frac{15}{2}O_2(g) \rightarrow 6CO_2(g) + 3H_2O(\ell)$	-780.98	-3267.6

9-7 STANDARD ENTHALPY OF FORMATION

The heat involved in the formation of 1 mol of a compound from the elements at 25°C, each substance being in its standard state, is the *standard enthalpy of formation* of the compound. The standard enthalpy of formation of the elements in their standard states at 25°C has been assigned the value of zero.

Examples

	$\Delta H°_{298}$ kcal/mol	kJ/mol
Methane:		
$C(\text{graphite}) + 2H_2(g) \rightarrow CH_4(g)$	-17.89	-74.85
Ethyl alcohol:		
$2C(\text{graphite}) + 3H_2(g) + \frac{1}{2}O_2(g) \rightarrow C_2H_6O$	-66.356	-277.63

The only product in the reaction is one molecule of the compound whose standard enthalpy of formation is to be determined.

9-8 HESS' LAW

The enthalpy (heat content) of a substance is a function of the physical state of a substance. Changes in enthalpy are functions of the initial and final states only and are independent of the path. One illustration of this fundamental principle is Hess' law;

> The amount of heat involved in a chemical reaction is the same whether reaction takes place in one step or in several steps.

Example 9-5

Calculate the enthalpy of formation of cyanamide, CH_2N_2, from its enthalpy of combustion, -177.20 kcal/mol.

[D. J. Solley and J. B. Gray, *J. Am. Chem. Soc.*, *70*, 2650 (1948)]

From the definition of enthalpy of formation we have

$$C(s) + H_2(g) + N_2(g) \rightarrow CH_2N_2(s) \qquad \Delta H^\circ_f = ?$$

The combustion of cyanamide is given by

$$CH_2N_2(s) + \frac{3}{2}O_2(g) \rightarrow CO_2(g) + H_2O(\ell) + N_2(g) \qquad \Delta H^\circ = -177.20 \text{ kcal/mol}$$

The other data required (from Table 9-3) are

$$C(s) + O_2(g) \rightarrow CO_2(g) \qquad \Delta H^\circ_f = -93.97 \frac{\text{kcal}}{\text{mol}}$$

$$H_2(g) + \frac{1}{2}O_2(g) \rightarrow H_2O(\ell) \qquad \Delta H^\circ_f = -68.32 \frac{\text{kcal}}{\text{mol}}$$

Applying Hess' law, changing the sign of ΔH when the reaction is reversed, we have

$$N_2(g) + H_2O(\ell) + CO_2(g) \rightarrow CH_2N_2(s) + \frac{3}{2}O_2(g) \qquad \Delta H^\circ = +177.20 \text{ kcal/mol}$$

$$C(s) + O_2(g) \rightarrow CO_2(g) \qquad \Delta H^\circ_f = -93.97 \text{ kcal/mol}$$

$$H_2(g) + \frac{1}{2}O_2(g) \rightarrow H_2O(\ell) \qquad \Delta H^\circ_f = -68.32 \text{ kcal/mol}$$

$$C(s) + H_2(g) + N_2(g) \rightarrow CH_2N_2(s) \qquad \Delta H^\circ_f = 14.91 \text{ kcal/mol}$$

Example 9-6

Calculate the standard enthalpy of formation of boron carbide from the enthalpy of combustion, -683.3 kcal/mol, as determined by Smith, Dworkin, and Van Artsdalen.

[*J. Am. Chem. Soc.*, *77*, 2655 (1955)]

The formation of B_4C from the elements is given by

$$4B(c) + C(s) \rightarrow B_4C(s) \qquad \Delta H^\circ_f = ?$$

TABLE 9-3 Some Standard Enthalpies of Formation

	$\Delta H^\circ_{f,298}$			$\Delta H^\circ_{f,298}$	
Substance	kcal/mol	kJ/mol	Substance	kcal/mol	kJ/mol
$Au(OH)_3$	-100.0	-418.4	HCl	-22.019	-92.127
B_2H_6	7.5	31.38	NF_3	-30.4	-127
B_2O_3	-305.3	-1277	N_2H_4	12.05	50.42
BF_3	-265.4	-1110	NH_3	-11.04	-46.19
CH_4	-15.99	-66.90	Nb_2O_5	-455.2	-1905
CF_4	-216.6	-906.3	$PrCl_3$	-257.8	-1079
CO_2	-94.05	-393.5	PH_3	5.5	23
Cr_2O_3	-272.7	-1141	TaN	-58.2	-244
$H_2O(\ell)$	-68.32	-285.9	Ta_2O_5	-488.8	-2045
$H_2O(g)$	-57.11	-238.9	$Tl(OH)_3$	-122.6	-513.0
HBr	-8.66	-36.2	WO_3	-201.46	-842.91
HF	-64.2	-269	ZrC	-44.1	-185
HF(aq)	-78.66	-329.1	ZrO_2	-261.5	-1094
H_2S	-4.815	-20.15			

The combustion of B_4C is given by:

$$B_4C(s) + 4O_2 \rightarrow 2B_2O_3(c) + CO_2(g) \quad \Delta H^\circ_f = -683.3\ \frac{\text{kcal}}{\text{mol}}$$

Applying Hess' law:

$$2B_2O_3(s) + CO_2(g) \rightarrow B_4C(s) + 4O_2(g) \quad \Delta H^\circ_f = +683.3 \text{ kcal/mol}$$

$$C(s) + O_2(g) \rightarrow CO_2(g) \quad \Delta H^\circ_f = -93.97 \text{ kcal/mol}$$

$$4B(c) + 3O_2(g) \rightarrow 2B_2O_3(c) \quad \Delta H^\circ_f = 2(-304.1) \text{kcal/mol}$$

$$C(s) + 4B(c) \rightarrow B_4C(s) \quad \Delta H^\circ_f = -18.9 \text{ kcal/mol}$$

The standard enthalpies of formation of a large number of compounds have been determined, and they are available as Technical Notes 270-3, 270-4, 270-5, and 270-6 from the National Bureau of Standards titled, *Selected Values of Chemical Thermodynamic Properties*. Most of the values in Table 9-3 are from Lewis and Randall's *Thermodynamics*, 2nd ed., revised by K. S. Pitzer and L. Brewer, McGraw-Hill, New York (1961): and *Handbook of Chemistry and Physics*, 52nd ed., the Chemical Rubber Co., Cleveland (1971).

9-9 CALORIMETRY

Measurements of heats of reaction are often made in a calorimeter. The masses of the reactants and products are determined as accurately as possible. The quantity of heat evolved is equal to the heat absorbed by a known quantity of water surrounding the reaction vessel (bomb) and the other parts of the calorimeter. The *water equivalent* of a calorimeter is the quantity of water that will absorb the same quantity of heat as the calorimeter with all of its parts. The *energy equivalent* of a calorimeter is the quantity of heat equivalent to a 1° rise in temperature of the calorimeter assembly.

Example 9-7

The combustion of 2.650 g of indium raised the temperature of a calorimeter 1.055°C. The energy equivalent of the calorimeter was 2412.81 cal/°C. Calculate the molar energy of combustion of indium.

[C. E. Holly, Jr., E. J. Huber, Jr., and E. H. Meierkord, *J. Am. Chem. Soc.*, *74*, 1084 (1952)]

$$2413\ \frac{\text{cal}}{°\text{C}} \times 1.055°\text{C} = 2546\ \text{cal}$$

$$\frac{2546\ \text{cal}}{2.650\ \not{\text{g}}} \times 114.8\ \frac{\not{\text{g}}}{\text{mol}} = 110,300\ \frac{\text{cal}}{\text{mol}}$$

Reactions in a calorimeter occur at constant volume. The heat evolved or absorbed at constant volume is equal to the change in energy of the reaction ΔE. To obtain the value of the change in enthalpy of the reaction ΔH, a correction must be made.

$$\Delta H = \Delta E + \Delta(PV)$$

Since V is constant in a calorimeter measurement

$$\Delta H = \Delta E + \Delta P(V)$$

$$\Delta H = \Delta E + (P_2 - P_1)V$$

From PV = nRT

$$P_2 = n_2 \frac{RT}{V}$$

$$P_1 = n_1 \frac{RT}{V}$$

$$\Delta H = \Delta E = (n_2 - n_1) \frac{RT}{V} V$$

$$= \Delta E + \Delta nRT$$

$$R = 1.9872 \frac{\text{cal}}{\text{K-mol}}$$

Δn is found by subtracting the number of gaseous reactants from the number of gaseous products based on a balanced equation. Any solids or liquids present are ignored.

Example 9-8

The energy of combustion of ZrC as measured in a bomb calorimeter was -310,863 cal/mol at 25°C. Calculate the enthalpy of combustion.

[A. D. Mah and B. J. Boyle, *J. Am. Chem. Soc.*, *77*, 6513 (1955)]

$$ZrC(s) + 2O_2(g) \rightarrow ZrO_2(s) + CO_2(g)$$

$$\Delta H = \Delta E + \Delta nRT$$

Since the change in volume of the solids may be neglected

$$\Delta n = \text{moles } CO_2 - \text{moles } O_2 = 1 - 2 = -1$$

$$\Delta H = -310,863 + -1 \times 1.9872 \times 298.15$$

$$= -310,863 - 592 = -311,455 \text{ cal}$$

$$= -311.5 \frac{\text{kcal}}{\text{mol}}$$

Example 9-9

The energy of combustion of NbN, as measured in a bomb calorimeter, was -170.38 kcal/mol. Calculate the enthalpy of combustion. [A. D. Mah and N. L. Gellert, *J. Am. Chem. Soc.*, *78*, 3261 (1956)]

$$NbN(s) + \frac{5}{4}O_2(g) \rightarrow \frac{1}{2}Nb_2O_5(s) + \frac{1}{2}N_2(g)$$

$$\Delta H = \Delta E + \Delta nRT$$

$$\Delta n = \frac{1}{2} - \frac{5}{4} = -\frac{3}{4}$$

$$\Delta H = -170{,}380 + \left(-\frac{3}{4} \times 1.9872 \times 298.15\right)$$

$$= -170{,}380 - 444 = -170{,}824$$

$$= -170.82 \frac{\text{kcal}}{\text{mol}} \text{ NbN}$$

9-10 BOND ENERGIES

The *standard bond-dissociation energy* (DH°) [S. W. Benson, *J. Chem. Educ.*, *42*, 502 (1965)] is defined as the change in enthalpy in the breaking of 1 mol of a given bond, reactants and products being in their standard states of hypothetical ideal gas at 1 atm and 25°C. The bond-dissociation energy of a *diatomic substance* is the energy required to cause the dissociation of 1 mol of the substance, in the gas phase, into its constituent atoms, also in the gas phase. The bond-dissociation energy of hydrogen, for example, is 104.2 kcal/mol for the reaction $H_2(g) \rightarrow 2H(g)$.

Example 9-10

Calculate the bond-dissociation energy of O_2.

$$O_2 \rightarrow 2O$$

$$\Delta H^\circ_f(O_2) = 0$$

$$\Delta H^\circ_f(O) = 59.6 \frac{\text{kcal}}{\text{mol}}$$

$$\Delta H^\circ_{react} = 2\Delta H^\circ_f(O) - \Delta H^\circ_f(O_2)$$

$$\Delta H^\circ_{react} = 2 \times 59.6 - 0 = 119.2 \text{ kcal}$$

$$DH^\circ(O-O) = 119.2 \frac{\text{kcal}}{\text{mol}}$$

The bond energies of polyatomic substances may be determined and recorded as *average bond-dissociation energies*. The average bond-dissociation energy DH° of CH_4 is one-fourth of the total enthalpy needed to break all the bonds in CH_4.

$$CH_4(g) \rightarrow C(g) + 4H(g)$$

$$DF^\circ(C-H) = \frac{1}{4} H^\circ_{react}$$

Average bond-dissociation energies have limited use, especially from the point of view of reactivity. The average bond-dissociation energy of the O—H bond in water is 110.5 kcal. However, the energy required to break the first O—H bond in H--O--H is 118 kcal, whereas the energy required to break the second is 103 kcal. This is a significant difference.

Example 9-11

Calculate the average C—H bond energy in CH_4 from the standard enthalpy of formation of CH_4 and the atomic species resulting from the dissociation.

$$CH_4(g) \rightarrow C(g) + 4H(g)$$

$$\Delta H^\circ_f(CH_4) = -17.89 \frac{\text{kcal}}{\text{mol}}$$

$$\Delta H^\circ_f(C) = 170.9 \frac{\text{cal}}{\text{mol}}$$

$$\Delta H^\circ_f(H) = 52.10 \frac{\text{cal}}{\text{mol}}$$

$$\Delta H^\circ_{react} = \Delta H^\circ_f(C) + 4\Delta H^\circ_f(H) - \Delta H^\circ_f(CH_4)$$

$$\Delta H^\circ_{react} = 170.9 + 4(52.10) - (-17.89) = 397.2 \text{ kcal}$$

$$DH^\circ(CH_4) = \frac{1}{4}\Delta H^\circ_{react} = \frac{397.2}{4} = 99.3 \text{ kcal}$$

The DH° values of the bonds in CH_4 are 104 kcal/mol for DH°(CH_3–H), 106 kcal/mol for DH°(CH_2–H) and DH°(CH–H), and 81 kcal/mol for DH°(C–H). The average of the four values is 99.3 kcal/mol.

PROBLEMS

In the following problems, those marked with an asterisk involve calculations using the joule as a unit of energy.

9-1 Calculate the quantity of heat required to change 4.13×10^{-3} mol of solid aluminum, specific heat 0.274 cal/g, from 600°C to the liquid state at the melting point, 658°C. The heat of fusion of aluminum is 76.8 cal/g. (See Table 9-4.)

9-2 A sample of sulfur containing 2.00 mol was heated from 20 to 150°C. Calculate the heat absorbed.

Specific heat: S_r, 0.176 cal/g; S_m, 0.181 cal/g; liquid, 0.220 cal/g.

$$\Delta H_{fus} = 13.2\ \frac{cal}{g};$$

$$\Delta H_{trans} = 0.096\ \frac{kcal}{mol};$$

mp = 119°; transition temp. = 95.5°.

9-3 According to J. C. Southard, 4360 cal of heat was required to change 1 mol of crystalline boron oxide to the glass form at 298 K. Calculate the difference in heat content between 10.0 g of B_2O_3 (glass) and 10.0 g of B_2O_3(c) at 25°C.
[*J. Am. Chem. Soc.*, *63*, 3147 (1941)]

9-4 On a basis of energy evolved per kilogram of fuel, which of the following substances offers the best possibility as a rocket fuel? Assume that oxygen is the oxidizer.

a. Diborane b. Methane c. Hydrazine

9-5 Isooctane is a hydrocarbon found in gasoline. It has assigned to it an octane number of 100. Ethanol (ethyl alcohol) is added to

TABLE 9-4 Standard Enthalpies of Formation

Element	ΔH°_f kcal/mol	ΔH°_f kJ/mol	Element	ΔH°_f kcal/mol	ΔH°_f kJ/mol
H	52.1	218	Cl	28.9	121
D	53.0	222	K	21.3	89.1
Be	78.3	328	Ca	46.0	192
B	133(4)	556	Cu	81.5	341
C	170.9	715.0	Zn	31.2	131
N	113.0	472.8	Ge	78.4	328
O	59.6	249	Br	26.7	112
F	18.9	79.1	Ag	69.1	289
Na	25.8	108	As	72.9	305
Mg	35.3	148	Sn	72	300
Al	78.0	326	Sb	60.8	254
Si	106(4)	443	I	25.5	107
P	79.8	334	Hg	14.7	61.5
S	65.7	275	Pb	46.7	195

Source: S. W. Benson, *J. Chem. Educ.*, *42*, 517 (1965).

gasoline to increase the octane rating of the gasoline. The heat of formation of isooctane $(C_8H_{18})(\ell)$ is -61.97 kcal/mol and that of ethanol $(C_2H_6O)(\ell)$ is -66.36 kcal/mol. Determine the heat of combustion of each of these.

*9-6 A sample of tungsten weighing 9.5886 g was burned in a bomb calorimeter and 43.549 J of heat was evolved. Calculate the energy of combustion of tungsten (at the temperature of the experiment) in kilocalories per gram atom.

[G. Huff, E. Squitieri, and P. E. Snyder, *J. Am. Chem. Soc.*, *70*, 3380 (1948)]

*9-7 The heat of formation of curium sesquioxide (Cm_2O_3) was recently measured by reactions in 6 M HCl. From the following heats of reaction, determine the heat of formation of Cm_2O_3.

	ΔH_r kJ/mol
$Cm_2O_3(s) + 6HCl(aq) \rightarrow 2CmCl_3(aq) + 3H_2O(\ell)$	-400
$Cm(s) + 6HCl(aq) \rightarrow CmCl_3(aq) + 3H_2(g)$	-611
$H_2(g) + O_2(g) \rightarrow H_2O(\ell)$ (in HCl)	-286.6

[L. R. Morse, J. Fuger, J. Goffort, and R. G. Haire, *Inorg. Chem.*, *22*, 1993 (1983)]

9-8 What is the standard enthalpy of formation of chromium hexacarbonyl, $Cr(CO)_6$, if the standard enthalpy of combustion was found to be -443.09 kcal/mol?
[F. A. Cotton, A. K. Fischer, and G. Wilkinson, *J. Am. Chem. Soc.*, *78*, 5169 (1956)]

9-9 Humphrey and O'Brien determined the standard enthalpy of combustion of tin to tin(IV) oxide to be -138,820 cal/mol. The heat of combustion of SnO(s) at 25°C was found to be -70,470 cal/mol. Calculate the standard heat of formation of tin(II) oxide.
[*J. Am. Chem. Soc.*, *75*, 2805 (1953)]

9-10 The energy of combustion ($\Delta E°_c$) for solid cubane (C_8H_8) was found to be -1156.0 kcal/mol. What is $\Delta H°_f$ of cubane?
[Kybett et al., *J. Am. Chem. Soc.*, *88*, 626 (1966)]

9-11 Walker et al. determined the enthalpy of formation of CF_3CN by two separate measurements. The heat of reaction of

$$2CF_3CN(g) \rightarrow C_2F_6(g) + C_2N_2(g)$$

was -10.54 kcal/mol at 25°C, and the heat of reaction of

$$CF_3CN(g) + \frac{5}{3}NF_3(g) \rightarrow 2CF_4(g) + \frac{4}{3}N_2(g)$$

was -274.8 kcal/mol at 25°C. Given the following data, calculate the enthalpy of formation of CF_3CN from each reaction.

Compound	ΔH°_f (kcal/mol)
$CF_4(g)$	-223.05
$NF_3(g)$	-31.43
$C_2F_6(g)$	-321.2
$C_2N_2(g)$	+73.87

[Walker, Sinke, Perette, Jong, *J. Am. Chem. Soc.*, *92*, 4525 (1970)]

9-12 Calculate the heat of combustion at constant pressure of TaN. Assume that nitrogen gas and tantalum(V) oxide are the products. The measured value was -184.38 kcal/mol, according to Mah and Gellert.

[A. D. Mah and N. L. Gellert, *J. Am. Chem. Soc.*, *78*, 3261 (1956)]

9-13 The heat of reaction for the atomization process

$$UCl_4(g) \rightarrow U(g) + 4Cl(g)$$

was found to be 437.5 kcal/mol. If the heat of formation of UCl_4 is -194.5 kcal/mol, what is the heat of formation of U(g)? Compare this result to that in Problem 9-64.

[K. Lau and D. L. Hildebrand, *J. Chem. Phys.*, *80*(3), 1312 (1984)]

9-14 Calculate the heat of combustion of ZrC from the data in Table 9-3 and the equation

$$ZrC(s) + 2O_2(g) \rightarrow ZrO_2(s) + CO_2(g).$$

[A. D. Mah and B. J. Boyle, *J. Am. Chem. Soc.*, *77*, 6513 (1955)]

9-15 Calculate the standard heat of formation of ONF from the following data:

$$NO + \frac{1}{2}F = ONF \qquad \Delta H^\circ = -37.4 \text{ kcal/mol}$$

$$\frac{1}{2}N_2 + \frac{1}{2}O_2 = NO \qquad \Delta H^\circ_f = +21.6 \text{ kcal/mol}$$

[H. S. Johnston and H. J. Bertin, Jr., *J. Am. Chem. Soc.*, *81*, 6404 (1959)]

9-16 The heat of combustion of B_4C, as measured in a bomb calorimeter, was -682.0 kcal/mol at 298 K. Calculate the standard heat of formation of B_4C. (This paper is recommended for student reading.)

[D. Smith, A. S. Dworkin, and E. R. Van Artsdalen, *J. Am. Chem. Soc.*, *77*, 2655 (1955)]

9-17 A compound prepared by the action of bromine trifluoride on gold had a mass ratio of compound to gold of 1.290. A sample of the compound was hydrolyzed, and the heat of the reaction with water was -39.4 kcal/mol. Calculate the standard enthalpy of formation of the compound using HF(aq) in the equation.

[A. A. Woolf, *J. Chem. Soc.*, *1954*, 4694]

9-18 The heats of combustion of hafnium metal and hafnium nitride, HfN, at 25°C were -265,463 cal/mol and -177,516 cal/mol, respectively, as measured in a bomb calorimeter. The other product of the combustion of HfN was nitrogen gas. What is the standard enthalpy of formation of HfN as calculated from these data?

[G. L. Humphrey, *J. Am. Chem. Soc.*, *75*, 2806 (1953)]

9-19 The energy of formation of arsenic pentafluoride, as measured by direct combination of the elements in a bomb calorimeter, was -3933.4 cal/g of arsenic at 25°. Calculate the standard enthalpy of formation of arsenic pentafluoride and the average bond-dissociation energy of the As–F bond in the compound.

[P. A. G. O'Hare and W. N. Hubbard, *J. Phys. Chem.*, *69*, 4360 (1965)]

9-20 The value of $\Delta E°_f$ of ruthenium pentafluoride, as measured in a fluorine bomb calorimeter by direct combination, was -2096.8 cal/g. Calculate the standard enthalpy of formation of RuF_5.

[H. A. Porter, E. Greenberg, and W. N. Hubbard, *J. Phys. Chem.*, *69*, 2308 (1965)]

9-21 The heat of solution of 0.3086 g of $ThCl_4$ was 36.49 cal. What was the heat of solution, in kilocalories per mole?

[L. Eyring and E. F. Westrum, Jr., *J. Am. Chem. Soc.*, *72*, 5555 (1950)]

9-22 Calculate the heat of formation of $B_3N_3H_3Cl_3(s)$ from the heat of solution, -113.8 kcal, as represented by the equation

$$B_3N_3H_3Cl_3(s) + 9H_2O(\ell) + aq \rightarrow 3H_3BO_3(aq) + 3NH_4Cl(aq)$$

The standard heats of formation of $NH_4Cl(aq)$ and $H_3BO_3(aq)$ are -71.76 kcal/mol and -255.2 kcal/mol, respectively.
[E. R. Van Artsdalen and A. S. Dworkin, *J. Am. Chem. Soc.*, *74*, 3402 (1952)]

*9-23 The combustion of 2.724 g of indium raised the temperature of a calorimeter 1.083°C. The energy equivalent of the calorimeter was 10,095.2 J/°C. Calculate the molar energy of combustion of indium in kilojoules and kilocalories.
[C. E. Hally, E. J. Huber, Jr., and E. H. Meierkord, *J. Am. Chem. Soc.*, *74*, 1084 (1952)]

*9-24 In one of a series of measurements to calibrate a calorimeter, the combustion of 1.08907 g of benzoic acid (heat of combustion 26.4284 kJ/g) led to a temperature rise of 2.15266°C. Heat from other sources amounted to 21.6 cal. Calculate the energy equivalent of the calorimeter, in calories per degree Celsius.
[D. J. Solley and J. B. Gray, *J. Am. Chem. Soc.*, *70*, 2651 (1948)]

9-25 Calculate the bond-dissociation energy of fluorine from the value of ΔH°_f of F, which is 18.9 kcal/mol.

9-26 Lau and Hildebrand measured the four bond-dissociation energies of UCl_4. They obtained the following values:

	DH°_{298} kcal/mol
UCl	108
UCl-Cl	117
UCl_2-Cl	117
UCl_3-Cl	101

What is the average bond-dissociation energy for UCl_4?
[*J. Chem. Phys.*, *80*(3), 1312 (1984)]

9-27 The atomization reaction is the enthalpy needed to completely dissociate a molecule into its component atoms. Lau and Hildebrand measured an enthalpy of reaction for UCl_4 of 437.5 kcal/mol. How does this compare to the result obtained from the DH° in Problem 9-26?
[*J. Chem. Phys.*, *80*(3), 1312 (1984)]

9-28 The enthalpy of atomization for the following reactions was recently measured by Ho, Binkley, Coltrin, and Melios:

	$\Delta H_{rx,298}$ kcal/mol
$SiH_3(g) \rightarrow Si(g) + 3H(g)$	216.20
$SiH_4(g) \rightarrow Si(g) + 4H(g)$	307.92

What is the first bond-dissociation energy of SiH_3-H from these data?
[*J. Phys. Chem.*, *89*(21), 4654 (1985)]

9-29 Calculate the N—H bond-dissociation energy in NH_2 and compare it with the average N—H bond-dissociation energy in ammonia. The standard enthalpy of formation of $NH_2(g)$ is 40 kcal/mol.

9-30 What is the bond-dissociation energy of the bonds in hydrogen bromide?

Supplementary Problems

9-31 The energy of a waterfall was converted into useful work. If 20 kJ of work was done by the waterfall per unit volume of water, and 30 kJ of heat was lost to the water per unit volume of water, what is the energy change of this unit volume of water?

9-32 Calculate the heat evolved if 8.60 mol of Bi is cooled from 300 to 200°C. The heat of fusion of bismuth is 12.64 cal/g; its melting point is 268°. The specific heat of the liquid is 0.0363 cal/g, and that of the solid is 0.0338 cal/g.

9-33 The standard enthalpy of formation of amorphous boron oxide was determined as -299.74 kcal/mol, and that of crystalline boron oxide was -304.1 kcal/mol. Calculate the standard enthalpy of transition from amorphous boron oxide to crystalline boron oxide.
[W. D. Good and M. Mansson, *J. Phys. Chem.*, *70*, 97 (1966)]

9-34 The standard heat of formation of U_3O_8 is -853.5 kcal/mol. The standard heat of the reaction $3UO_2 + O_2 \rightarrow U_3C_8$ is -76.01 kcal according to E. J. Huber, Jr., C. E. Hally, and E. H. Meierkord. Calculate the standard heat of formation of UO_2.
[*J. Am. Chem. Soc.*, *74*, 3406 (1952)]

9-35 Calculate the standard enthalpy of formation of PrOCl from the heat of hydrolysis of $PrCl_3$, 22.98 kcal. The equation is

$$PrCl_3(s) + H_2O(g) \rightarrow PrOCl(s) + 2HCl(g)$$

[C. W. Koch and B. B. Cunningham, *J. Am. Chem. Soc.*, *76*, 1473 (1954)]

9-36 Calculate the standard heat of formation of melamine, $C_3H_6N_6$, from the heat of combustion at constant pressure, -471.76 kcal/mol.

[D. J. Solley and J. B. Gray, *J. Am. Chem. Soc.*, *70*, 2650 (1948)]

9-37 What is the standard enthalpy of formation of NbN if the heat of combustion at 25° (corrected for change in pressure) was -170.83 kcal/mol?

[A. D. Mah and N. L. Gellert, *J. Am. Chem. Soc.*, *78*, 3261 (1956)]

9-38 Calculate the heat of formation of WC(s) from the heat of reaction with oxygen to yield WO_3 and CO_2 (-285.80 kcal/mol) and the heat of formation of CO_2 from graphite.

[G. Huff, E. Squitieri, and P. E. Snyder, *J. Am. Chem. Soc.*, *70*, 3381 (1948)]

9-39 The energy evolved during the combustion in a bomb calorimeter of 1 g of dicyandiamide, $C_2H_4N_4$, was 3943.1 cal. Calculate the heat of combustion at constant pressure and the standard enthalpy of formation of dicyandiamide.

[D. J. Solley and J. B. Gray, *J. Am. Chem. Soc.*, *70*, 2650 (1948)]

9-40 The energy evolved during combustion of indium to In_2O_3 in a bomb calorimeter was 959.8 cal/g at 25°C. Calculate the standard enthalpy of formation of In_2O_3.

[C. E. Hally, E. J. Huber, Jr., and E. H. Meierkord, *J. Am. Chem. Soc.*, *74*, 1084 (1952)]

9-41 Calculate the standard enthalpy of formation of cadmium oxide from the heat of combustion of cadmium at 25°C obtained in a bomb calorimeter, 540.2 cal/g.

[A. D. Mah, *J. Am. Chem. Soc.*, *76*, 3363 (1954)]

9-42 Calculate the standard enthalpy of formation of TlF_3 from the heat of hydrolysis, -8.4 kcal. The reaction was

$$TlF_3 + 3H_2O \rightarrow Tl(OH)_3 + 3HF(aq)$$

[A. A. Woolf, *J. Chem. Soc.*, *1954*, 4694]

9-43 The heat of combustion of ZrB_2 at 25°C was found to be -486.1 kcal/mol. Calculate the standard enthalpy of formation of ZrB_2.

[E. J. Huber, E. L. Head, and C. E. Halley, *J. Phys. Chem.*, *68*, 3040 (1964)]

9-44 The standard enthalpy of formation of boron nitride was calculated from its measured heat of reaction with fluorine, which was -211.68 kcal/mol (after correction for pressure change). Make the calculation, assuming that the products were boron trifluoride and nitrogen.

[S. S. Wise, J. L. Margrave, H. M. Feder, and W. N. Hubbard, *J. Phys. Chem.*, *70*, 7 (1966)]

9-45 Calculate the average bond-dissociation energy of the B—F bonds in boron trifluoride from the fact that 25,083 cal/g of B was evolved during combination of the elements in a bomb calorimeter.

[G. K. Johnson, H. M. Feder, and W. N. Hubbard, *J. Phys. Chem.*, *70*, 1 (1966)]

9-46 From the heat of formation of SiF_4 (-385.98 kcal/mol) and the heat of reaction of the fluorination of α-quartz, calculate the standard heat of formation of α-quartz.

$$SiO_2(\alpha\text{-quartz}) + 2F_2(g) \rightarrow SiF_4(g) + O_2(g)$$

$$\Delta H^\circ = -168.26 \text{ kcal/mol}$$

[S. S. Wise, J. L. Margrave, H. M. Feder, and W. N. Hubbard, *J. Phys. Chem.*, *67*, 819 (1963)]

9-47 The bond dissociation energy of SiH was measured from its atomization reaction. The atomization enthalpy corrected for systematic errors was found to be 68.1 kcal/mol. What is the enthalpy of formation of SiH based on this data? The atomization reaction is

$$SiH(g) \rightarrow Si(g) + H(g)$$

[P. Ho., N. Coltrin, J. Binkley, and C. Melios, *J. Phys. Chem.*, *89*(21), 4654 (1985)]

9-48 Calculate the heat of formation of hydrogen fluoride from the following data and any other required data from Table 9-3.

$$SiO_2(\alpha\text{-quartz}) + 2F_2(g) \rightarrow SiF_4(g) + O_2 \qquad \Delta H° = -168.26 \text{ kcal/mol}$$

$$SiO_2(\text{crystobalite}) \rightarrow SiO_2(\alpha\text{-quartz}) \qquad \Delta H° = -0.35 \text{ kcal/mol}$$

$$SiO_2(\text{crystobalite}) + 4HF(g) \rightarrow SiF_4(g) + 2H_2O(g) \qquad \Delta H° = -24.53 \text{ kcal/mol}$$

[H. M. Feder, W. N. Hubbard, S. S. Wise, and J. L. Margrave, *J. Phys. Chem.*, *67*, 1148 (1963)]

9-49 The energy of formation of yttrium trifluoride was measured by direct combination of the elements in a fluorine bomb calorimeter. The value at 25°C was -409.8 kcal/mol. Calculate the standard enthalpy of formation of $YF_3(s)$ at 25°C.

[E. Rudzitis, H. M. Feder, and W. N. Hubbard, *J. Phys. Chem.*, *69*, 2305 (1965)]

9-50 Calculate the heat of formation of lithium hydride from its heat of solution (-31.76 kcal/mol) and the heat of reaction of lithium metal with a large excess of water (-53.10 kcal/mol).

[C. E. Messer, L. G. Fasolino, and C. E. Thalmayer, *J. Am. Chem. Soc.*, *77*, 4525 (1955)]

9-51 Calculate the heat of solution of potassium hydride $KH(s) + H_2O(\ell) + aq \rightarrow KOH(aq) + H_2(g)$. The heat of reaction of potassium metal with a large excess of water is -47.05 kcal/mol, and the heat of formation of KH(s) is -15.16 kcal/mol.

[C. E. Messer, L. G. Fasolino, and C. E. Thalmayer, *J. Am. Chem. Soc.*, *77*, 4525 (1955)]

9-52 The combustion of 2.863 g of indium raised the temperature of a calorimeter 1.140°C. The energy equivalent of the calorimeter was 2412.81 cal/°C. Calculate the molar energy of combustion of indium.

[C. E. Halley, E. J. Huber, Jr., and E. H. Meierkord, *J. Am. Chem. Soc.*, *74*, 1084 (1952)]

9-53 The heat of combustion of liquid bicyclo[2,2,2]octene (C_8H_{12}) was determined recently by Wong and Westrum. $\Delta E°_c$ was -1193.42 kcal/mol. What is the enthalpy of formation of this compound?

[*J. Am. Chem. Soc.*, *93*, 5317 (1971)]

9-54 Calculate the enthalpy of formation of In_2O_3 from the following calorimetric data:

mass of In, 3.116 g
temperature increase, 1.241°C
energy equivalent of calorimeter, 2412.81 cal/°C

[C. E. Hally, E. J. Huber, Jr., and E. H. Meierkord, *J. Am. Chem. Soc.*, *74*, 1084 (1952)]

9-55 The energy equivalent of a calorimeter was determined from heats of combustion of benzoic acid as 3207.1 cal/°C. Calculate the water equivalent of the calorimeter. The specific heat of water is 0.99828 cal/g°C.
[D. J. Solley and J. B. Gray, *J. Am. Chem. Soc.*, *70*, 2651 (1948)]

9-56 The combustion of 1.000 g of salicylic acid, $C_7H_6O_3$, in a calorimeter whose water equivalent was 2,343.8 g produced a temperature rise of 2.240°C. Calculate the molar heat of combustion of salicylic acid. (The specific heat of water is 0.99828 cal/g°C.)
[F. A. Cotton, A. K. Fischer, and G. Wilkinson, *J. Am. Chem. Soc.*, *78*, 5168 (1956)]

9-57 Calculate the H—Cl bond-dissociation energy.

9-58 The successive bond-dissociation energies of SiH_4 were recently measured by Ho, Binkley, et al. The following data were obtained:

	DH°_{298} kcal/mol
$SiH_3\text{-}H \rightarrow SiH_3 + H$	91.7
$SiH_2\text{-}H \rightarrow SiH_2 + H$	72.4
$SiH\text{-}H \rightarrow SiH + H$	75.7
$Si\text{-}H \rightarrow Si + H$	68.1

What is the average bond-dissociation energy in SiH_4?
[*J. Phys. Chem.*, *89*(21), 4654 (1985)]

9-59 Compare the average N—F bond-dissociation energy in NF_3, calculated from the data in Tables 9-3 and 9-4, with the F_2N—F bond-dissociation energy. The standard enthalpy of formation of $NF_2(g)$ is 10.0 kcal/mol.

9-60 Calculate the average bond-dissociation energy in hydrogen sulfide.

9-61 Calculate the average bond-dissociation energy of the bonds in carbon tetrachloride. $\Delta H^\circ_f(CCl_4)_g = -25.50$ kcal/mol.

9-62 Calculate the average bond-dissociation energy of Kr—F. The standard enthalpy of formation of KrF_2(g) is +13.7 kcal/mol.
[S. Gunn, *J. Am. Chem. Soc.*, *88*, 5924 (1966)]

*9-63 The heat of formation of $CuGe_2$(g) was determined based on the reaction

$$CuGe_2(g) \rightarrow Cu(g) + 2Ge(g)$$

The enthalpy of this reaction was calculated to be 501.5 kJ/mol, assuming the molecule is Ge-Ge-Cu and bent. The heat of vaporization of Ge is 374.5 kJ/mol, and that of Cu is 336.8 kJ/mol. Estimate the enthalpy of formation of $CuGe_2$ from these data and compare this to the best value of 580 ± 28 kJ/mol determined by the authors.
[J. Kincade, K. Gingerich, and U. Choudary, *J. Phys. Chem.*, *82*(1), 49 (1978)]

9-64 Armentrout and Beauchamp found the enthalpy of formation of UF(g) at 298°C to be 22 kcal/mol, and that of U(g) was 128 kcal/mol. What is the bond-dissociation energy for UF(g) from these data?
[*J. Phys. Chem.*, *85*(26), 4103 (1981)]

10
SOLUTIONS OF ELECTROLYTES

Electrolytes are substances that conduct electricity in aqueous solutions. Typical electrolytes, such as sodium chloride, potassium sulfate, and sodium hydroxide, are ionic in the solid state. When they dissolve, the solution contains hydrated cations and anions that were present as unhydrated ions in the crystals. Other electrolytes, such as acids like hydrogen chloride, are molecular in nature until they dissolve in water. The molecules react with water to form ions.

10-1 COLLIGATIVE PROPERTIES

Whether the electrolyte is ionic or molecular in nature, the number of solute particles is greater than if no ions were present, and hence the colligative properties are greater.

Example 10-1

$$NaCl(s) \rightarrow Na^+(aq) + Cl^-(aq)$$

Example 10-2

$$K_2SO_4(s) \rightarrow 2K^+(aq) + SO_4^{2-}(aq)$$

Example 10-3

$$NaOH(s) \rightarrow Na^+(aq) + OH^-(aq)$$

Example 10-4

$$HCl(g) + H_2O(\ell) \rightarrow H_3O^+(aq) + Cl^-(aq)$$

In many cases an equilibrium between un-ionized molecules and ions is established. Acetic acid and ammonia solutions are typical examples.

Example 10-5

$$CH_3COOH(aq) + H_2O(\ell) \rightleftharpoons H_3O^+(aq) + CH_3COO^-(aq)$$

Example 10-6

$$NH_3(aq) + H_2O(\ell) \rightleftharpoons NH_4^+(aq) + OH^-(aq)$$

The concentration of ions is small here because the equilibrium lies far to the left. These reactions are considered in Chapter 11.

Example 10-7

The vapor pressures of a series of aqueous solutions of potassium chloride at 20.0°C were measured by B. F. Lovelace, J. C. W. Frazer, and E. Miller. In one experiment 0.20 mol of KCl lowered the vapor pressure of 1000 g of water in 0.110 mm. If KCl were molecular and did not dissociate in solution, the vapor-pressure lowering (depression) expected from Raoult's law could be calculated.

[*J. Am. Chem. Soc.*, *38*, 515 (1916)]

$$P = P^\circ X$$

$$P^\circ = 17.535 \text{ mm at } 20.0^\circ\text{C (from table)}$$

$$X = \frac{\text{moles of solvent}}{\text{moles of solvent + moles of solute}}$$

$$H_2O = 18.0 \frac{\text{g}}{\text{mol}}$$

$$\text{Moles of solvent} = 1000 \text{ g} \times \frac{1 \text{ mol}}{18.0 \text{ g}} = 55.56$$

$$X = \frac{55.56}{55.56 + 0.20}$$

$$P = 17.54 \frac{55.56}{55.76} = 17.48 \text{ mm}$$

$$\begin{array}{r} 17.54 \\ -17.48 \\ \hline 0.06 \end{array}$$

0.06 mm calculated depression

0.110 mm observed depression

The observed depression is nearly twice as great, indicating that KCl exists as $K^+(aq)$ and $Cl^-(aq)$ ions in the solution.

Example 10-8

A solution containing 1.195 g of NaCl in 100 g of H_2O exhibited a freezing-point depression of 0.693°. Compare this result with the values expected if no dissociation occurred, and if complete dissociation occurred.

[L. Kahlenberg, *J. Phys. Chem.*, *5*, 353 (1901)]

$$\Delta t_f = K_b m$$

$$m = \frac{\text{mol NaCl}}{\text{kg } H_2O} \qquad 1 \text{ kg} = 1000 \text{ g}$$

$$1.195 \text{ g} \times \frac{1000 \text{ g}}{100 \text{ g}} = \frac{11.95 \text{ g NaCl}}{\text{kg } H_2O}$$

$$\begin{array}{ll} \text{NaCl} & 22.99 \\ & \underline{35.45} \\ & 58.44 \text{ g/mol} \end{array}$$

$$11.95 \text{ g} \times \frac{1 \text{ mol}}{58.44 \text{ g}} = 0.2045 \text{ mol}$$

$$\Delta t_f = 1.86 \frac{°\text{C}}{\text{mol/kg}} \times 0.204 \frac{\text{mol}}{\text{kg}} = 0.379°\text{C}$$

This is the freezing-point depression expected if sodium chloride were not dissociated. A depression of 2 × 0.379° = 0.758° would be expected if NaCl were completely dissociated and no other complicating effects were present. The observed value indicates that NaCl is indeed highly dissociated, but other effects must be involved.

Example 10-9

The boiling-point elevation of a 0.0500 m solution of sodium chloride was observed to be 0.0479° by R. P. Smith. Compare this result with the values predicted assuming no dissociation, and assuming complete dissociation.

[*J. Am. Chem. Soc.*, *61*, 500 (1939)]

$$\Delta t_b = k_b m$$

$$\Delta t_b = 0.52 \frac{°C}{mol/kg} \times 0.050 \frac{mol}{kg} = 0.026°C$$

This is the elevation expected if no dissociation occurred. If complete dissociation occurred and there were no other complicating effects, the elevation would be 0.052°.

10-2 CONCENTRATION OF IONS

It is common practice to describe a solution by designating the concentration of one of the ions present. Chemical reactions involve ions, and the particular ion in question may be the participant in a chemical reaction.

Example 10-10

A 1.60 M solution of sodium perchlorate was used in a kinetics investigation. The sodium ion concentration and the perchlorate ion concentration were each 1.60 M as well.

[R. C. Thompson and G. Gordon, *Inorg. Chem.*, 5, 563 (1966)]

$$NaClO_4(s) \rightarrow Na^+(aq) + ClO_4^-(aq)$$

Example 10-11

In the same study as in Example 10-10, a 1.69 M zinc perchlorate solution was used. In this case the zinc ion concentration was 1.69 M, but the perchlorate ion concentration was 3.38 M.

$$Zn(ClO_4)_2(s) \rightarrow Zn^{2+}(aq) + 2ClO_4^-(aq)$$

Example 10-12

Calculate the quantity of sodium bromide required to make 100 mL of a 0.033 M Br^- solution.

$$NaBr(s) \rightarrow Na^+(aq) + Br^-(aq)$$

Since 1 mol of NaBr yields 1 mol of $Br^-(aq)$, 0.033 mol of NaBr is required to yield 0.033 mol of $Br^-(aq)$. A 0.033 M solution contains 0.033 mol of solute in 1000 mL of solution. Since 100 mL of solution is required here, only 0.0033 mol of NaBr is needed.

$$0.033 \text{ mol} \times \frac{100 \text{ mL}}{1000 \text{ mL}} = 0.0033 = 3.3 \times 10^{-3} \text{ mol}$$

NaBr	23.0
	79.9
	102.9 g/mol

$$3.3 \times 10^{-3} \text{ mol} \times 103 \frac{\text{g}}{\text{mol}} = 0.34 \text{ g NaBr}$$

Example 10-13

If the 0.0033 M Br^- solution was prepared from magnesium bromide, how many grams of salt were required?

$$MgBr_2 \rightarrow Mg^{2+}(aq) + 2Br^-(aq)$$

Since 1 mol of $Br^-(aq)$ requires 0.5 mol $MgBr_2$, 0.033 mol $Br^-(aq)$ will require only 0.0165 mol $MgBr_2$. For 100 ml of a 0.0165 M solution, 0.00165 mol $MgBr_2$ is required.

$MgBr_2$		24.3
	2 × 79.9	159.8
		184.1 g/mol

$$1.65 \times 10^{-3} \text{ mol} \times 184 \frac{\text{g}}{\text{mol}} = 0.304 \text{ g } MgBr_2$$

Example 10-14

A 3.0 M solution of perchloric acid will yield 3.0 M H_3O^+ and 3.0 M ClO_4^-.

$$HClO_4(\ell) + H_2O(\ell) \rightarrow H_3O^+(aq) + ClO_4^-(aq)$$

10-3 ANALYTICAL CONCENTRATION

The *analytical concentration* of a substance is the total number of moles of a pure substance dissolved in 1 L of solution. The analytical concentration of electrolytes is different from the concentration of the species actually present in the solution. A solution prepared by dissolving 7.45 g of KCl in sufficient water to make a liter of solution has an analytical concentration of KCl of 0.1 M. For all practical purposes the solution contains no KCl, only hydrated K^+ and Cl^-. A 0.01 M solution of acetic acid has an analytical concentration of 0.0100. The actual concentration of acetic acid molecules present in the solution is 0.00987 M. The remaining acetic acid is present as hydronium ions and acetate ions. The symbol C is often used to designate the analytical concentration. The analytical concentration of $NaClO_4$ in Example 10-10 is 1.60 M. The analytical concentration of $HClO_4$ in Example 10-14 is 3.0 M.

10-4 pH AND pOH

Very dilute solutions of electrolytes are in common use in the laboratory. Solutions of acids in which the hydrogen ion concentration (or, more accurately, the concentration of the hydronium ion: $[H_3O^+]$) has values of 1.0×10^{-4} to 1.0×10^{-10} mol/L are frequently encountered. Such concentrations are often expressed on the pH scale. The pH of a solution is defined by

$$pH = -\log[H_3O^+]$$

The following examples have been solved using the table of logarithms (Table D-2). They may be solved more easily by using a hand calculator, with which values of logarithms may be obtained directly.

Example 10-15

Express the hydrogen ion concentration of a 0.0217 M H_3O^+ solution on the pH scale.

$$[H_3O^+] = 0.0217 \text{ M}$$

$$pH = -\log 0.0217 = -(-2 + 0.337)$$

$$pH = 1.66$$

Remember, a pH is a *logarithm* and is the sum of a characteristic and a mantissa. The characteristic denotes the location of the decimal point. Since the pH scale is limited to dilute solutions, the characteristic is always negative. The mantissa is unique to a particular combination of digits and is always positive. Here the characteristic is -2 and the mantissa from Table D-2 is 0.337.

Example 10-16

Express the hydrogen ion concentration of the following solution on the pH scale.

Method 1

$[H_3O^+] = 6.02 \times 10^{-4}$ M

pH $= -\log 6.02 \times 10^{-4}$

$= -(-4 + .780) = 3.22$

In the example, -4 is the characteristic and 0.780 is the mantissa obtained from Table D-2.

Method 2

$pH = -\log 6.02 \times 10^{-4} = -(\log 6.02 + \log 10^{-4})$

The logarithm of a product (6.02×10^{-4}) is the sum of the logarithms of the terms.

$pH = -[0.780 + (-4.000)] = 3.22$

Example 10-17

$[H_3O^+]$ mol/L		Logarithm	pH
(a)	3.71×10^{-8}	$pH = -\log 3.71 \times 10^{-8}$	7.43
(b)	0.00765	$pH = -\log 0.00765$	2.12
(c)	1.00×10^{-5}	$pH = -\log 1.00 \times 10^{-5}$	5.00
(d)	21.2×10^{-9}	$pH = -\log 21.2 \times 10^{-9}$	7.67

Note in (c) that the pH is equal to the absolute magnitude of the exponent 5 because the logarithm of 1.00 is zero. In (d) it is necessary to express the concentration as 2.12×10^{-8} in order to use the exponent as the characteristic of the logarithm in method 1. Method 2 will lead directly to the answer, however. When the hydrogen ion concentration is expressed as a power of 10, for example, 3.21×10^{-8}, the pH is always less than the absolute value of the exponent but greater than the absolute value of the exponent minus 1.

Example 10-18

If the pH of a solution was 5.34 what was the hydrogen ion concentration in moles per liter?

$pH = 5.34$

$-\log[H_3O^+] = 5.34$

$\log[H_3O^+] = -5.34$

Because this logarithm is *negative*, it may be expressed as a negative characteristic and a positive mantissa.

$$-6.00 + 0.66 = -5.34$$

$$\log[H_3O^+] = -6.00 + 0.66$$

$$[H_3O^+] = 4.57 \times 10^{-6} \text{ mol/L}$$

Example 10-19

pH		Logarithm	$[H_3O^+]$ mol/L
(a)	5.00	$-\log[H_3O^+] = 5.00$	1.00×10^{-5}
(b)	4.38	$-\log[H_3O^+] = 4.38$	4.17×10^{-5}
(c)	8.09	$-\log[H_3O^+] = 8.09$	8.13×10^{-9}
(d)	11.18	$-\log[H_3O^+] = 11.18$	6.61×10^{-12}

In (a) the exponent is equal to the characteristic of the logarithm, since the mantissa is 0. In (b) to (d) the exponent is equal to the characteristic plus 1.

The hydroxide ion concentration of aqueous solutions may be expressed as pOH, which is analogous to pH.

$$pOH = -\log [OH^-]$$

Example 10-20

The pOH of a 0.00200 M solution of sodium hydroxide may be calculated as follows.

Sodium hydroxide is a strong electrolyte, i.e., completely ionized.

$$NaOH(aq) \rightarrow Na^+(aq) + OH^-(aq)$$

The $[OH^-]$ is, therefore, 0.00200 mol/L, since 1 mol of OH^- is formed for each mole of NaOH dissolved.

$$pOH = -\log 0.00200 = -\log 2.00 \times 10^{-3}$$

$$pOH = -[0.301 + (-3.000)] = 2.70$$

PROBLEMS

10-1 What is the pH of the following solutions whose hydronium ion concentrations are listed?

a. 6.91×10^{-8} M
b. 8.71×10^{-10} M
c. 1.21×10^{-3} M
d. 0.0111 M
e. 3.71×10^{-4} M

10-2 What is the pH of the solutions of strong acids listed below?

a. 3.81×10^{-3} M $HClO_4$
b. 6.72×10^{-4} M HBr
c. 0.000179 M HCl
d. 5.42×10^{-4} M HNO_3
e. 9.81×10^{-3} M HCl

10-3 What is the hydrogen ion concentration of the solutions whose pHs are listed?

a. 9.95
b. 1.97
c. 5.75
d. 10.08
e. 2.42

10-4 What is the pH of each of the solutions that have been prepared by diluting the concentrated strong acid?

a. 1.53 mL of 12.0 M HCl diluted to 1.00 L
b. 2.12 mL of 5.43 M HBr diluted to 1.00 L
c. 0.862 mL of 3.17 M $HClO_4$ diluted to 2.50 L

10-5 What was the pH of a solution of nitric acid prepared by diluting 1.00 mL of a 70.0% solution to 3.25 L? The specific gravity of the concentrated acid was 1.42.

10-6 What was the pH of a solution made by diluting 8.50 mL of 38.0% HCl to 1.50 L? The density of the concentrated acid was 1.19 g/mL.

10-7 What volume of 70.0% perchloric acid (sp. gr. 1.172) would be required to make 1.00 L of a solution whose pH was 4.11?

10-8 What volume of 36.5% hydrochloric acid (sp. gr. 1.18) would be required to make 1.00 L of a solution whose pH was 3.00?

10-9 Calculate the potassium ion concentration, in moles per liter, of a solution prepared by dissolving 0.217 g of potassium nitrate in sufficient water to make 100 mL of solution.

10-10 What is the concentration, in moles per liter, of sodium ions in a solution containing 17.3 g of sodium sulfate in 1.00 L of solution?

10-11 A solution containing 0.200 mol of sodium ions in 100 mL of solution was required for an experiment. What mass, in grams, of sodium perchlorate was required to prepare the solution?

10-12 How many grams of potassium perchlorate would be required to prepare 100 mL of a solution containing 0.200 mol of potassium ion? After you have made this calculation, consult a handbook for the solubility of potassium perchlorate in water.

10-13 Is it possible to prepare a solution of lithium fluoride in which the lithium ion concentration is 0.0100 M? If it is possible, how many grams of lithium fluoride would be required to make 50.0 mL of such a solution?

10-14 Compare the observed vapor-pressure lowering of 0.217 mm for a 0.40 m aqueous solution of KCl at 20.0°C with that calculated from Raoult's law.

[B. F. Lovelace, J. C. W. Frazer, and E. Miller, *J. Am. Chem. Soc.*, *38*, 515 (1916)]

10-15 Calculate the boiling-point elevation expected for a 0.079 m aqueous solution of sodium chloride assuming complete dissociation. Compare your answer with the observed elevation of R. P. Smith: 0.0751°.

[*J. Am. Chem. Soc.*, *61*, 500 (1939)]

10-16 Calculate the freezing-point depression expected for a 0.00443 m solution of potassium nitrate in water.

[L. D. Adams, *J. Am. Chem. Soc.*, *37*, 494 (1915)]

10-17 Calculate the predicted boiling-point elevation of a solution of 1.9501 g of cadmium iodide in 44.69 g of water.

[E. Beckmann, *Z. Physik. Chem. (Leipzig)*, *6*, 460 (1890)]

10-18 Compare the results of a freezing-point depression experiment, in which 0.699 g of magnesium sulfate in 100 g of water had a freezing point of -0.154°C, with the value calculated from theory.

[L. Kahlenberg, *J. Phys. Chem.*, *5*, 353 (1901)]

10-19 Plot the vapor pressures *calculated* for each of the solutions listed versus the molal concentration. Plot the observed vapor pressure for each solution on the same graph. Discuss the results.

KCl (m)	ΔP (observed[a])
0.2	0.110
0.4	0.217
0.6	0.329
0.8	0.438
0.9	0.493
1.0	0.547
1.2	0.663
1.5	0.826
2.0	1.102

[a]Temperature, 20.0°C

10-20 Calculate the molecular mass of cadmium iodide from the fact that 5.6977 g in 44.69 g of water boiled at 100.181°C.
[E. Beckmann, *Z. Physik. Chem. (Leipzig)*, *6*, 460 (1890)]

10-21 Calculate the molecular mass of potassium nitrate from the fact that a 0.00817 m aqueous solution had a freezing point of -0.0295°C. Account for the difference in results in this problem and Problem 10-20.
[L. D. Adams, *J. Am. Chem. Soc.*, *37*, 494 (1915)]

10-22 Calculate the theoretical freezing-point depression of each of the solutions of KCl listed and plot them versus the molal concentration of KCl. On the same graph plot the observed depressions. Discuss the resulting graph.
[L. D. Adams, *J. Am. Chem. Soc.*, *37*, 494 (1915)]

m	Δt
0.00506	0.0184
0.00963	0.0348
0.01648	0.0590
0.03170	0.1122
0.05818	0.2031
0.11679	0.4014

10-23 The analytical concentration of a solution of strontium iodate was 0.101 M. What was the concentration of the various species in the solution?

10-24 What was the analytical concentration of a solution of magnesium bromide that contained 0.216 mol of bromide ions per liter of solution?

10-25 What was the pH of a solution whose analytical concentration of trifluoroacetic acid was 0.00382 M?

10-26 Many experiments require a relatively high concentration of an ionic compound to maintain the ionic strength of a solution at a fairly constant value. In the past $NaClO_4$ was often used for this purpose. Calculate the mass, in grams, of $NaClO_4$ required to prepare 500 mL of a solution in which the total ion concentration was 0.125 M.

10-27 In recent nears $NaBF_4$ has been used instead of $NaClO_4$. Calculate the mass, in grams, of $NaBF_4$ required to make 500 mL of a solution with the same total ion concentration as the solution in Problem 10-26.

10-28 Calculate the PF_6^- concentration in a solution prepared by dissolving 2.34 g of $Ba(PF_6)_2$ in 250 mL of water.

10-29 Calculate the pH of a solution containing 0.00467 g of HBF_4 in 25.0 mL of water.

10-30 Which solution will have the lower freezing point, 10.0 g of KPF_6 in 100 g of water or 4.50 g of $Ba(PF_6)_2$ in 100 g of water?

10-31 Many experiments are conducted in nonaqueous solvents in which the common salts are not very soluble. Salts like $(CH_3)_4NCl$, which are called tetraalkylammonium salts, are soluble in nonaqueous solvents. How many grams of $(CH_3)_4NBr$ would be required to prepare 100 mL of a solution in methanol in which the bromide ion concentration is 0.0937 M?

11
CHEMICAL EQUILIBRIUM, I

11-1 REVERSIBLE REACTIONS—EQUILIBRIUM

In principle, all chemical reactions are reversible. However, in many instances the reaction proceeds in one direction to such an extent that it is practically complete. Thus fluorine reacts with ammonia, but the reverse reaction is negligible.

$$3F_2(g) + 2NH_3(g) \rightarrow N_2(g) + 6HF(g)$$

Such a reaction is said to be *quantitative.*

In other reactions, a state of measurable equilibrium is established. The *concentration* of each *reactant decreases* until a constant minimum concentration is reached. The concentration of each *product increases* from zero to a constant maximum concentration. At equilibrium the concentration of each reactant and product remains constant.

11-2 THE EQUILIBRIUM CONSTANT

The relationship of the concentrations of products and reactants may be represented by the equilibrium constant K_c. In the generalized equilibrium reaction

$$aA + bB \rightleftharpoons cC + dD$$

$$K_c = \frac{[C]^c[D]^d}{[A]^a[B]^b}$$

The brackets represent the concentration of the particular substance in moles per liter. Note that:

1. The substances on the *right* side of the chemical equation are placed in the *numerator*, those on the *left* in the *denominator*.

2. The value of the concentration of each substance is raised to a *power* that is the respective coefficient in the balanced chemical equation.

3. The numerical value of the equilibrium constant changes with change in *temperature*.

Example 11-1

In an appropriate temperature range, dinitrogen tetroxide is in equilibrium with nitrogen dioxide.

$$N_2O_4(g) \rightleftharpoons 2NO_2(g)$$

Write the equilibrium-constant expression for this equilibrium.

$$K_c = \frac{[NO_2]^2}{[N_4O_4]}$$

Example 11-2

Write the equilibrium-constant expression for the reaction

$$4HCl(g) + O_2(g) \rightleftharpoons 2H_2O(g) + 2Cl_2(g)$$

$$K_c = \frac{[H_2O]^2[Cl_2]^2}{[HCl]^4[O_2]}$$

The form of the equilibrium-constant expression must represent the chemical equation for the equilibrium as it is written. The chemical equation in Example 11-1 may be written

$$\frac{1}{2}N_2O_4(g) \rightleftharpoons NO_2(g)$$

The equilibrium-constant expression would be

$$K_c = \frac{[NO_2]}{[N_2O_4]^{1/2}}$$

The numerical value of K_c will be quite different in this case.

11-3 THE DYNAMIC NATURE OF CHEMICAL EQUILIBRIA

The fact that the concentrations of the substances at equilibrium are not changing does not mean that changes are not occurring.

At equilibrium two opposing reactions are occurring *at the same rate*. Since reactions are occurring, the system is said to be in a state of *dynamic equilibrium*. Despite the fact that the reactions are occurring, the concentrations of substances present remain constant; the substances are formed at the same rate they are consumed.

11-4 HOMOGENEOUS AND HETEROGENEOUS EQUILIBRIA

Equilibria involving substances in the same phase, mixtures of gases or substances in solution, are called *homogeneous* equilibria. *Heterogeneous* equilibria involve substances in more than one phase. Examples are precipitates in equilibrium with ions in solution, water vapor in equilibrium with hydrates, and decomposition products in equilibrium with solids. The important consideration here is that the effective concentration of the solid substance is constant regardless of the concentration of the other species present.

Example 11-3

Calcium carbonate decomposes on heating to form carbon dioxide according to the equation

$$CaCO_3(s) \rightleftharpoons CaO(s) + CO_2(g)$$

In a closed container at constant temperature, an equilibrium is established. Write an equilibrium expression for the system.

$$K_c = [CO_2]$$

Both $CaCO_3$ and CaO are solids.

11-5 K_c AND K_p

Equilibrium constants in general are represented by K. The symbol K_c has been used to emphasize the fact that the equilibrium expression contains the *concentrations* of the substances involved.

When the substances at equilibrium are gases, the *partial pressure* of each substance may be used in place of the concentration. The equilibrium is then represented by the constant K_p. In the equilibrium $N_2O_4(g) \rightleftharpoons 2NO_2(g)$

$$K_p = \frac{P^2_{NO_2}}{P_{N_2O_4}}$$

The decomposition of $CaCO_3$ is represented by $K_p = P_{CO_2}$.

Example 11-4

What is the K_p expression for the equilibrium $4HCl(g) + O_2(g) \rightleftharpoons 2H_2O(g) + 2Cl_2(g)$?

$$K_p = \frac{P^2_{H_2O} \cdot P^2_{Cl_2}}{P^4_{HCl} \cdot P_{O_2}}$$

Example 11-5

Write a K_p expression for the equilibrium

$$Na_2SO_3(s) \rightleftharpoons Na_2O(s) + SO_2(g)$$

$$K_p = P_{SO_2}$$

11-6 CALCULATION OF EQUILIBRIUM CONSTANTS

The units of K are often omitted because they can easily be determined from the equilibrium expression if necessary. In a more rigorous treatment, the concentrations are referred to standard states and are represented by dimensionless ratios. For these reasons, units will usually not be included here, except for a few illustrations.

Example 11-6

At 8.2°C, 1.00 L of a chloroform solution was found to contain 0.129 mol of N_2O_4 and 1.17×10^{-3} mol of NO_2 at equilibrium. Calculate K_c for the dissociation of N_2O_4 under these conditions.
[J. T. Cundell, *J. Chem. Soc.*, *67*, 808 (1895)]

$$N_2O_4 \rightleftharpoons 2NO_2$$

$$K_c = \frac{[NO_2]^2}{[N_2O_4]} = \frac{(1.17 \times 10^{-3})^2}{1.29 \times 10^{-1}} = 1.07 \times 10^{-5}$$

If the units of K_c are to be required, the units of concentration are substituted in the expression for K_c.

$$K_c = \frac{(\text{mole/liter})^2}{\text{mole/liter}} = \text{mole/liter}$$

In this example, the units of K_c are moles per liter.

Example 11-7

The decomposition of hydrogen iodide was investigated at 457.6°C. The concentrations of the species present at equilibrium were 5.62×10^{-3} mol/L of H_2, 5.94×10^{-4} mol/L of I_2, and 1.27×10^{-2} mol/L of HI. Calculate K_c.

[A. H. Taylor and R. H. Crist, *J. Am. Chem. Soc.*, *63*, 1381 (1941)]

$$2HI(g) \rightleftharpoons H_2(g) + I_2(g)$$

$$K_c = \frac{[H_2][I_2]}{[HI]^2}$$

$$K_c = \frac{(5.62 \times 10^{-3}\ \text{mol/L})(5.94 \times 10^{-4}\ \text{mol/L})}{(1.27 \times 10^{-2}\ \text{mol/L})^2} = 2.06 \times 10^{-2}$$

Example 11-8

K. V. von Falckenstein investigated the decomposition of HBr at elevated temperatures. In one experiment at 1108°, the partial pressures of the components of the mixture were HBr, 0.998 atm; H_2, 3.82×10^{-3} atm; Br_2, 3.82×10^{-3} atm. Calculate the K_p for the equilibrium.

[*Z. Physik. Chem. (Leipzig)*, *68*, 270 (1910)]

$$2HBr(g) \rightleftharpoons H_2(g) + Br_2(g)$$

$$K_p = \frac{P_{H_2} \cdot P_{Br_2}}{P^2_{HBr}}$$

$$K_p = \frac{(3.82 \times 10^{-3}\ \text{atm})^2}{(0.998\ \text{atm})^2} = 1.47 \times 10^{-5}$$

Example 11-9

Hydrogen iodide was found to be 22.3% dissociated at 730.8 K. Calculate K_c.

[A. H. Taylor and R. H. Crist, *J. Am. Chem. Soc.*, *63*, 1377 (1941)]

$$2HI(g) \rightleftharpoons H_2(g) + I_2(g)$$

Assume that the initial concentration of HI was 1 mol/L.

$1.000 \times 0.223 = 0.223$ mol/L HI dissociated

$1.000 - 0.223 = 0.777$ mol/L HI remaining at equilibrium

$\frac{0.223}{2} = 0.112$ mol/L H_2 and I_2 formed.

From the equation, for each mole of HI dissociated, 1/2 mol H_2 and 1/2 mol of I_2 are formed.

$$K_c = \frac{[H_2][I_2]}{[HI]^2}$$

$$K_c = \frac{0.112 \times 0.112}{0.777^2} = \frac{0.0125}{0.604} = 2.07 \times 10^{-2}$$

Example 11-10

At 160°C and 1 atm, $PCl_5(g)$ is 13.5% dissociated into $PCl_3(g)$ and $Cl_2(g)$. Calculate K_c for the equilibrium.

[W. Fischer and O. Jubermann, *Z. Anorg. Allgem. Chem.*, *235*, 337 (1938)]

$$PCl_5(g) \rightleftharpoons PCl_3(g) + Cl_2(g)$$

$$K_c = \frac{[PCl_3][Cl_2]}{[PCl_5]}$$

Assume that the initial concentration of $PCl_5(g)$ was 1.00 mol/L. The number of moles per liter of PCl_5 that dissociated was $1.00 \times 0.135 = 0.135$. According to the chemical equation, 1.00 mol of PCl_3 and 1.00 mol of Cl_2 are formed for each mole of PCl_5 that dissociates. Therefore, 0.135 mol/L of PCl_3 and 0.135 mol/L of Cl_2 were formed. The concentration of PCl_5 remaining is $1.000 - 0.135 = 0.865$ mol/L.

$$K_c = \frac{0.135 \times 0.135}{0.865} = 0.0211$$

11-7 CALCULATION OF CONCENTRATIONS FROM EQUILIBRIUM CONSTANTS

The calculation of the concentration or partial pressure of a species from an equilibrium constant requires careful consideration of the relationships among the components of the equilibrium system.

Example 11-11

Calculate the concentration of NO_2 present at equilibrium in a chloroform solution that contained 0.129 mol/L of N_2O_4. K_c for the dissociation of $N_2O_4 = 1.07 \times 10^{-5}$.

$$N_2O_4 \rightleftharpoons 2NO_2$$

$$K_c = \frac{[NO_2]^2}{[N_2O_4]}$$

Let $X = [NO_2]$.

$$[N_2O_4] = 0.129 \frac{\text{mol}}{\text{L}}$$

$$\frac{X^2}{0.129} = 1.07 \times 10^{-5}$$

$$X^2 = 0.129 \times (1.07 \times 10^{-5}) = 1.38 \times 10^{-6}$$

$$X = \sqrt{1.38 \times 10^{-6}} = 1.17 \times 10^{-3} \frac{\text{mol}}{\text{L}}$$

Example 11-12

Calculate the equilibrium partial pressure of hydrogen formed by the decomposition of hydrogen bromide if the partial pressure of the remaining HBr was 0.998 atm. The K_p for the decomposition of HBr is 3.83×10^{-3} at the temperature of the experiment.

$$2HBr \rightleftharpoons H_2 + Br_2$$

$$K_p = \frac{P_{H_2} \cdot P_{Br_2}}{P^2_{HBr}} = 3.83 \times 10^{-3}$$

Let X = partial pressure of H_2, P_{H_2}; then $X = P_{Br_2}$, since 1 mol of Br_2 is formed for each mole of H_2 formed according to the equation.

$$\frac{X^2}{0.998^2} = 3.83 \times 10^{-3}$$

$$X^2 = 0.996 \times 3.83 \times 10^{-3} \text{ atm}$$

$$X = 6.17 \times 10^{-2} \text{ atm}$$

Example 11-13

Calculate the percent decomposition of PCl_5 at 160° from the equilibrium constant K_c.

$$PCl_5 \rightleftharpoons PCl_3 + Cl_2$$

From Example 11-10

$$K_c = \frac{[PCl_3][Cl_2]}{[PCl_5]} = 0.0211$$

Assume initial concentration of PCl_5 = 1.00 mol/L. Let X = $[PCl_3]$ at equilibrium; then $[Cl_2]$ = X. Since 1 mol of Cl_2 is formed for each mole of PCl_3 formed, X moles of Cl_2 are formed for each X moles of PCl_3 formed. Also $[PCl_5]$ = 1.00 - X, since 1 mol of PCl_5 must dissociate to produce 1 mol of PCl_3.

$$\frac{X^2}{1.00 - X} = 0.0211$$

$$X^2 = 0.0211 - 0.0211X$$

$$X^2 + 0.0211X - 0.0211 = 0$$

To solve a quadratic equation ($AX^2 + BX + C$), we can use the quadratic formula

$$X = \frac{-B \pm \sqrt{B^2 - 4AC}}{2A}$$

Here A = 1, B = 0.0211, and C = -0.0211.

$$X = \frac{-0.0211 \pm \sqrt{(0.0211)^2 + (4)(1)(.0211)}}{2}$$

$$X = \frac{-0.0211 \pm 0.2913}{2}$$

Since the negative root has no physical meaning

$$X = 0.135 \frac{\text{mol}}{\text{L}}$$

Example 11-14

Calculate the iodine concentration in an equilibrium mixture of hydrogen iodide and its constituents. The hydrogen concentration was 5.62×10^{-3} mol/L, and the HI concentration was 1.27×10^{-2} mol/L. $K_c = 2.06 \times 10^{-2}$.

[A. H. Taylor and R. H. Crist, *J. Am. Chem. Soc.*, *63*, 1377 (1941)]

$$2HI(g) \rightleftharpoons H_2(g) + I_2(g)$$

$$K_c = \frac{[H_2][I_2]}{[HI]^2} = 2.06 \times 10^{-2}$$

$$\frac{(5.62 \times 10^{-3})[I_2]}{(1.27 \times 10^{-2})^2} = 2.06 \times 10^{-2}$$

$$[I_2] = \frac{(1.61 \times 10^{-4}) \times (2.06 \times 10^{-2})}{5.62 \times 10^{-3}}$$

$$= 5.90 \times 10^{-4} \frac{\text{mol}}{\text{L}}$$

Example 11-15

A 1.00-mol sample of HI, contained in a 1.00-L flask, was heated to 730.8 K. Calculate the concentration of each component of the mixture at equilibrium. $K_c = 2.06 \times 10^{-2}$.

$$2HI \rightleftharpoons H_2 + I_2$$

$$K_c = \frac{[H_2][I_2]}{[HI]^2} = 2.06 \times 10^{-2}$$

Let X = moles HI that decomposed; then [HI] = (1.00 - X).

$$[H_2] = [I_2] = \frac{X}{2}$$

since 2 mol of HI gives 1 mol of H_2 and 1 mol of I_2 in the equation

$$\frac{X/2 \times X/2}{(1 - X)^2} = 2.06 \times 10^{-2}$$

Take the square root of both sides.

$$\frac{X/2}{1 - X} = \frac{X}{2 - 2X} = 0.144$$

$$X = 0.288 - 0.288X$$

$$1.288X = 0.288$$

$$X = \frac{0.288}{1.288} = 0.223\ \frac{\text{mol}}{\text{L}}$$

$$[HI] = 1.00 - 0.223 = 0.78\ \frac{\text{mol}}{\text{L}}$$

$$[H_2] = [I_2] = \frac{0.223}{2} = 0.112\ \frac{\text{mol}}{\text{L}}$$

PROBLEMS

11.1 Write concentration equilibrium-constant expressions for the following equilibria. Balance the equations when necessary.

a. $2NO(g) \rightleftharpoons N_2(g) + O_2(g)$

b. $N_2(g) + 3H_2(g) \rightleftharpoons 2NH_3(g)$

c. $NO_2(g) + H_2(g) \rightleftharpoons NH_3(g) + H_2O(g)$

d. $C(s) + O_2(g) \rightleftharpoons CO(g)$

e. $H_2(g) + Fe_3O_4(s) \rightleftharpoons Fe(s) + H_2O(g)$

f. $CO_2(g) + H_2(g) \rightleftharpoons CO(g) + H_2O(g)$

g. $HgO(s) \rightleftharpoons Hg(\ell) + 1/2\ O_2(g)$

h. $N_2O_3(g) \rightleftharpoons NO_2(g) + NO(g)$

11-2 Write K_p expressions for the equilibria in Problem 11-1.

11-3 The value of K_c for the reaction

$$2HI \rightleftharpoons H_2 + I_2$$

is 2.06×10^{-2}. From this calculate the value of K_c for the reaction

$$HI \rightleftharpoons \frac{1}{2}H_2 + \frac{1}{2}I_2$$

[A. H. Taylor and R. H. Crist, *J. Am. Chem. Soc.*, *63*, 1381 (1941)]

11-4 In another experiment, J. T. Cundell found that at 8.2°C, a chloroform solution contained 0.778 mol/L of N_2O_4 and 2.84×10^{-3} mol/L of NO_2. Calculate K_c for the reaction $2NO_2 \rightleftharpoons N_2O_4$.
[J. T. Cundell, *J. Chem. Soc.*, *67*, 808 (1895)]

11-5 Calculate the equilibrium constant for the formation of hydrogen iodide from the elements at 763.8 K. At equilibrium, the concentration of iodine was 0.406 mol/L and the hydrogen and hydrogen iodide concentrations were 0.172 and 1.78 mol/L, respectively.

[A. H. Taylor and R. H. Crist, *J. Am. Chem. Soc.*, *63*, 1381 (1941)]

11-6 The equilibrium constant for the decomposition of hydrogen iodide at 763.8 K was also determined by A. H. Taylor and R. H. Crist. In one experiment the concentrations at equilibrium were iodine, 0.242; hydrogen, 0.242; and hydrogen iodide, 1.64 mol/L. Calculate the value of K_c for the decomposition of HI.

[*J. Am. Chem. Soc.*, *63*, 1381 (1941)]

11-7 The partial pressure of $PCl_3(g)$ and $Cl_2(g)$ in equilibrium with $PCl_5(g)$ at 422.6 K was 6.08×10^{-2} atm. The partial pressure of $PCl_5(g)$ was 0.453 atm. What was the value of K_p for the system?

[W. Fischer and O. Jubemann, *Z. Anorg. Allgem. Chem.*, *235*, 337 (1938)]

11-8 Would the numerical value of K_p in Problem 11-7 be the same if the data had been given in mm of Hg, as recorded in the original paper? $PCl_5(g)$, 344.4 mm; $PCl_3(g)$, 46.2 mm; $Cl_2(g)$, 46.2 mm.

11-9 Determine K_p for the equilibrium represented by

$$1/6B_2H_6(g) + 2/3BF_3(g) \rightleftharpoons HBF_2(g)$$

The partial pressures of the components of the equilibrium mixture were B_2H_6, 2.1×10^{-2}; BF_3, 1.4×10^{-2}; HBF_2, 1.5×10^{-2} atm at 296 K.

[R. F. Porter and S. K. Wason, *J. Phys. Chem.*, *69*, 2211 (1955)]

11-10 The hydrolysis of $SmCl_3(s)$ at 893 K was investigated by C. W. Koch and B. B. Cunningham, who found that the partial pressures of HCl and H_2O at equilibrium were 418.8 mm and 1.60 mm, respectively, in one experiment. The other product of the reaction was SmOCl(s). They recorded an equilibrium constant of 144 for this experiment. Calculate the numerical value of K and give the appropriate units.

[*J. Am. Chem. Soc.*, *75*, 796 (1953)]

11-11 Calculate K_p for the equilibrium

$$LaCl_3(s) + H_2O(g) \rightleftharpoons LaOCl(s) + 2HCl(g) \quad \text{at } 892\text{ K}$$

The partial pressure of HCl was 80.9 mm, and the partial pressure of H_2O was 2.92 mm.

[C. W. Koch, A. Broido, and B. B. Cunningham, *J. Am. Chem. Soc.*, *74*, 2349 (1952)]

11-12 Sudborough and Miller found that NOCl was 8.22% dissociated at 796°C into NO and Cl_2. Calculate the equilibrium constant for the dissociation.

[*J. Chem. Soc.*, *59*, 271 (1891)]

11-13 A sample of N_2O_4 gas, initial concentration 0.02968 mol/L, was allowed to reach equilibrium at 25°C. The total pressure at equilibrium was 0.862 atm. Calculate the equilibrium constant for the dissociation of N_2O_4.

[F. H. Verhoek and F. Daniels, *J. Am. Chem. Soc.*, *53*, 1250 (1931)]

11-14 The compound OPF(g) has recently been prepared by Schnockel et al. at high temperatures and low pressures by the following reaction:

$$P(O)FBr_2(g) + 2Ag(s) \rightleftharpoons OPF(g) + 2AgBr(g)$$

Initially only pure $P(O)FBr_2$ and Ag were present at 1021 K. At equilibrium the partial pressure of OPF(g) was 7.94×10^{-6} bar, and that of $P(O)FBr_2(g)$ was 1.44×10^{-6} bar.

a. What is the expression for K_p for the reaction?

b. What is the value for the equilibrium constant based on these data?

[*J. Am. Chem. Soc.*, *108*, 7905 (1986)]

11-15 When Bertholet and St. Giles mixed 1.00 mol/L of acetic acid and 0.33 mol/L of alcohol, 0.293 mol/L of ester was observed. Compare their observation with the quantity of ester expected from the equilibrium constant calculated in Problem 11-14.

11-16 Calculate the equilibrium constant for the equilibrium $2NO(g) + Br_2(g) = 2NOBr(g)$ from the observation that the total pressure at 296.9 K was 110.5 mm at equilibrium. The initial partial pressures of NO and Br_2 were 98.4 mm and 41.3 mm, respectively.

[C. M. Blair, Jr., P. D. Brass, and D. M. Yost, *J. Am. Chem. Soc.*, *56*, 1916 (1934)]

11-17 The pressure of bromine in equilibrium with $FeBr_3$ at 138°C was 748 mm. What are the values of K_p (in atm) for the two equilibria?

$$FeBr_3(s) \rightleftharpoons FeBr_2(s) + \frac{1}{2}Br_2(g)$$

$$2FeBr_3(s) \rightleftharpoons 2FeBr_2(s) + Br_2(g)$$

[N. W. Gregory and B. A. Thackrey, *J. Am. Chem. Soc.*, *72*, 3176 (1950)]

11-18 Calculate the pressure of bromine, in torr, in equilibrium with $FeBr_3$ at 108°C. K_p for the reaction $FeBr_3(s) \rightleftharpoons FeBr_2(s) + 1/2Br_2(g)$ is 0.456 atm.

11-19 The density of CF_3COOH vapor at 118.1°C and 468.5 torr was 2.784 g/L. Calculate K_c for the association of the acid $2CF_3COOH \rightleftharpoons (CF_3COOH)_2$.

[R. E. Lundin, F. E. Harris, and L. K. Nash, *J. Am. Chem. Soc.*, *74*, 4654 (1952)]

11-20 Calculate the equilibrium partial pressures of the monomer of acetic acid at 121.1°C. K_p (in torr) for the dissociation is 270, and the partial pressure of the dimer was 26.74 torr.

[M. D. Taylor, *J. Am. Chem. Soc.*, *73*, 315 (1951)]

11-21 Acetone and iodine react at 371°C to form the compound iodoacetone according to the following reaction:

$$CH_3COCH_3(g) + I_2(g) \rightleftharpoons CH_3COCH_2I(g) + HI(g)$$

A mixture of acetone, originally at 276.5 torr, and iodine, at 4.52 torr, had pressures of 275.9 and 3.97 torr, respectively, at equilibrium. What is K_p for this reaction?

[Solly, Golden, and Benson, *J. Am. Chem. Soc.*, *92*, 4653 (1970)]

11-22 When a mixture of tetrafluorohydrazine (N_2F_4) and Cl_2 is exposed to uv light, chlorodifluoramine (NF_2Cl) is formed. In

one experiment, chlorine at 406 mm pressure and chlorodifluoramine at 100 mm pressure were mixed at 49.9°C. At equilibrium the pressure of each gas was as follows: N_2F_4, 45.1 mm; Cl_2, 451.1 mm; and NF_2Cl, 9.86 mm. What is K_p for this reaction?
[R. Petry, *J. Am. Chem. Soc.*, *89*, 4600 (1967)]

11-23 In a series of experiments, the equilibrium constant for the reaction of iodine with propene according to the following equation was determined.

$$I_2(g) + C_3H_6(g) \rightleftharpoons C_3H_5I(g) + HI(g)$$

Some of the results obtained at 545 K were as follows:

Initial pressures			Equilibrium pressures	
I_2 [P_0 (mm)]	C_3H_6 [P_0 (mm)]	HI [P_0 (mm)]	HI [P_{eq} (mm)]	C_3H_5I [P_{eq} (mm)]
23.9	505.8	0	1.80	1.80
11.9	512.2	0	1.25	1.25
24.0	81.6	0	0.756	0.756
16.1	355.3	1.62	2.27	0.645

Determine K_p for each of the above results and the average value of K_p.
[A. Rodgers, D. Golden, and S. Benson, *J. Am. Chem. Soc.*, *88*, 3194 (1966)]

11-24 Find the concentration of monomeric dichloroacetic acid in a carbon tetrachloride solution that contains 0.0129 g of the acid in 100 mL of solution. The dissociation constant of the dimeric acid is 9.3×10^{-4}.
[J. T. Harris, Jr., and M. E. Hobbs, *J. Am. Chem. Soc.*, *76*, 1419 (1954)]

11-25 If the equilibrium composition of the system $CO_2(g) + H_2(g) \rightleftharpoons CO(g) + H_2O(g)$ was CO_2, 7.15; H_2, 46.93; and CO = H_2O, 22.96 mol%, what was the equilibrium constant? W. Nernst and F. Haber calculated a value of 1.57 for K_p. Is this value consistent with your calculated K?
[F. Haber and F. Richardt, *Z. Anorg. Chem.*, *38*, 5 (1904)]

11-26 The equilibrium constant for the formation of ammonia from the elements is 0.0266 atm^{-1} at 350°. Calculate K_c, $1/2N_2(g)$ + $3/2N_2(g) \rightleftharpoons NH_3(g)$.
[A. T. Larson and R. L. Dodge, *J. Am. Chem. Soc.*, *45*, 2918 (1923)]

11-27 Hydrogen iodide gas (1.69×10^{-2} mol/L) was in equilibrium with its constituents at 457.6°C. Calculate the hydrogen concentration if the iodine concentration was 1.52×10^{-3} mol/L. $K_c = 2.06 \times 10^{-2}$.
[A. H. Taylor and R. H. Crist, *J. Am. Chem. Soc.*, *63*, 1377 (1941)]

11-28 A chloroform solution of dinitrogen tetroxide was allowed to come to equilibrium with nitrogen dioxide at 8.2°C. Calculate the concentration of nitrogen dioxide present if the concentration of dinitrogen tetroxide at equilibrium was 1.29×10^{-1} mol/L. $K_c = 1.07 \times 10^{-5}$.
[J. T. Cundell, *J. Chem. Soc.*, *67*, 808 (1895)]

11-29 Find the equilibrium concentrations of the gases in the equilibrium $CO_2 + H_2 \rightleftharpoons CO + H_2O$ at 1000°C assuming that 1.00 mol of CO_2 and 1 mol of H_2 were heated to that temperature. K_c = 1.7 at 1000°C.

11-30 Calculate the equilibrium concentration of the monomer of CF_3COOH at 103.3°C if 1.00 mol of the dimer was present. The equilibrium constant for the dimerization at that temperature is 2.29.
[R. E. Lundin, F. E. Harris, and L. K. Nash, *J. Am. Chem. Soc.*, *74*, 4654 (1952)]

Supplementary Problems

11-31 In the gaseous state, selenium reacts with $SeOCl_2$ and HCl in two different ways:

a. $3Se(s) + SeOCl_2(g) + 2HCl(\ell) \rightleftharpoons 2Se_2Cl_2(g) + H_2O(\ell)$

b. $Se(s) + SeOCl_2(g) + 2HCl(\ell) \rightleftharpoons 2SeCl_2(g) + H_2O(\ell)$

Write the equilibrium constant expression (K_p) for both reactions.
[M. Mahadevan and J. Milne, *Inorg. Chem.*, *22*, 1648 (1983)]

11-32 Calculate the value of K_p for the dissociation of HBr at 1024° from the following data: HBr, 0.997 atm; H_2, 3.81×10^{-3} atm; Br_2, 3.81×10^{-3} atm.
[K. V. Falckenstein, *Z. Physik. Chem. (Leipzig)*, *68*, 270 (1910)]

11-33 The K_p for the reaction

$$2SO_3(g) \rightleftharpoons 2SO_2(g) + O_2(g)$$

is 0.625 atm. Find the value of K_p for each of the following reactions from the above data.

a. $SO_3(g) \rightleftharpoons SO_2(g) + 1/2O_2(g)$

b. $O_2(g) + 2SO_2(g) \rightleftharpoons 2SO_3(g)$

11-34 Use the data in Problem 11-32 to calculate K_p for the *formation* of HBr from the elements.

11-35 Calculate the value of K_p (include units) for the dissociation of PCl_5 at 464.5 K from the data of W. Fischer and O. Jubermann: $PCl_5(g)$, 274.6 mm; $Cl_2(g)$, 153.3 mm.
[*Z. Anorg. Allgem. Chem.*, *235*, 337 (1938)]

11-36 Calculate K_p for the formation of PCl_5 from PCl_3. At equilibrium the partial pressure of $PCl_3(g)$ was 37.4 mm and the partial pressure of $PCl_5(g)$ was 52.5 mm at 445.3 K.
[W. Fischer and O. Jubermann, *Z. Anorg. Allgem. Chem.*, *235*, 337 (1938)]

11-37 A gaseous mixture at equilibrium contained HBr (757.96 mm), H_2 (1.81 mm), and Br_2 (1.81 mm). Calculate K_p for the dissociation of HBr in this experiment.
[K. V. von Falckenstein, *Z. Physik. Chem. (Leipzig)*, *68*, 270 (1910)]

11-38 The equilibrium constants for a series of reactions whose general formula is

$$MX_2(s) + 2LX_3(g) \rightleftharpoons ML_2X_8(g)$$

were measured by F. P. Emmenegger. In the case where M is Mg, X is Cl, and L is Fe, the following reaction occurs:

$$MgCl_2(s) + 2FeCl_3(g) \rightleftharpoons MgFe_2Cl_8(g)$$

When the partial pressure of $FeCl_3$ was 1.22×10^{-3} atm, an equilibrium constant of 22.0 atm^{-1} was obtained at 700 K. What would be the partial pressure of $MgFe_2Cl_8$ under these conditions?
[*Inorg. Chem.*, *16*, 343 (1977)]

11-39 At 1024°, hydrogen bromide was found to be 0.467% dissociated. Determine the equilibrium constant for the dissociation.
[K. V. von Falckenstein, *Z. Physik. Chem. (Leipzig)*, *68*, 270 (1910)]

11-40 Calculate the equilibrium constant for the combination of deuterium and iodine at 730.8 K. The equilibrium mixture contained 0.10945×10^{-5} mol/L of I_2, 0.44150×10^{-5} mol/L of D_2, and 1.4160 mol/L of DI.
[A. H. Taylor, Jr., and R. H. Crist, *J. Am. Chem. Soc.*, *63*, 1382 (1941)]

11-41 Determine the equilibrium constant for the dissociation of N_2O_3 at 25°. At a total pressure of 1 atm the gas mixture contained 10.5% N_2O_3.
[M. Bodenstein et al., *Z. Physik. Chem. (Leipzig)*, *100*, 75 (1922)]

11-42 The equilibrium partial pressures of PCl_5 and PCl_3 were found to be 312.5 mm and 99.3 mm, respectively, at 446.4 K. Find K_p for the dissociation of PCl_5.
[W. Fischer and O. Jubermann, *Z. Anorg. Chem.*, *235*, 337 (1938)]

11-43 Calculate the value of K_p for the equilibrium $CO_2(g) + H_2(g) \rightleftharpoons CO(g) + H_2O(g)$ from the data of F. Haber and F. Richardt. The equilibrium composition observed was CO_2, 0.69; H_2, 80.52; CO, 9.40; H_2O, 9.40 mole%.
[*Z. Anorg. Chem.*, *38*, 5 (1904)]

11-44 Iron(III) bromide was heated to 128°C. The pressure due to the decomposition, with the formation of iron(II) bromide and bromine vapor, was 454 mm. Find K_p in atmospheres.
[N. W. Gregory and B. A. Thackrey, *J. Am. Chem. Soc.*, *72*, 3176 (1950)]

11-45 Determine the equilibrium constant for the reaction $CO_2(g) + C(s) \rightleftharpoons 2CO(g)$ at 700°. The gas mixture contains 42.3% CO_2 and 57.7% CO.

11-46 Calculate the value of K_p for the reaction $2CO_2 \rightleftharpoons 2CO + O_2$ at 1205°C. $K_c = 1.41 \times 10^{-13}$.

11-47 Calculate the equilibrium constant for the association of CH_3COOH vapor at 81.4°C. The density was 2.444 g/L at a pressure of 310.2 torr.

[R. E. Lundin, F. E. Harris, and L. K. Nash, *J. Am. Chem. Soc.*, *74*, 4654 (1952)]

11-48 At 1345.7°, bromine was found to be 4.77% dissociated. Calculate K_p.

[M. Bodenstein and F. Cramer, *Z. Elektrochem.*, *22*, 336 (1916)]

11-49 Calculate K_p for the reaction $CS_2(g) + 4H_2(g) \rightleftharpoons CH_4(g) + 2H_2S(g)$ at 650°C. The partial pressures of CS_2, H_2, CH_4, and H_2S were 407.0, 270.6, 67.9, and 14.4 mm, respectively.

[E. Terres and E. Sasse, *Angew. Chem.*, *47*, 241 (1934)]

11-50 Schnockel et al. recently measured the equilibrium constant for the reaction

$$P(O)FBr_2(g) + 2Ag(s) \rightleftharpoons OPF(g) + 2AgBr(g)$$

using a mass spectrometer. The following partial pressures were measured at 1084 K:

$$p\,(OPF) = 9.63 \times 10^{-6}\ \text{bar}$$

$$p\,(P(O)FBr_2) = 5.30 \times 10^{-7}\ \text{bar}$$

$$p\,(AgBr) = 3.14 \times 10^{-6}\ \text{bar}$$

What is the value of K_p from these data?

[*J. Am. Chem. Soc.*, *108*, 7905 (1986)]

11-51 Calculate the equilibrium constant K_c for the reaction $CS_2(g) + 4H_2(g) \rightleftharpoons CH_4(g) + 2H_2S(g)$ at 900°. $K_p = 3.0 \times 10^{-5}$.

[E. Terres and E. Sasse, *Angew. Chem.*, *47*, 241 (1934)]

11-52 The vapor-phase dissociation of dimeric acetic acid was studied by M. D. Taylor. At 100.6°C the pressure of 359.8 mL of the vapor was 40.68 mm. The pressure that would have been observed if no dissociation had occurred is 25.33 mm. Find K_p for the dissociation.

[*J. Am. Chem. Soc.*, *73*, 315 (1951)]

11-53 Determine the partial pressure of $N_2O_4(g)$ in equilibrium with $NO_2(g)$ if the total pressure was 0.157 atm. $K_p = 0.142$ for the dissociation.

[F. H. Verhoeck and F. Daniels, *J. Am. Chem. Soc.*, *53*, 1250 (1931)]

11-54 Calculate the equilibrium partial pressures of the components of the reaction $CS_2 + 4H_2 \rightleftharpoons CH_4 + 2H_2S$ if the total pressure was 2.00 atm at 900°. $K_p = 3.0 \times 10^{-5}$.

[E. Terres and E. Sasse, *Angew. Chem.*, *47*, 238 (1934)]

11-55 The equilibrium constant for the formation of ammonia from the elements $1/2N_2(g) + 3/2H_2(g) \rightleftharpoons NH_3(g)$ at a given temperature is 6.59×10^{-3} atm^{-1}. Calculate the partial pressure of each gas at equilibrium if the initial partial pressure of N_2 and H_2 was 5.00 atm each.

11-56 A sample of iodine vapor was observed to exert a pressure of 0.11216 atm at 1274 K. Calculate the value of the pressure due to iodine atoms if the equilibrium constant for the decomposition was 0.1692 atm.

[M. L. Periman and G. K. Rollefson, *J. Chem. Phys.*, *9*, 362 (1941)]

11-57 Determine the partial pressure of F_2 molecules at equilibrium with F atoms at 1115 K if K_p for the dissociation was 7.55×10^{-2}. The total pressure was 0.831 atm.

[R. N. Doescher, *J. Chem. Phys.*, *20* (1952)]

11-58 Use the equilibrium constant calculated in Problem 11-3 to calculate the concentration of N_2O_4 in equilibrium with 2.13×10^{-3} M NO_2 in chloroform.

[J. T. Cundell, *J. Chem. Soc.*, *67*, 808 (1895)]

11-59 Determine the percent monomeric form of benzoic acid present in equilibrium with the dimeric form in CCl_4 solution. $K_{diss} = 7 \times 10^{-5}$ at 25°.

[J. T. Harris, Jr., and M. E. Hobbs, *J. Am. Chem. Soc.*, *76*, 1419 (1954)]

11-60 Calculate the percent monomeric acetic acid in equilibrium with the dimeric form at 71.3°C. K_p for the dissociation is 16.4 mm.

[M. D. Taylor, *J. Am. Chem. Soc.*, *73*, 315 (1951)]

11-61 Use the equilibrium constant calculated in Problem 11-43 to calculate the mole percent of CO_2 present at equilibrium when the equilibrium mixture contained 22.85 mol% H_2 and 27.86 mol% each of CO and H_2O.

11-62 In another experiment K_c was found to be 2.01×10^{-2} for the decomposition of hydrogen iodide. Determine the hydrogen and iodine concentrations if the hydrogen iodide concentration was 1.00×10^{-2} mol/L after equilibrium was established.

[A. H. Taylor and R. H. Crist, *J. Am. Chem. Soc.*, *63*, 1377 (1941)]

11-63 A solution was prepared by dissolving N_2O_4 in chloroform until the concentration was 0.325 mol/L. Calculate the NO_2 concentration of the solution after equilibrium was established at 8.2°C. $K_c = 1.07 \times 10^{-5}$.

[J. T. Cundell, *J. Chem. Soc.*, *67*, 808 (1895)]

11-64 Find the equilibrium concentrations of the monomer and dimer of chloroacetic acid in carbon tetrachloride solution if the analytical concentration of $CH_2ClCOOH$ was 1.00 mol/L. K_c for the dissociation of the dimeric form is 6.2×10^{-4}.

[J. T. Harris and M. E. Hobbs, *J. Am. Chem. Soc.*, *76*, 1419 (1954)]

12
CHEMICAL EQUILIBRIUM, II

In Chapter 10, problems connected with solutions of strong electrolytes were discussed. In this chapter, problems concerned with solutions of electrolytes that are slightly ionized are considered.

12-1 WEAK ACIDS AND BASES

Some acids that react with water to form hydronium ions establish an equilibrium which lies far to the left. Hydrocyanic acid is such an acid.

$$HCN + H_2O \rightleftharpoons H_3O^+ + CN^-$$

This equilibrium can be represented by

$$K = \frac{[H_3O^+][CN^-]}{[HCN][H_2O]}$$

In dilute solutions, the quantity of water present is practically constant and can be included in the constant.

$$K \times [H_2O] = \frac{[H_3O^+][CN^-]}{[HCN]}$$

Such an equilibrium constant may be called the *ionization constant*, K_{ion} or K_a, or the *dissociation constant*.

$$K_a = \frac{[H_3O^+][CN^-]}{[HCN]}$$

A similar equilibrium involves the base NH_3, and the constant is K_b.

$$NH_3 + H_2O \rightleftharpoons NH_4^+ + OH^-$$

$$K_b = \frac{[NH_4^+][OH^-]}{[NH_3]}$$

Always write the balanced equation for the equilibrium and the dissociation-constant expression before attempting to calculate a numerical answer to a problem involving ionic equilibria. These constants are often expressed as pK_a and pK_b.

$$pK_a = -\log K_a \qquad pK_b = -\log K_b$$

12-2 CALCULATION OF DISSOCIATION CONSTANTS

Dissociation constants of slightly ionized acids and bases have been calculated from the results of conductance measurements, potentiometric titrations, and spectrometry.

Example 12-1

By conductance measurements, a 0.0500 M solution of acetic acid was observed to be 1.90% dissociated. Calculate K_a for acetic acid.

[D. A. MacInnes and T. Shedlovsky, *J. Am. Chem. Soc.*, *54*, 1435 (1932)]

$$CH_3COOH + H_2O \rightleftharpoons H_3O^+ + CH_3COO^-$$

$$K_a = \frac{[H_3O^+][CH_3COO^-]}{[CH_3COOH]}$$

$0.0500 \times 0.0190 = 0.000950$ mole of CH_3COOH ionized

$[CH_3COOH] = 0.0500 - 0.000950 = 0.0490$ mol CH_3COOH unionized

$[CH_3COO^-] = [H_3O^+] = 0.000950$ mol

since 1 mol of CH_3COO^- and 1 mol of H_3O^+ are formed for each mole of CH_3COOH that ionizes.

$$K_a = \frac{(9.50 \times 10^{-4})^2}{4.90 \times 10^{-2}} = \frac{9.02 \times 10^{-7}}{4.90 \times 10^{-2}} = 1.84 \times 10^{-5}$$

12.3 CALCULATIONS FROM DISSOCIATION CONSTANTS

Experimentally determined dissociation constants (Table 12-1) may be used to calculate the approximate concentrations of species in solution.

TABLE 12-1 Ionization Constants of Weak Electrolytes

	Formula		K_a	pK_a
Acid				
Acetic acid	CH_3COOH		1.76×10^{-5}	4.75
Benzoic acid	C_6H_5COOH		6.46×10^{-5}	4.19
Butyric acid	C_3H_7COOH		1.52×10^{-5}	4.81
Carbonic acid	H_2CO_3	(1)	4.30×10^{-7}	6.37
		(2)	5.61×10^{-11}	10.25
Formic acid	$HCOOH$		1.77×10^{-4}	3.75
Hydrocyanic acid	HCN		4.93×10^{-10}	9.31
Nitrous acid	HNO_2		4.6×10^{-4}	3.33
Phosphoric acid	H_3PO_4	(1)	7.52×10^{-3}	2.12
		(2)	6.23×10^{-8}	7.21
		(3)	2.2×10^{-13}	12.66
Propionic acid	C_2H_5COOH		1.34×10^{-5}	4.87
Malonic acid	$CH_2(COOH)_2$	(1)	1.49×10^{-3}	2.83
		(2)	2.03×10^{-6}	5.69
o-Phthalic acid	$C_6H_4(COOH)_2$	(1)	1.49×10^{-3}	2.83
		(2)	2.03×10^{-6}	5.69
Salicylic acid	$C_6H_4OH(COOH)$	(1)	1.0×10^{-3}	3.00
		(2)	3.98×10^{-15}	14.40
			K_b	pK_b
Base				
Ammonia	$NH_3(aq)$		1.79×10^{-5}	4.75
Pyridine	C_5H_5N		5.62×10^{-6}	5.25
Trimethylamine	$(CH_3)_3N$		1.55×10^{-10}	9.81

Example 12-2

Calculate the hydronium ion concentration in a 0.200 M acetic acid solution.

$$CH_3COOH + H_2O \rightleftharpoons H_3O^+ + CH_3COO^-$$

$$K_a = \frac{[H_3O^+][CH_3COO^-]}{[CH_3COOH]}$$

At equilibrium, let $X = [H_3O^+]$; then

$$X = [CH_3COO^-]$$

(1 mol of CH_3COO^- is formed for each mole of H_3O^+.)

$$0.200 - X = [CH_3COOH]$$

(1 mol of CH_3COOH is consumed for each mole of H_3O^+ formed.)

$$\frac{X^2}{0.200 - X} = 1.76 \times 10^{-5}$$

$$X^2 = 3.52 \times 10^{-6} - 1.76 \times 10^{-5}X$$

$$X^2 + 1.76 \times 10^{-5}X - 3.52 \times 10^{-6} = 0$$

$$X = \frac{-1.76 \times 10^{-5} \pm \sqrt{3.10 \times 10^{-10} + 14.1 \times 10^{-6}}}{2}$$

$$X = \frac{-1.76 \times 10^{-5} + 3.75 \times 10^{-3}}{2} = 1.86 \times 10^{-3} \frac{\text{mol}}{\text{L}}$$

The quantity of H_3O^+ ions is very small compared to the quantity of undissolved acid CH_3COOH present. A simplified, approximate calculation may be made.

$$\frac{X^2}{0.200 - X} = 1.76 \times 10^{-5}$$

Assume that X is very small compared to 0.200 and neglect it in the denominator.

$$\frac{X^2}{0.200} = 1.76 \times 10^{-5}$$

$$X^2 = 3.52 \times 10^{-6}$$

$$X = 1.88 \times 10^{-3} \frac{\text{mol}}{\text{L}}$$

The result obtained by this shortcut method is usually sufficiently accurate, especially when the value of K_a is less than 10^{-3} and the analytical concentration of the acid is 0.100 or less.

Example 12-3

Calculate the percent dissociation of a 0.10 M HCN solution.

$$HCN + H_2O \rightleftharpoons H_3O^+ + CN^-$$

$$K_a = \frac{[H_3O^+][CN^-]}{[HCN]}$$

let $X = [H_3O^+] = [CN^-]$

then $0.100 - X = [HCN]$

$$\frac{X^2}{0.100 - X} = 4.93 \times 10^{-10}$$

Assume that X is very small compared to 0.100 and neglect it in the denominator

$$X^2 = 0.10 \times 4.93 \times 10^{-10} = 4.93 \times 10^{-11}$$

$$X = 7.0 \times 10^{-6} \frac{\text{mol}}{\text{L}}$$

$$\text{Percent diss} = \frac{\text{moles HCN dissociated}}{\text{total moles HCN}} \times 100$$

$$\text{Percent diss} = \frac{7.0 \times 10^{-6}}{0.10} \times 100 = 7.0 \times 10^{-3}\%\text{, or } 0.0070\%$$

Example 12-4

What is the pH of a 0.200 M formic acid solution?

$$HCOOH + H_2O \rightleftharpoons H_3O^+ + HCOO^-$$

$$K_a = \frac{[H_3O^+][HCOO^-]}{[HCOOH]} = 1.77 \times 10^{-4}$$

let $X = [H_3O^+] = [HCOO^-]$

then $0.200 - X = [HCOOH]$

$$\frac{X^2}{0.200 - X} = 1.77 \times 10^{-4}$$

Assume X is very small compared to 0.2 and neglect it in the denominator.

$$X^2 = 3.54 \times 10^{-5}$$

$$X = 5.95 \times 10^{-3} \frac{\text{mol}}{\text{L}}$$

$$pH = -\log[H_3O^+]$$
$$= -\log 5.95 \times 10^{-3} = -(-3 + 0.774)$$
$$= 2.23$$

Example 12-5

What is the approximate OH^- concentration of a 0.150 M NH_3 solution?

$$NH_3 + H_2O \rightleftharpoons NH_4^+ + OH^-$$

$$K_b = \frac{[NH_4^+][OH^-]}{[NH_3]}$$

let $X = [OH^-] = [NH_4^+]$

then $0.150 - X = [NH_3]$

$$\frac{X^2}{0.150 - X} = 1.79 \times 10^{-5}$$

Neglecting X in the denominator

$$X^2 = 0.150 \times 1.79 \times 10^{-5} = 2.68 \times 10^{-6}$$

$$X = 1.64 \times 10^{-3} \frac{\text{mol}}{\text{L}}$$

12-4 THE ION PRODUCT OF WATER

Water is the most common self-ionizing solvent.

$$2H_2O \rightleftharpoons H_3O^+ + OH^-$$

The equilibrium concentrations of the ions are related by the ion product of water K_w.

$$K_w = [H_3O^+][OH^-] = 1.0 \times 10^{-14}$$

In pure water, $[H_3O^+] = [OH^-]$

Example 12-6

Calculate the hydronium ion concentration and hydroxide ion concentration in pure water at 25°C.

$$2H_2O \rightleftharpoons H_3O^+ + OH^-$$

$$K_w = [H_3O^+][OH^-] = 1.0 \times 10^{-14}$$

let $X = [H_3O^+] = [OH^-]$

$$X^2 = 1.0 \times 10^{-14}$$

$$X = 1.0 \times 10^{-7} \frac{\text{mol}}{\text{L}}$$

12-5 pK_W

In Chapter 10 the pH notation for the concentrations of H_3O^+ was described.

$$pH = -\log[H_3O^+]$$

and similarly

$$pOH = -\log[OH^-]$$

Using the same notation, we may define pK_W.

$$pK_W = -\log K_W$$

since

$$K_W = [H_3O^+][OH^-] = 1 \times 10^{-14}$$

$$\log K_W = \log[H_3O^+] + \log[OH^-]$$

$$-\log K_W = -\log[H_3O^+] + (-\log[OH^-])$$

or

$$pK_W = pH + pOH = 14.0$$

Example 12-7

What is the pOH of a solution of pH 6.0?

$$pH + pOH = 14.0$$

$$pOH = 14.0 - 6.0 = 8.0$$

12-6 THE COMMON-ION EFFECT

The addition of a salt containing an ion, which is also present in a solution of a slightly ionized acid or base, will repress the dissociation of the acid or base. This is expected from Le Chatelier's principle. The addition of sodium acetate to a solution of acetic acid will increase the concentration of undissociated acid and decrease the concentration of hydronium ions.

$$CH_3COOH + H_2O \rightleftharpoons H_3O^+ + CH_3COO^-$$

Since sodium acetate is completely ionized, the addition of this salt increases the CH_3COO^- ion concentration and shifts the equilibrium to the left. From the point of view of the dissociation-constant expression

$$K_a = \frac{[H_3O^+][CH_3COO^-]}{[CH_3COOH]}$$

an increase in $[CH_3COO^-]$ requires that $[CH_3COOH]$ increase and $[H_3O^+]$ decrease if K_a is to remain constant.

Example 12-8

Calculate the hydronium ion concentration in a 0.200 M solution of acetic acid that also contains 0.100 M sodium acetate.

$$CH_3COOH + H_2O \rightleftharpoons H_3O^+ + CH_3COO^-$$

$$CH_3COONa(s) \rightarrow CH_3COO^-(aq) + Na^+(aq)$$

$$K = \frac{[H_3O^+][CH_3COO^-]}{[CH_3COOH]} = 1.76 \times 10^{-5}$$

let $x = [H_3O^+] = [CH_3COO^-]$ from acetic acid

$[CH_3COO^-]$ from $CH_3COONa = 0.100$ M

$[CH_3COOH] = 0.200 - x$

$$\frac{x(0.100 + x)}{0.200 - x} = 1.76 \times 10^{-5}$$

Since x is so much smaller than 0.200 and 0.100, it may be neglected.

$$\frac{0.100\,x}{0.200} = 1.76 \times 10^{-5}$$

$$x = 3.52 \times 10^{-5} \frac{\text{mol}}{\text{L}}$$

Example 12-9

Calculate the hydroxide concentration of a 0.175 M solution of ammonia that also contains 0.125 M ammonium nitrate.

$$NH_3(aq) + H_2O(\ell) \rightleftharpoons NH_4^+(aq) + OH^-(aq)$$

$$NH_4NO_3(s) \rightarrow NH_4^+(aq) + NO_3^-(aq)$$

$$K = \frac{[NH_4^+][OH^-]}{[NH_3]} = 1.79 \times 10^{-5}$$

let $x = [OH^-] = [NH_4^+]$ from $NH_3(aq)$

$[NH_4^+]$ from $NH_4NO_3(s) = 0.125$ M

$[NH_3] = 0.175 - x$

$$\frac{(0.125 + x)x}{0.175 - x} = 1.79 \times 10^{-5}$$

Since x is very small compared to 0.175 and 0.125, it may be neglected.

$$\frac{0.125x}{0.175} = 1.79 \times 10^{-5}$$

$$0.714x = 1.79 \times 10^{-5}$$

$$x = 1.25 \times 10^{-5} \frac{mol}{L}$$

12-7 BUFFER SOLUTIONS

One of the practical applications of the common-ion effect is the *buffer solution*, which resists changes in pH on the addition of acid or base. A buffer solution contains a weak acid and one of its salts (common anion) or a weak base and one of its salts (common cation, or using Brønsted acid-based nomenclature, a buffer solution contains a weak acid and its conjugate base or a weak base and its conjugate acid.

A solution of acetic acid and sodium acetate will resist change in hydronium ion concentration (pH) when a small quantity of strong acid is added. The hydronium ions added will combine with the acetic ions present to form un-ionized acetic acid molecules. When a strong base is added, the hydroxide ions will react with un-ionized acetic acid.

Example 12-10

Calculate the pH of a buffer solution containing 0.100 mol/L of sodium acetate in 0.100 M acetic acid.

$$CH_3COOH + H_2O \rightleftharpoons H_3O^+ + CH_3COO^-$$

$$CH_3COONa \rightarrow CH_3COO^- + Na^-$$

$$K = \frac{[H_3O^+][CH_3COO^-]}{[CH_3COOH]}$$

let $X = [H_3O^+]$

then $0.100 + X = [CH_3COO^-]$

and

$$0.100 - X = [CH_3COOH]$$

$$K = \frac{X \times (0.100 + X)}{0.100 - X} = 1.76 \times 10^{-5}$$

Since X is negligible compared to 0.100

$$K = \frac{0.100X}{0.100} = 1.76 \times 10^{-5} = X$$

$$pH = -\log[H_3O^+]$$

$$= -\log 1.76 \times 10^{-5} = -(-5 + 0.2455)$$

$$= 4.76$$

12-8 THE HENDERSON-HASSELBALCH EQUATION

A limiting law that is useful in making buffer calculations is the Henderson-Hasselbalch equation. It is used quite often by biochemists. This equation is given as

$$pH = pK_a + \log \frac{[\text{salt}]}{[\text{acid}]}$$

for an acid buffer, or in general as

$$pH = pK_a + \log \frac{[\text{conjugate base}]}{[\text{conjugate acid}]}$$

Example 12-11

Calculate the pH of a buffer solution containing 0.100 mol/L of sodium acetate in 0.100 M acetic acid. (Compare this to Example 12-10.)

[conjugate base] = 0.1 M acetate ion (from 0.1 M sodium acetate)

[conjugate acid] = 0.1 M acetic acid

$$pH = 4.75 + \log \frac{[0.1]}{[0.1]}$$

$$pH = 4.75$$

Example 12-12

What is the pH of a 0.1 M ammonium hydroxide solution containing 1 mol/L of ammonium chloride?

[conjugate base] = 0.1 M NH_3 in water

[conjugate acid] = 1.0 M NH_4^+

$pK_b = 4.75$

For a basic buffer, the pK_a taken is that of the conjugate acid of the base. This is given by

$$pK_a + pK_b = 14.00$$

so

$$pK_a(NH_4^+) = 9.25$$

$$pH = 9.25 + \log \frac{[0.1]}{[1.0]}$$

$$pH = 9.25 - 1.00$$

$$pH = 8.25$$

12-9 POLYPROTIC ACIDS

Polyprotic acids ionize in two or more steps. Each ionization step reaches an equilibrium.

$$H_2S + H_2O \rightleftharpoons H_3O^+ + HS^-$$

$$HS^- + H_2O \rightleftharpoons H_3O^+ + S^{2-}$$

Each ionization step is represented by an equilibrium constant.

$$K_1 = \frac{[H_3O^+][HS^-]}{[H_2S]} = 9.1 \times 10^{-8}$$

$$K_2 = \frac{[H_3O^+][S^{2-}]}{[HS^-]} = 1.2 \times 10^{-15}$$

Hydrogen sulfide is chosen as the example of a polyprotic acid because of its importance in the development of wet chemical analyses. A saturated solution of H_2S is approximately 0.10 M.

The $[H_3O^+]$ and $[S^{2-}]$ of this solution are of interest.

Example 12-13

Calculate the $[S^{2-}]$ of 0.10 M H_2S solution.

$$H_2S + H_2O \rightleftharpoons H_3O^+ + HS^-$$

$$K_1 = \frac{[H_3O^+][HS^-]}{[H_2S]} = 9.1 \times 10^{-8}$$

$$HS^- + H_2O \rightleftharpoons H_3O^+ + S^{2-}$$

$$K_2 = \frac{[H_3O^+][S^{2-}]}{[HS^-]} = 1.2 \times 10^{-15}$$

Because $K_1 \gg K_2$ practically all of the H_3O^+ comes from the dissociation of H_2S and $[H_3O^+] \approx [HS^-]$ ($\approx$ means approximately equal to).

Let $X = [H_3O^+]$

then $X \approx [HS^-]$

$$K_2 = \frac{X[S^{2-}]}{X} = 1.2 \times 10^{-15}$$

Sometimes the overall ionization constant is useful in a calculation.

$$H_2S + 2H_2O \rightleftharpoons 2H_3O^+ + S^{2-}$$

$$K_{12} = \frac{[H_3O^+]^2[S^{2-}]}{[H_2S]} = K_1 \times K_2$$

$$K_{12} = 9.1 \times 10^{-3} \times 1.2 \times 10^{-15} = 1.09 \times 10^{-22}$$

Example 12-14

What is the sulfide ion concentration in a saturated H_2S solution, if the pH is adjusted to 1.0?

$$pH = 1.0$$

$$[H_3O^+] = 1 \times 10^{-1}$$

$$[H_2S] = 0.1\ M$$

$$K_{12} = 1.09 \times 10^{-22} = \frac{[H_3O^+]^2[S^{2-}]}{[H_2S]}$$

$$1.09 \times 10^{-22} = \frac{[10^{-1}]^2[S^{2-}]}{[0.1]}$$

$$[S^{2-}] = 1.09 \times 10^{-21} \frac{mol}{L}$$

12-10 ADDITIONAL TYPES OF CALCULATIONS INVOLVING DISSOCIATION CONSTANTS

Example 12-15

Calculate the pH of a solution prepared by mixing 40.0 mL of 0.100 M acetic acid with 60.0 mL of 1.00 M sodium acetate.

$$CH_3COOH + H_2O \rightleftharpoons H_3O^+ + CH_3COO^-$$

$$CH_3COONa \rightarrow CH_3COO^- + Na^+$$

$$\frac{[H_3O^+][CH_3COO^-]}{[CH_3COOH]} = 1.76 \times 10^{-5}$$

$$M(mol/\cancel{L}) \times V(\cancel{L}) = mol\ solute$$

$$M(mmol/\cancel{mL}) \times V(\cancel{mL}) = mmol\ solute \qquad \text{(Section 7-4)}$$

$$40.0\ \cancel{mL} \times \frac{1\ L}{1000\ \cancel{mL}} = 0.0400\ L$$

$$60.0\ \cancel{mL} \times \frac{1\ L}{1000\ \cancel{mL}} = 0.0600\ L$$

$$0.100\ mol/\cancel{L} \times 0.0400\ \cancel{L} = 0.00400\ mol\ CH_3COOH$$

$$1.00\ mol/\cancel{L} \times 0.600\ \cancel{L} = 0.0600\ mol\ CH_3COONa$$

$$0.00400\ mol \times \frac{1000\ \cancel{mL}/\cancel{L}}{100\ \cancel{mL}} = 0.0400 \frac{mol}{L}$$

$$0.600 \text{ mol} \times \frac{1000 \text{ mL/L}}{100 \text{ mL}} = 0.600 \frac{\text{mol}}{\text{L}}$$

$$\frac{600}{1000} = 0.600 \text{ mol/L}$$

Let $X = [H_3O^+]$

$$[CH_3COO^-] = 0.600 + X$$

$$[CH_3COOH] = 0.0400 - X$$

X is negligible compared to 0.600

$$\frac{X \times 0.600}{0.0400} = 1.76 \times 10^{-5}$$

$$X = \frac{(1.76 \times 10^{-5}) \times (4.00 \times 10^{-2})}{6.00 \times 10^{-1}} = 1.17 \times 10^{-6} \text{ mol/L}$$

$$pH = -\log(1.17 \times 10^{-6}) = 5.93$$

Example 12-16

How many grams of sodium cyanide must be added to 200 mL of a 0.100 M solution of HCN to give a solution whose pH is 6.22?

$$HCN + H_2O \rightleftharpoons H_3O^+ + CN^-$$

$$NaCN \rightarrow Na^+ + CN^-$$

$$\frac{[H_3O^+][CN^-]}{[HCN]} = 4.93 \times 10^{-10}$$

First calculate mass of NaCN on the basis of 1 L of solution.

$$pH = 6.22 = -\log[H_3O^+]$$

$$[H_3O^+] = 6.03 \times 10^{-7} \frac{\text{mol}}{\text{L}}$$

Let $[CN^-] = x$

$$\frac{(6.03 \times 10^{-7})X}{0.100} = 4.93 \times 10^{-10}$$

$$X = \frac{4.93 \times 10^{-11}}{6.03 \times 10^{-7}}$$

$$= 8.18 \times 10^{-5} \frac{\text{mol}}{\text{L}}$$

$$8.18 \times 10^{-5} \frac{\text{mol}}{\text{L}} \times \frac{200 \text{ mL}}{1000 \text{ mL}} = 1.64 \times \frac{10^{-5} \text{ mol}}{200 \text{ mL}}$$

NaCN

$$\begin{array}{l} 1 \times 23.0 = 23.0 \\ 1 \times 12.0 = 12.0 \\ 1 \times 14.0 = \underline{14.0} \\ \qquad\qquad 49.0 \text{ g/mol} \end{array}$$

$$1.64 \times 10^{-5} \cancel{\text{mol}} \times 49.0 \frac{\text{g}}{\cancel{\text{mol}}} = 8.04 \times 10^{-4} \text{ g}$$

PROBLEMS

12-1 Calculate the value of the dissociation constant of 0.0200 M acetic acid which was observed to be 2.99% dissociated.
[D. A. MacInnes and T. Shedlovsky, *J. Am. Chem. Soc.*, *54*, 1435 (1932)]

12-2 A 0.01173 M solution of n-butyric acid was found to be 3.53% ionized by conductance measurements. Calculate the ionization constant of the acid, C_3H_7COOH.
[J. F. P. Dippy, *J. Chem. Soc.*, *1938*, 1226]

12-3 What is the value of K_a for benzoic acid, C_6H_5COOH, based on the observation that a 0.00823 M solution was 8.65% ionized?
[D. J. G. Ives, *J. Chem. Soc.*, *1933*, 734]

12-4 Calculate the hydronium concentration of 0.025 M acetic acid.

12-5 Calculate the pOH of a 0.200 M solution of ammonia.

12-6 Calculate the concentration of a benzoic acid solution whose pH is 2.12.

12-7 Determine the hydronium ion concentration of a 0.0500 M formic acid solution.

12-8 A. Klemenc and E. Hayek determined the dissociation constants of nitrous acid at 0, 12.5, and 30°C as 3.2, 4.6, and

6.0×10^{-4}, respectively. Find the hydronium concentrations of a 0.100 M HNO_2 solution at each of the three temperatures.
[*Monatsh. Chem.*, *54*, 407 (1929)]

12-9 What is the cyanide ion concentration of a 0.0120 M HCN solution?

12-10 Calculate the pOH of a 0.0040 M butyric acid solution.

12-11 Calculate the pH of a 1.02×10^{-3} M cyanoacetic acid solution. $K = 3.34 \times 10^{-3}$.
[B. Saxton and L. S. Dacken, *J. Am. Chem. Soc.*, *62*, 848 (1962)]

12-12 Find the dissociation constant of an acid if a 0.100 M solution of the acid has a pH of 3.41.

12-13 Calculate the dissociation constant of an acid if a 0.125 M solution has a pH of 4.671.

12-14 Determine the pH of a solution of formic acid that contains 10.02 g/L.

12-15 What is the pH of a 2.00% solution of acetic acid? The specific gravity of the solution is 1.0012.

12-16 What is the pH of a 1.00% solution of ammonia, sp. gr. 0.9939?

12-17 Assuming all the $[H^+]$ comes from the first ionization of H_3PO_4, what is the pH of a 0.1 M H_3PO_4 solution?

12-18 Calculate the hydronium concentration of a 0.120 M acetic acid solution that also contains 0.200 mol of sodium acetate per liter.

12-19 What is the pH of a 0.225 M formic acid solution that also contains 0.300 mol of potassium formate per liter?

12-20 How many grams of sodium acetate must be added to 200 mL of 0.100 M acetic acid to give a solution of pH 4.20?

12-21 Under normal conditions the pH of blood is 7.4. Assuming blood is buffered by the carbonic acid bicarbonate equilibrium,

What is the ratio of bicarbonate ion to carbonic acid in blood? (Assume all the CO_2 is present as H_2CO_3.)

12-22 What is the pH of a buffer solution containing 0.750 M acetic acid and 0.45 M sodium acetate?

12-23 How many moles of NaH_2PO_4 would you use to make 100 mL of a buffer of pH 3.0 containing 0.1 M H_3PO_4?

12-24 Use the Henderson-Hasselbalch equation to calculate the ratio of the number of moles of salt to acid necessary to make each of the following buffers:

a. sodium acetate, acetic acid, pH = 3.5
b. sodium formate, formic acid, pH = 5.0
c. sodium dihydrogen phosphate, disodium hydrogen phosphate, pH = 7.0

12-25 A 0.433 M solution of an acid HX also contained 0.623 mol/L of the salt KX. Find the dissociation constant of the acid if the pH of the solution was 4.882.

12-26 A student in a freshman lab was given 0.01 mol of a weak organic acid and told to determine the K_a of the acid. He prepared 100 mL of an aqueous solution containing the sample; 50 mL of this solution was then titrated with NaOH to the equivalence point. The titrated solution was then mixed with the other 50 mL of solution and the pH determined. A value of 4.80 was obtained for the pH of the mixed solutions. What is the approximate K_a value of the acid?

12-27 Determine the $[S^{2-}]$ of a 0.100 M H_2S solution that is also 0.300 M in HCl.

12-28 Calculate the sulfide ion concentration of a 0.10 M H_2S solution that is also 1.00 M in acetic acid.

12-29 Determine the pH of a solution prepared by mixing 30 mL of 0.015 M propionic acid with 20 mL of 0.10 M sodium propionate.

12-30 Calculate the pH of a solution prepared by mixing 150 mL of 0.020 M aqueous ammonia with 80.0 mL of 0.50 M ammonium chloride solution.

12-31 Calculate the pH of a solution made by mixing 500 mL of 0.100 M acetic acid with 500 mL of 0.200 M sodium hydroxide.

12-32 What is the pH of 750 mL of 0.0200 M acetic acid solution to which 250 mL of 0.50 M calcium acetate solution has been added?

Supplementary Problems

12-33 J. F. P. Dippy and F. R. Williams found that a 0.002907 M solution of benzoic acid was 13.8% dissociated. Calculate pK_a for the acid.

[*J. Chem. Soc.*, *1934*(1888)]

12-34 In an early determination of the ionization constant of acetic acid, J. Kendall found that a 0.03685 M solution was 2.22% ionized. Calculate K_a and pK_a for acetic acid from this measurement.

[*J. Chem. Soc.*, *101*, 1279 (1912)]

12-35 What is the pH of a 0.0250 M solution of propionic acid?

12-36 Calculate the pOH of a 0.0110 M benzoic acid solution.

12-37 Calculate the pH of a 0.0300 M benzoic acid solution.

12-38 What is the pH of a 0.030 M ammonia solution?

12-39 What fraction of propionic acid is dissociated in the following solutions?

a. 0.100 M
b. 0.0100 M
c. 0.00100 M

12-40 Calculate the pH of a solution prepared by diluting 100 mL of 0.216 M formic acid solution to 2.00 L.

12-41 What is the hydrogen ion concentration of a 0.043 M H_2S solution?

12-42 Calculate the pH of a 2.8×10^{-2} M H_2S solution.

12-43 What is the pH of 2×10^{-2} M o-phthalic acid solution?

12-44 Calculate the pH of a 0.100 M malonic acid solution.

12-45 The solubility of salicylic acid in water is 0.16 g/100 g. Calculate the approximate pH of a saturated solution of the acid.

12-46 How many grams of trimethylamine must be present in 250 mL of an aqueous solution to give a pH of 11.91?

12-47 What is the pOH of a 0.0212 M solution of pyridine?

12-48 What concentration of nitrous acid is required to give a solution whose pH is 2.23?

12-49 What is the pH of a 0.208 M phosphoric acid solution?

12-50 Calculate the pH of a 0.125 M hydrogen sulfide solution.

12-51 What is the pH of a 0.309 M sulfurous acid solution? $K_1 = 1.54 \times 10^{-2}$.

12-52 Find the pH of 0.100 M HCN solution.

12-53 What is the pH of a 1.00% solution of formic acid? The density of the solution is 1.0019 g/mL.

12-54 Calculate the pH of a solution of acetic acid that contains 20.02 g/L.

12-55 What is the hydroxide ion concentration of a 0.125 M ammonia solution?

12-56 What is the concentration of a propionic acid solution whose hydronium concentration is 1.2×10^{-3} M?

12-57 What is the ammonium ion concentration of a 0.0200 M solution of ammonia that is also 0.100 M in potassium hydroxide?

12-58 What is the malonate ion concentration of a 0.100 M malonic acid solution that also contains 0.100 M HCl?

12-59 What mass of potassium nitrite must be included in the solution in Problem 12-48 to raise the pH from 2.23 to 3.16?

13
CHEMICAL EQUILIBRIUM, III

13-1 THE SOLUBILITY PRODUCT

The equilibrium between a solid ionic salt and its saturated aqueous solution is described by the solubility-product expression. This relationship was first observed by Nernst in 1899. The usual example of this principle is the equilibrium between solid silver chloride and its saturated solution.

$$AgCl(s) \rightleftharpoons Ag^+ + Cl^-$$

The solubility-product expression is

$$K_S = [Ag^+][Cl^-]$$

where K_S is the solubility-product constant of silver chloride. It has a value of 1.8×10^{-10}.

The solubility-product constant may also be recorded as pK_S, where p serves the same function as in pH.

$$pK_S = -\log K_S$$

For silver chloride, $pK_S = 9.75$.

Table 13-1 is a compilation of K_S and pK_S values obtained by a critical examination of experimental data. The result of a single experiment in the examples and problems here may differ from the values in Table 13-1.

13-2 CALCULATION OF K_S FROM SOLUBILITY

The solubility of a slightly soluble salt in water is its analytical concentration. If we assume that the dissolved salt is completely ionized, the value of K_S can be calculated from the solubility.

TABLE 13-1 Solubility Products

Substance	pK_S	K_S
$AgBr$	12.28	5.2×10^{-13}
$AgBrO_3$	4.28	5.2×10^{-5}
$AgCN$	15.92	1.2×10^{-16}
$AgCl$	9.75	1.8×10^{-10}
Ag_2CrO_4	11.95	1.1×10^{-12}
AgI	16.08	8.3×10^{-17}
$AgIO_3$	7.52	3.0×10^{-8}
$AgSCN$	12.00	1.0×10^{-12}
Ag_2SeO_4	7.25	5.6×10^{-8}
Ag_2SO_4	4.80	1.6×10^{-5}
$Ba(BrO_3)_2$	5.50	3.2×10^{-6}
$BaCrO_4$	9.93	1.2×10^{-10}
BaF_2	5.98	1.0×10^{-6}
$BaSO_4$	9.87	1.3×10^{-10}
CaF_2	10.31	4.9×10^{-11}
$CaSO_4$	5.92	1.2×10^{-6}
$Cd(IO_3)_2$	9.50	3.2×10^{-10}
CuI	11.96	1.1×10^{-12}
$Cu(IO_3)_2$	7.13	7.4×10^{-8}
Hg_2Cl_2	17.88	1.3×10^{-18}
$KClO_4$	1.97	1.1×10^{-2}
KIO_3	1.70	5.0×10^{-2}
K_2PtCl_6	4.96	1.1×10^{-5}
$La(IO_3)_2$	11.21	6.2×10^{-12}
MgF_2	8.19	6.5×10^{-9}

(continued)

TABLE 13-1 (continued)

Substance	pK_S	K_S
$PbBr_2$	4.41	3.9×10^{-5}
$PbCl_2$	4.79	1.6×10^{-5}
$PbCrO_4$	13.75	1.8×10^{-14}
$Pb(IO_3)_2$	12.49	3.2×10^{-13}
$PbSO_4$	7.79	1.6×10^{-8}
$PbSeO_4$	6.84	1.4×10^{-7}
$SrCrO_4$	4.44	3.6×10^{-5}
$Sr(IO_3)_2$	6.48	3.3×10^{-7}
$SrSO_4$	6.49	3.2×10^{-7}
TlBr	5.47	3.4×10^{-6}
$TlBrO_3$	4.07	8.5×10^{-5}
TlCl	3.76	1.7×10^{-4}
Tl_2CrO_4	12.01	9.8×10^{-13}
TlI	7.19	6.5×10^{-8}
$TlIO_3$	5.51	3.1×10^{-6}

Source: L. Meites (ed.), *Handbook of Analytical Chemistry*, pp. 1-13, McGraw-Hill, New York (1963).

Example 13-1

The solubility of silver bromide in water was determined by C. Bedell as 1.86×10^{-6} mol/L at 20°. Calculate the corresponding value of K_S.

[*Compt. Rend. 207*, 632-634 (1938)]

$$AgBr(s) \rightleftharpoons Ag^+ + Br^-$$

$$K_S = [Ag^+][Br^-]$$

$$[Ag^+] = [Br^-] = C = 1.86 \times 10^{-6}$$

where C is the analytical concentration of AgBr.

$[Ag^+] = [Br^-] = C$, because one Ag^+ and one Br^- are formed for each AgBr that dissolves.

$$K_S = (1.86 \times 10^{-6})^2 = 3.46 \times 10^{-12}$$

Example 13-2

What is the value of the K_S of $Cu(IO_3)_2$ at 25°? The solubility of $Cu(IO_3)_2$ in water was found to be 3.245×10^{-3} mol/L.
[R. M. Keefer, *J. Am. Chem. Soc.*, *70*, 476-479 (1948)]

$$Cu(IO_3)_2(s) \rightleftharpoons Cu^{2+} + 2IO_3^-$$

$$K_S = [Cu^{2+}][IO_3^-]^2$$

$$[Cu^{2+}] = C = 3.245 \times 10^{-3} \frac{mol}{L}$$

One $Cu(IO_3)_2$ forms one Cu^{2+} ion.

$$[IO_3^-] = 2C = 6.490 \times 10^{-3}$$

One $Cu(IO_3)_2$ forms two IO_3^- ions.

$$K_S = (3.245 \times 10^{-3}) \times (6.490 \times 10^{-3})^2$$

$$= 136.6 \times 10^{-9} = 1.37 \times 10^{-7}$$

It is important to realize that the K_S expression describes an experimental observation. When $[A]^n$ appears in the expression, the total concentration of A, regardless of its source, must be raised to the power n. In Example 13-2 the *total concentration* of IO_3^- $[IO_3^-]$ was raised to the power of 2 because the K_S expression demanded it. The total concentration of IO_3^- was equal to 2C because the only source of IO_3^- ions was $Cu(IO_3)_2$ and *one* $Cu(IO_3)_2$ formed *two* IO_3^- ions.

Example 13-3

The average of several determinations of the solubility of silver chloride in water at 25°C was 0.00194 g/L. Calculate the pK_S of AgCl by using this value.
[A. Seidell, *Solubilities of Inorganic and Metal Organic Compounds*, 3rd ed., p. 32, Van Nostrand, Princeton, New Jersey (1940)]

$$AgCl(s) \rightleftharpoons Ag^+ + Cl^-$$

$$K_S = [Ag^+][Cl^-]$$

$[Ag^+] = [Cl^-] = C$

AgCl 107.870
35.453

143.323 g/mol

$$C = \frac{1.94 \times 10^{-3}\ \not{g}/L}{1.43 \times 10^{+2}\ \not{g}/mol} = 1.36 \times 10^{-5}\ \frac{mol}{L}$$

$K_S = (1.36 \times 10^{-5})^2 = 1.85 \times 10^{-10}$

$pK_S = -\log_{10}K_S = -\log(1.85 \times 10^{-10})$

$pK_S = 9.73$

13-3 CALCULATION OF SOLUBILITIES FROM K_S

Example 13-4

Calculate the solubility of thallous chloride in water at 20°C from its K_S, 1.7×10^{-4}. The experimental value obtained by Viktorin and Sirucek was 0.01376 mol/L.
[*Collection Czech. Chem. Commun.*, *11*, 474-493 (1939)]

$TlCl(s) \rightleftharpoons Tl^+ + Cl^-$

$K_S = [Tl^+][Cl^-] = 1.7 \times 10^{-4}$

Solubility of TlCl = C

$[Tl^+] = [Cl^-] = C$

$C^2 = 1.7 \times 10^{-4}$

$$C = \sqrt{1.7 \times 10^{-4}} = 1.3 \times 10^{-2}\ \frac{mol}{L}$$

Example 13-5

Compare the aqueous solubility of lead iodate found by R. M. Keefer and H. G. Reiber, 3.61×10^{-5} mol/L, with the value calculated from K_S, 3.2×10^{-13}.
[*J. Am. Chem. Soc.*, *63*, 689-692 (1941)]

$Pb(IO_3)_2 \rightleftharpoons Pb^{2+} + 2IO_3^-$

$K_S = [Pb^{2+}][IO_3^-]^2 = 3.2 \times 10^{-13}$

Solubility of $Pb(IO_3)_2$ = C

$[Pb^{2+}]$ = C (one $Pb(IO_3)_2$ gives *one* Pb^{2+})

$[IO_3^-]$ = 2C (one $Pb(IO_3)_2$ gives *two* IO_3^-)

$C \times (2C)^2 = 3.2 \times 10^{-13}$

$4C^3 = 3.2 \times 10^{-13}$

$C^3 = 0.80 \times 10^{-13} = 8.0 \times 10^{-14}$

$C = \sqrt[3]{8.0 \times 10^{-14}} = \sqrt[3]{80 \times 10^{-15}}$

$C = 4.3 \times 10^{-5} \frac{mol}{L}$

Example 13-6

Calculate the solubility of silver iodide in water from the K_S.

$AgI(s) \rightleftharpoons Ag^+ + I^-$

$K_S = [Ag^+][I^-] = 8.3 \times 10^{-17}$

Let C = solubility of AgI; then

$C = [Ag^+] = [I^-]$ (one AgI forms one Ag^+ and one I^-)

$C^2 = 8.3 \times 10^{-17} = 83 \times 10^{-18}$

$C = 9.1 \times 10^{-9} \frac{mol}{L}$

Example 13-7

Calculate the concentration of barium in a saturated solution of barium iodate at 25°C. The pK_S of barium iodate is 8.82.

[G. Macdougall and C. W. Davies, *J. Chem. Soc.*, *1935*, 1416]

$Ba(IO_3)_2 \rightleftharpoons Ba^{2+} + 2IO_3^-$

$K_S = [Ba^{2+}][IO_3^-]^2$

$-\log K_S = 8.82$

$\log K_S = -8.82 = -9. + .18$

$K_S = 1.51 \times 10^{-9}$

Let X = $[Ba^{2+}]$; then

$2X = [IO_3^-]$

$(X)(2X)^2 = 1.51 \times 10^{-9}$

$4X^3 = 1.51 \times 10^{-9}$

$X^3 = 0.38 \times 10^{-9}$

$X = \sqrt[3]{.38 \times 10^{-3}}$

$X = 7.24 \times 10^{-2} \frac{mol}{L}$

13-4 COMMON-ION EFFECT APPLIED TO SOLUBILITY

The common-ion effect applies to equilibria involving slightly soluble salts. The addition of a soluble salt with an ion in common with the slightly soluble salt diminishes the concentration of the other ion. The addition of potassium sulfate to a saturated solution of strontium sulfate will diminish the concentration of strontium ions.

$SrSO_4(s) \rightleftharpoons Sr^{2+} + SO_4^{2-}$

$K_S = [Sr^{2+}][SO_4^{2-}] = 3.2 \times 10^{-7}$

Example 13-8

Calculate the strontium ion concentration in a 0.200 M K_2SO_4 solution. Let $X = [Sr^{2+}]$; then

$0.200 + X = [SO_4^{2-}]$

$X \times (0.200 + X) = 3.2 \times 10^{-7}$

X is negligibly small compared to 0.200

$0.200X = 3.2 \times 10^{-7}$

$X = 1.6 \times 10^{-6} \frac{mol}{L}$

In this example the excess sulfate ion was very large. If the excess is small, X cannot be neglected and the quadratic equation must be involved.

Example 13-9

Calculate the solubility of lead iodate in a 0.020 M KIO_3 solution.

$$Pb(IO_3)_2(s) \rightleftharpoons Pb^{2+} + 2IO_3^-$$

$$K_S = [Pb^{2+}][IO_3^-]^2 = 3.2 \times 10^{-13}$$

Let C = solubility of $Pb(IO_3)_2$ (analytical concentration); then

$$[Pb^{2+}] = C$$

$$[IO_3^-] = 0.020 + 2C$$

$$C(0.020 + 2C)^2 = 3.2 \times 10^{-13}$$

2C << 0.020 and can be neglected

$$4.0 \times 10^{-4} C = 3.2 \times 10^{-13}$$

$$C = 0.80 \times 10^{-9} = 8.0 \times 10^{-10} \frac{\text{mol}}{\text{L}}$$

Example 13-10

50.0 mL of 2.50×10^{-2} M silver nitrate solution was added to 50.0 mL of 5.00×10^{-2} M sodium chloride solution. Calculate the silver ion concentration of the resulting solution.

$$Ag^+ + NO_3^- + Na^+ + Cl^- \rightleftharpoons AgCl(s) + Na^+ + NO_3^-$$

$$AgCl(s) \rightleftharpoons [Ag^+][Cl^-]$$

$$50.0 \times 2.50 \times 10^{-2} = 1.25 \text{ mmol } AgNO_3$$

$$50.0 \times 5.00 \times 10^{-2} = 2.50 \text{ mmol NaCl}$$

(Twice the number of mmol of NaCl required to precipitate the $AgNO_3$ has been added.)

1.25 mmol of NaCl remains in 100 mL of solution. The concentration of NaCl is 12.5 mmol/liter or 0.0125 mol/L. Let $X = [Ag^+]$; then

$$X + 0.0125 = [Cl^-]$$

$$(X)(X + 0.0125) = 1.8 \times 10^{-10}$$

$$1.25 \times 10^{-2} X = 1.8 \times 10^{-10}$$

X << 0.0125 and can be neglected in (X + 0.0125)

$$X = \frac{1.8 \times 10^{-10}}{1.25 \times 10^{-2}} = 1.44 \times 10^{-8} \frac{\text{mol}}{\text{L}}$$

13-5 COMPLEX IONS

Complex ions exist in equilibrium with their constituents in solution. The constituents are metal cations and neutral molecules or anions. The neutral molecules and anions are called *ligands*.

The most familiar example to general-chemistry students is the silver amine complex $Ag(NH_3)_2{}^+$. The equilibrium may be represented by

$$Ag(NH_3)_2{}^+ \rightleftharpoons Ag^+ + 2NH_3$$

and the equilibrium constant

$$K_{diss} = \frac{[Ag^+][NH_3]^2}{[Ag(NH_3)_2{}^+]}$$

13-6 DISSOCIATION CONSTANT

K_{diss} is called the *dissociation constant* of the complex ion (see Table 13-2). Some tables list formation constants, which are, of course, reciprocals of the dissociation constants.

TABLE 13-2 Dissociation Constants of Complex Ions

Equilibrium	K_{diss}
$Ag(CN)_2{}^- \rightleftharpoons Ag^+ + 2CN^-$	1.4×10^{-20}
$Ag(NH_3)_2{}^+ \rightleftharpoons Ag^+ + 2NH_3$	6.8×10^{-8}
$Ag(S_2O_3)_2{}^{3-} \rightleftharpoons Ag^+ + 2S_2O_3{}^{2-}$	6×10^{-14}
$Cd(NH_3)_4{}^{2+} \rightleftharpoons Cd^{2+} + 4NH_3$	2.8×10^{-7}
$Co(NH_3)_6{}^{2+} \rightleftharpoons Co^{2+} + 6NH_3$	1.2×10^{-5}
$Co(NH_3)_6{}^{3+} \rightleftharpoons Co^{3+} + 6NH_3$	2.2×10^{-34}
$Cu(NH_3)_4{}^{2+} \rightleftharpoons Cu^{2+} + 4NH_3$	4.7×10^{-15}
$Hg(CN)_4{}^{2-} \rightleftharpoons Hg^{2+} + 4CN^-$	4×10^{-42}
$Ni(NH_3)_6{}^{2+} \rightleftharpoons Ni^{2+} + 6NH_3$	2×10^{-9}
$Zn(NH_3)_4{}^{2+} \rightleftharpoons Zn^{2+} + 4NH_3$	3×10^{-10}
$Zn(OH)_4{}^{2-} \rightleftharpoons Zn^{2+} + 4OH^-$	3×10^{-16}

The dissociation of a complex ion, such as $Ag(NH_3)_2^+$, is a stepwise process similar to the ionization of polyprotic acids. Each step constitutes an equilibrium with an equilibrium constant.

$$Ag(NH_3)_2^+ \rightleftharpoons AgNH_3^+ + NH_3$$

$$K_1 = \frac{[AgNH_3^+][NH_3]}{[Ag(NH_3)_2^+]}$$

$$AgNH_3^+ \rightleftharpoons Ag^+ + NH_3$$

$$K_2 = \frac{[Ag^+][NH_3]}{[AgNH_3^+]}$$

The equilibria involved in the stepwise dissociation of complex ions may become extremely complicated. The problems in this treatment will be confined to those in which the ligand concentration is sufficiently high to eliminate the necessity for considering the intermediate steps in the dissociation. Only the overall dissociation constant need be considered.

$$K_{diss} = K_1 \times K_2 = \frac{[Ag(NH_3)^+][NH_3]}{[Ag(NH_3)_2^+]} \times \frac{[Ag^+][NH_3]}{[Ag(NH_3)^+]}$$

Example 13-11

Determine the silver ion concentration in a 0.200 M solution of $Ag(NH_3)_2NO_3$.

$$Ag(NH_3)_2^+ \rightleftharpoons Ag^+ + 2NH_3$$

$$K_{diss} = \frac{[Ag^+][NH_3]^2}{[Ag(NH_3)_2^+]} = 6.8 \times 10^{-8}$$

Let $X = [Ag^+]$; then

$$2X = [NH_3] \text{ and } 0.200 - X = [Ag(NH_3)_2^+]$$

$$\frac{4X^3}{0.200 - X} = 6.8 \times 10^{-8}$$

$$4X^3 = 1.36 \times 10^{-8}$$

$$X = \sqrt[3]{0.34 \times 10^{-8}} = \sqrt[3]{3.4 \times 10^{-9}}$$

$$X = 1.5 \times 10^{-3} \frac{\text{mol}}{\text{L}}$$

Example 13-12

How many moles of ammonia must be added to 1 L of a 0.100 M silver nitrate solution to reduce the silver ion concentration to 2.00×10^{-7} M?

$$Ag(NH_3)_2^+ \rightleftharpoons Ag^+ + 2NH_3$$

$$K_{diss} = \frac{[Ag^+][NH_3]^2}{[Ag(NH_3)_2^+]} = 6.8 \times 10^{-8}$$

(1) Sufficient ammonia must be added to combine with practically all of the silver ions present. (2) An additional amount to further reduce the silver ion concentration is also necessary.

1. $Ag^+ + 2NH_3 \rightarrow Ag(NH_3)_2^+$

$0.100 \text{ mol} + 0.200 \text{ mol} \rightarrow 0.100 \text{ mol}$

2. $Ag(NH_3)_2^+ \rightleftharpoons Ag^+ + 2NH_3$

$$K_{diss} = \frac{[Ag^+][NH_3]^2}{[Ag(NH_3)_2^+]}$$

Let $[NH_3] = X$

$[Ag^+] = 2.0 \times 10^{-7}$

$[Ag(NH_3)_2^+] = 0.100 - 2 \times 10^{-7}$

$$\frac{2.0 \times 10^{-7} X^2}{0.100} = 6.8 \times 10^{-8}$$

$$X^2 = \frac{6.8 \times 10^{-9}}{2.0 \times 10^{-7}} = 3.4 \times 10^{-2}$$

$$X = 0.184 \frac{\text{mol}}{\text{L}}$$

$$0.200 + 0.184 = 0.384 \frac{\text{mol}}{\text{L}}$$

13-7 HYDROLYSIS

In Chapter 10, solutions of salts whose ions did not react with the solvent water were considered. Since anions are bases and some hydrated cations are acids, reactions with water may be expected. Reactions of this type are called *hydrolytic reactions*. The phenomenon is called *hydrolysis*.

Hydrolysis occurs when salts of strong bases and weak acids, weak bases and strong acids, and weak bases and weak acids are dissolved in water. Examples are sodium acetate, ammonium chloride, and ammonium acetate.

13-8 SODIUM ACETATE

This will serve as an illustration of the hydrolysis of a salt of a strong base and a weak acid. When CH_3COONa dissolves in water, the ions dissociate, $CH_3COO^- + Na^+$. The solution is basic to indicators. The following reversible reactions are involved.

$$CH_3COO^- + H_3O^+ \rightleftharpoons CH_3COOH + H_2O$$

$$2H_2O \rightleftharpoons H_3O^+ + OH^-$$

They may be combined:

$$CH_3COO^- + H_2O \rightleftharpoons CH_3COOH + OH^-$$

The equilibrium constant expression has the form

$$K_h = \frac{[CH_3COOH][OH^-]}{[CH_3COO^-]}$$

and is called the *hydrolysis constant*. The numerical value of K_h may be obtained by making suitable substitutions.

$$[CH_3COOH] = \frac{[H_3O^+][CH_3COO^-]}{K_a}$$

$$[OH^-] = \frac{K_w}{[H_3O^+]}$$

$$K_h = \frac{[H_3O^+][CH_3COO^-]/K_a \times K_w/[H_3O^+]}{[CH_3COO^-]}$$

$$K_h = \frac{[H_3O^+][CH_3COO^-]}{K_a} \times \frac{K_w}{[H_3O^+]} \times \frac{1}{[CH_3COO^-]} = \frac{K_w}{K_a}$$

Example 13-13

Calculate the hydroxide ion concentration of a 0.100 M sodium acetate solution.

$$CH_3COONa \rightarrow CH_3COO^- + Na^+$$

$$CH_3COO^- + H_2O \rightleftharpoons CH_3COOH + OH^-$$

$$K_h = \frac{[CH_3COOH][OH^-]}{[CH_3COO^-]} = \frac{K_w}{K_a} = \frac{1.00 \times 10^{-14}}{1.75 \times 10^{-5}}$$

$$K_h = 5.72 \times 10^{-10}$$

Let $X = [OH^-]$; then

$X = [CH_3COOH]$ (1 CH_3COOH is formed for each OH^-)

$0.100 - X = [CH_3COO^-]$ (1 CH_3COO^- is consumed for each OH^- formed)

$$\frac{X^2}{0.100 - X} = 5.72 \times 10^{-10}$$

X may be neglected in the denominator.

$$\frac{X^2}{1.00 \times 10^{-1}} = 5.72 \times 10^{-10}$$

$$X^2 = 5.72 \times 10^{-11}$$

$$X^2 = 57.2 \times 10^{-12}$$

$$X = 7.56 \times 10^{-6} \frac{\text{mol}}{\text{L}}$$

13-9 AMMONIUM CHLORIDE

Ammonium chloride is the salt of a weak base and a strong acid. When it dissolves in water, the ions dissociate: $NH_4^+ + Cl^-$. The solution is acidic to indicators. The reactions that occur in the solution are

$$NH_4^+ + OH^- \rightleftharpoons NH_3 + H_2O$$

$$2H_2O \rightleftharpoons H_3O^+ + OH^-$$

The overall equilibrium reaction is

$$NH_4^+ + H_2O \rightleftharpoons H_3O^+ + NH_3$$

$$K_h = \frac{[H_3O^+][NH_3]}{[NH_4^+]}$$

From

$$K_b = \frac{[NH_4^+][OH^-]}{[NH_4]}$$

we get

$$[NH_3] = \frac{[NH_4^+][OH^-]}{K_b}$$

and from

$$K_w = [H_3O^+][OH^-]$$

we get

$$[H_3O^+] = \frac{K_w}{[OH^-]}$$

Substituting

$$K_h = \frac{K_w/[OH^-] \times [NH_4^+][OH^-]/K_b}{[NH_4^+]}$$

$$= \frac{K_w}{[OH^-]} \times \frac{[NH_4^+][OH^-]}{K_b} \times \frac{1}{[NH_4^+]} = \frac{K_w}{K_b}$$

Example 13-14

What is the pH of a 0.100 M NH_4Cl solution?

$$NH_4^+ + H_2O \rightleftharpoons NH_3 + H_3O^+$$

$$K_h = \frac{[NH_3][H_3O^+]}{[NH_4^+]} = \frac{1.00 \times 10^{-14}}{1.75 \times 10^{-5}} = 5.72 \times 10^{-10}$$

Let $X = [H_3O^+]$; then

$$X = [NH_3]$$

$$0.100 - X = [NH_4^+]$$

$$\frac{X^2}{0.100 - X} = 5.72 \times 10^{-10}$$

X may be neglected in the denominator

$$\frac{X^2}{1.00 \times 10^{-1}} = 5.72 \times 10^{-10}$$

$$X^2 = 5.72 \times 10^{-11} = 57.2 \times 10^{-12}$$

$$X = 7.56 \times 10^{-6} \frac{\text{mol}}{\text{L}}$$

$$\text{pH} = -\log [H_3O^+] = -\log 7.56 \times 10^{-6}$$

$$\text{pH} = 5.12$$

13-10 AMMONIUM ACETATE

Ammonium acetate yields NH_4^+ and CH_3COO^- ions in solution. The ammonium ion is an acid and the acetate ion is a base.

$$NH_4^+ + H_2O \rightleftharpoons NH_3 + H_3O^+$$

$$CH_3COO^- + H_2O \rightleftharpoons CH_3COOH + OH^-$$

The H_3O^+ ions formed combine with the OH^- ions formed.

$$NH_4^+ + CH_3COO^- \rightleftharpoons NH_3 + CH_3COOH$$

$$K_h = \frac{[NH_3][CH_3COOH]}{[NH_4^+][CH_3COO^-]}$$

$$K_b = \frac{[NH_4^+][OH^-]}{[NH_3]}$$

$$K_a = \frac{[H_3O^+][CH_3COO^-]}{[CH_3COOH]}$$

$$[NH_3] = \frac{[NH_4^+][OH^-]}{K_b}$$

$$[CH_3COOH] = \frac{[H_3O^+][CH_3COO^-]}{K_a}$$

$$K_b = \frac{[NH_4^+][OH^-]/K_b \times [H_3O^+][CH_3COO^-]/K_a}{[NH_4^+][CH_3COO^-]}$$

$$K_h = \frac{K_w}{K_b \times K_a}$$

PROBLEMS

13-1 What is the value of K_S for $TlIO_3$ based on the solubility 0.001844 mol/L determined by V. K. La Mer and F. H. Goldman?
[*J. Am. Chem. Soc.*, *51*, 2632 (1929)]

13-2 Calculate the solubility-product constant of silver iodate at 75° from the solubility determined by W. P. Baxter, 0.8403 mmol/1000 g of water. The density of water at 25°C is 0.99707 g/mL.
[*J. Am. Chem. Soc.*, *48*, 615-621 (1926)]

13-3 M. P. Menon found the solubility of ErF_3 in water at 298 K to be 1.98×10^{-5} M. What is its K_S?
[*J. Radioanal. Nucl. Chem.*, *96*(3), 311 (1985)]

13-4 Determine a value of the K_S of the Ag_2SeO_4 from the solubility, 0.00242 mol/L at 25°.
[R. W. Gelbach and G. B. King, *J. Am. Chem. Soc.*, *64*, 1054 (1942)]

13-5 The solubility of $Ba(BrO_3)_2$ was determined by R. M. Keefer, H. G. Reiber, and C. S. Bisson as 7.895 g/1000 g H_2O. Calculate the value of K_S based on this solubility determination.
[*J. Am. Chem. Soc.*, *62*, 2951-2955 (1940)]

13-6 According to R. Dolique, 0.1361 g $SrSeO_4$ dissolves in 100 mL of water at 10°. Calculate a value of the K_S based on this solubility.
[*Bull. Soc., Chim. France*, *10*, 50 (1943)]

13-7 Three determinations of the solubility of silver sulfate in water at 25° gave 0.02677, 0.02675, and 0.02676 mol/1000 g water. Find a value of the K_S of silver sulfate from these data.
[G. Akerlof and H. C. Thomas, *J. Am. Chem. Soc.*, *56*, 596 (1934)]

13-8 Calculate the value of K_S for $La(IO_3)_3$ from the solubility in water and compare the solubilities calculated from K_S with those listed below.

Added salt	Concentration	Solubility of $La(IO_3)_3$ (mol/L)
None		9.426×10^{-4}
$La_2(SO_4)_3$	M/30	1.830×10^{-4}
$La_2(SO_4)_3$	M/60	1.532×10^{-4}
$La(NO_3)_3$	M/30	9.398×10^{-4}
$LaCl_3$	M/30	9.233×10^{-4}

Source: V. K. La Mer and F. H. Goldman, *J. Am. Chem. Soc.*, *51*, 2632 (1932).

13-9 What is the value of the solubility-product constant of praseodymium iodate, $Pr(IO_3)_3$, if the iodate ion concentration is 3.4×10^{-3} mol/L?

[F. H. Firsching and T. R. Paul, *J. Inorg. Nucl. Chem.*, *28*, 2415 (1966)]

13-10 Determine the solubility of silver cyanide from the K_S for AgCN.

13-11 Calculate the lead ion concentration in a saturated solution of lead selenate.

13-12 Calculate the concentration of calcium ions in a solution prepared by adding 25.0 mL of 0.125 M $Ca(NO_3)_2$ to 25.0 mL of 0.125 M Na_2SO_4.

13-13 P. Taylor and V. Lopata recently determined that the log K_S of BiI_3 at 298 K was -15.3.

a. What is the K_S of BiI_3 at that temperature?
b. What is the molality of a saturated solution of BiI_3?

[*Can. J. Chem.*, *64*(2), 290 (1986)]

13-14 A solution was prepared by mixing 30.0 mL of 0.250 M silver nitrate solution with 30.0 mL of 0.125 M sodium sulfate solution. What was the silver ion concentration of the solution after mixing?

13-15 Calculate the silver ion concentration in each of the solutions listed. Compare your calculated results with those observed by J. J. Renier and D. S. Martin, Jr.
[*J. Am. Chem. Soc.*, *78*, 1835 (1956)]

$[IO_3^-]$	$[Ag^+] \times 10^6$ (observed)
0.000940	90.2
0.00248	35.4
0.0103	9.05
0.0501	1.96
0.0634	1.68
0.492	1.33

13-16 A sample of sodium fluoride weighing 0.420 g was added to 100 mL of a saturated solution of calcium fluoride. Calculate the calcium ion concentration before and after addition of the sodium fluoride.

13-17 In the group I qualitative analysis scheme, $Pb(NO_3)_2$ is precipitated out of solution by the addition of HCl. If you start with a 0.1 M $Pb(NO_3)_2$ solution and bring the concentration of Cl^- to 0.1 M by the addition of HCl,

a. What will be the concentration of Pb^{2+} left in solution?
b. The lead also comes out in group II by the addition of H_2SO_4. If the SO_4^{-2} concentration is made 1 M, how much Pb^{2+} will now be left in solution?

13-18 Calculate the silver ion concentration of a solution prepared by dissolving 1.99 g of $KAg(CN)_2$ in sufficient water to make 100 mL of solution.

13-19 What is the Ag^+ concentration of a solution prepared by dissolving 0.100 mol of silver nitrate in 1 L of a solution in which the ammonia concentration is 0.0870 mol/L?

13-20 What is the Cu^{2+} concentration in a solution prepared by dissolving 2.00×10^{-3} mol of $CuSO_4$ in 100 mL of solution containing 0.0850 mol/L of ammonia?

13-21 A solution that contained 1.00 mol of potassium cyanide in 1.00 L of a 0.100 M silver nitrate solution was prepared. Determine the dissociation constant of the silver cyanide complex ion, $Ag(CN)_2^-$, if the silver ion concentration of the solution was 6.76×10^{-20} mol/L.

13-22 What is the pH of a 0.500 M CH_3COONa solution?

13-23 Calculate the hydroxide ion concentration of a 0.0125 M NH_4Cl solution.

13-24 Calculate the pH of a 5.00% potassium acetate solution, sp. gr. 1.0191.

13-25 What is the pH of 8.00% ammonium nitrate, sp. gr. 1.0313?

13-26 The pH of a saturated solution of calcium hydroxide is 12.58. Calculate K_S.

13-27 In a series of experiments the pH's of solutions of KCN that had hydrolyzed were determined colorimetrically. Find a value of pK_{HCN} from one of these experiments in which the KCN was 12.5% hydrolyzed and the pH was 10.34.

13-28 Calculate the pH of a solution prepared by adding 500 mL of 0.100 M sodium hydroxide to 500 mL of 0.100 M acetic acid.

13-29 What is the pH of a solution prepared by adding 250 mL of a 0.200 M solution of hydrochloric acid to 500 mL of 0.100 M ammonia?

13-30 A solution was prepared by mixing 200 mL of 0.125 M propionic acid with 100 mL of 0.250 potassium hydroxide. What was the pH of the solution?

Supplementary Problems

13-31 The solubility of lanthanum(III) iodate in water at 25°C as determined by V. K. La Mer and F. H. Goldman was 8.90×10^{-4} mol/L. Find the K_S of $La(IO_3)_3$.
[*J. Am. Chem. Soc.*, *51*, 2632 (1929)]

13-32 Calculate the K_S of a saturated solution of $PbCl_2$. The solubility is 0.03367 mol/L.

[A. Seidell and W. F. Linke, *Solubilities of Inorganic and Organic Compounds*, supplement to 3rd ed., p. 504, Van Nostrand Company, Princeton, NJ (1952)]

13-33 Richard Ramette measured the solubility of $Cd(IO_3)_2$ in $Cd(ClO_4)_2$ solutions using a polarograph. In one experiment, in which the concentration of $Cd(ClO_4)_2$ was 9.814 mmol, he found the solubility of $Cd(IO_3)_2$ to be 1.3016 mmol/L. What is the K_S for this salt?

[*Anal. Chem.*, *53*(14), 2244 (1981)]

13-34 What is the K_S of $TlIO_3$ at 25° from its solubility, 0.001844 mol/L?

[V. K. La Mer and F. H. Goldman, *J. Am. Chem. Soc.*, *51*, 2632 (1929)]

13-35 Calculate K_S for lanthanum iodate, $La(IO_3)_3$, from the iodate ion concentration of a saturated solution, 3.1×10^{-3} mol/L.

[F. H. Firsching and T. R. Paul, *J. Inorg. Nucl. Chem.*, *28*, 2415 (1966)]

13-36 The solubility of yttrium iodate was found by a determination of the iodate ion concentration of a saturated solution, 5.8×10^{-3} mol/L. Calculate the K_S for $Y(IO_3)_3$.

[F. H. Firsching and T. R. Paul, *J. Inorg. Nucl. Chem.*, *28*, 2415 (1966)]

13-37 M. P. Menon determined the solubility of TbF_3 to be 3.59×10^{-5} M. What is the K_S for this salt? (Note the error in *Chemical Abstracts*.)

[*C. A.*, 41729V (1982)] [*J. Chem. Eng. Data*, *27*(1), 81 (1982)]

13-38 In a polarographic determination of the solubility of $Cd(IO_3)_2$ in $Cd(ClO_4)_2$, Richard Ramette found 0.9796 mmol/L of $Cd(IO_3)_2$ were present when the $Cd(ClO_4)_2$ concentration was 19.839 mmol/L. What is the K_S for this salt under these conditions?

[*Anal. Chem.*, *53*(14), 2244 (1981)]

13-39 The solubility of calcium hydroxide at 0°C is 0.185 g/100 g H_2O. What is the molal concentration? Assume that the specific gravity is 1.00 and calculate K_S.

13-40 What is the solubility of copper iodate at 25°C?

13-41 What is the silver ion concentration of a saturated solution of silver chromate?

13-42 What is the solubility of europium iodate $Eu(IO_3)_3$ from the K_S, 1.1×10^{-11}?
[F. H. Firsching and T. R. Paul, *J. Inorg. Nucl. Chem.*, *28*, 2415 (1966)]

13-43 What is the chloride ion concentration in a saturated solution of mercurous chloride?

13-44 Calculate the copper(II) ion concentration of a solution prepared by adding 50.0 mL of 0.100 M sodium iodate to 25.0 mL of 0.100 M copper nitrate.

13-45 What is the thallium(I) ion concentration in a 0.0050 M potassium chromate solution?

13-46 Determine the mercurous ion concentration in a 0.100 M KCl solution. What is the mercurous ion concentration in a saturated solution of KCl? The solubility of KCl in water is 26.34 wt% at 25°. The density of the saturated solution is 1.178 g/mL.

13-47 What is the Gd^{3+} concentration of a saturated solution of $Gd(IO_3)_3$? $K_S = 1.8 \times 10^{-11}$.
[F. H. Firsching and T. R. Paul, *J. Inorg. Nucl. Chem.*, *28*, 2415 (1966)]

13-48 A saturated solution of barium chromate that also contained 0.020 mol/L of sodium chromate was prepared. What was the barium ion concentration of the solution?

13-49 Saturated solutions of silver iodide in aqueous potassium iodide that contained 0.118, 0.134, 0.277, 0.437, 0.584, 0.766, and 1.06 mol/L of iodide ion were prepared. Calculate the silver ion concentration of each of these solutions. Plot the silver ion concentration versus the iodide ion concentration.
[E. L. King, H. J. Krall, and M. L. Pandow, *J. Am. Chem. Soc.*, *74*, 3493 (1952)]

13-50 Calculate the solubility of lead iodate in a 0.0200 M calcium iodate solution.

13-51 Calculate the silver ion concentration in a solution made by dissolving 0.0200 mol of silver nitrate in 750 mL of 0.200 M sodium cyanide solution.

13-52 Calculate the cadmium ion concentration in a solution made by mixing 100 mL of 0.0100 M cadmium nitrate solution with 150 mL of 0.100 M aqueous ammonia.

13-53 What is the ammonia concentration required to reduce the silver ion concentration of a solution from 0.120 mol/L to 1.00×10^{-7} mol/L?

13-54 What concentration of sodium acetate is necessary to prepare an aqueous solution of pH 9.38?

13-55 Determine the mass, in grams, of ammonium chloride required to prepare 100 mL of a solution of pH 4.63.

13-56 A 0.0245 M KCN solution hydrolyzed to form HCN to the extent of 3.18%. Determine a value of K_{HCN} from this observation.

[H. T. S. Britton and E. N. Dodd, *J. Chem. Soc.*, *102*, 2333 (1931)]

13-57 The hydrolysis constant of KCN was reported to be 2.55×10^{-5} by H. T. S. Britton and E. N. Dodd. Calculate the percent hydrolysis of a 0.245 M KCN solution.

[*J. Chem. Soc.*, *102*, 2333 (1931)]

13-58 Calculate the pH of 2.00% $NaNO_2$, sp. gr. 1.0125.

13-59 A volume of 20.0 mL of 0.0992 M sodium hydroxide solution was run from a buret into 40.0 mL of 0.0496 M acetic acid. Calculate the pH of the solution.

13-60 Find the pH of a solution resulting from the titration of 32.6 mL of 0.104 M sodium hydroxide with 40.7 mL of 0.0833 M acetic acid. The solutions were titrated in a beaker that contained 50.0 mL of water.

14
OXIDATION-REDUCTION

Oxidation is the process in which an element increases its oxidation number; in reduction there is a decrease in oxidation number. A loss of electrons accompanies oxidation, and a gain of electrons accompanies reduction.

14-1 OXIDATION NUMBERS

Oxidation numbers are numbers assigned to atoms in a rather arbitrary fashion to designate electron transfers in oxidation-reduction reactions. They represent the charges that atoms would have if the electrons were assigned according to an arbitrary set of rules.

14-2 ASSIGNMENT OF OXIDATION NUMBERS

The assignment of oxidation numbers to elements is governed by the following set of rules.

1. The oxidation number of elements in the free (uncombined) state, for example, Li, H_2, P_4, S_8, and Tl, is 0.
2. The oxidation number of fluorine is -1 in all compounds, for example, HF, OF_2, and KF.
3. The oxidation number of oxygen in compounds, for example, H_2O, Li_2O, NO, and Cl_2O, usually is -2. Oxygen has an oxidation number of -1 in peroxides, for example, Na_2O_2, H_2O_2. It is +2 in OF_2.
4. The oxidation number of hydrogen in compounds, for example, H_2O, NH_3, HCl, and PH_3, is usually +1. In metal hydrides, for example, LiH and CaH_2, it is -1.

5. The oxidation numbers of the metals in groups IA and IIA (1 and 2) are equal to the group numbers.
6. The oxidation numbers of the remaining elements may vary, for example, C in CO and CO_2, N in NF_3 and N_2F_4, and Cr in Cr_2O_3 and Na_2CrO_4.
7. The sum of the oxidation numbers in compounds is equal to 0, for example, in H_2SO_4, $2(+1) + (+6) + 4(-2) = 0$.
8. The oxidation number of simple ions is equal to the charge on the ion, for example, Cl^-, F^-, O^{2-}, Na^+, Mg^{2+}.
9. The sum of the oxidation numbers in a polyatomic ion is equal to the charge on the ion, for example, NO_3^-, $+5 + 3(-2) = -1$.

Example 14-1

Assign oxidation numbers to each of the elements in the following substances: H_2Te, N_2, SF_4, Li_2SO_4, K_2O_2 (potassium peroxide).

H_2Te	$2(+1) + (-2)$
N_2	0
SF_4	$(+4) + 4(-1)$
Li_2SO_4	$2(+1) + (+6) + 4(-2)$
K_2O_2	$2(+1) + 2(-1)$

Example 14-2

Assign oxidation numbers to the elements in the following ions: I^-, Sc^{3+}, NO_2^-, HSO_3^-, BF_4^-.

I^-	-1
Sc^{3+}	$+3$
NO_2^-	$+3 + 2(-2) = -1$
HSO_3^-	$(+1) + (+4) + 3(-2) = -1$
BF_4^-	$(+3) + 4(-1) = -1$

Example 14-3

Indicate which elements are oxidized and which are reduced in the reaction:

$$HBr(aq) + HSO_4^- \rightarrow Br_2 + SO_2 + 2H_2O$$

reduction (HSO_4^- → SO_2)

$$3H^+ + 2Br^- + HSO_4^- \rightarrow Br_2 + SO_2 + 2H_2O$$

$$+1 \quad -1 \quad (+1)(+6)(-2) \quad 0 \quad (+4)(-2) \quad (+1)(-2)$$

oxidation (Br^- → Br_2)

14-3 OXIDATION-REDUCTION REACTIONS IN SOLUTION

Reactions that occur in the absence of polar solvents have been described in Section 4-3. Many useful and important oxidation-reduction reactions occur in aqueous solution. A typical example is the reaction of Fe^{2+} ions with MnO_4^-, the basis of a volumetric method of analysis for iron. Such reactions may be separated into two parts: the oxidation part and the reduction part. In the reaction of Fe^{2+} with MnO_4^- the parts are:

$$Fe^{2+} \rightarrow Fe^{3+} \qquad \text{(oxidation)}$$

$$8H^+ + MnO_4^- \rightarrow Mn^{2+} + 4H_2O \qquad \text{(reduction)}$$

14-4 BALANCING OXIDATION-REDUCTION REACTIONS BY THE ION-ELECTRON METHOD

Method

A. REACTIONS IN ACID SOLUTION
   1. Divide the reaction into half-reactions.
   2. Balance the *atoms* in each half-reaction.
      a. Add H^+ or H_2O where necessary to balance H and O.
   3. Balance the *charges* to each half-reaction.
      a. Add electrons to the side that is more positive.
   4. Make the *loss of electrons* equal to the *gain of electrons* by multiplying each half-reaction by the appropriate integer.
   5. Add the two balanced half-reactions.
   6. Check the balance of *atoms* and *charges*.

Example 14-4

Write a balanced equation for the reaction of Fe^{2+} with MnO_4^-. The products are Fe^{3+} and Mn^{2+}.

Step 1

$Fe^{2+} \rightarrow Fe^{3+}$

$MnO_4^- \rightarrow Mn^{2+}$

Step 2

$Fe^{2+} \rightarrow Fe^{3+}$

$MnO_4^- \rightarrow Mn^{2+} + 4H_2O$

to provide 4O's

$8H^+ + MnO_4^- \rightarrow Mn^{2+} + 4H_2O$

to provide 8H's

Step 3

$Fe^{2+} \rightarrow Fe^{3+} + e^-$

$+2 \quad = +3 \quad + (-1)$

An electron was added to the right side to provide a charge of +2, equal to the charge on the left side.

$8H^+ + MnO_4^- + 5e^- \rightarrow Mn^{2+} + 4H_2O$

$+7 \quad + (-5) = +2$

$5e^-$ must be added to the left side to give a charge of +2, equal to the charge of the right side.

Step 4

$5(Fe^{2+} \rightarrow Fe^{3+} + e^-)$

$5Fe^{2+} \rightarrow 5Fe^{3+} + 5e^-$

$8H^+ + MnO_4^- + 5e^- \rightarrow Mn^{2+} + 4H_2O$

Step 5

$$5Fe^{2+} \rightarrow 5Fe^{3+} + 5e^-$$

$$8H^+ + MnO_4^- + 5e^- \rightarrow Mn^{2+} + 4H_2O$$

$$5Fe^{2+} + 8H^+ + MnO_4^- \rightarrow 5Fe^{3+} + Mn^{2+} + 4H_2O$$

Step 6

$5Fe^{2+}$ $5Fe^{3+}$

$8H^+$ $4 \times 2H$

1 Mn 1 Mn

4 O 4 O

$5Fe^{2+} + 8H^+ + MnO_4^- \rightarrow 5Fe^{3+} + Mn^{2+} + 4H_2O$

$(+10) + (+8) + (-1) = +17 = (+15) + (+2) = +17$

Example 14-5

Chlorine is produced by the reaction of hydrochloric acid with potassium dichromate. Write a balanced ionic equation for the reaction. Dichromate ions are reduced to chromium(III) ions.

Step 1

$Cl^- \rightarrow Cl_2$

$Cr_2O_7^{2-} \rightarrow Cr^{3+}$

Step 2

$2Cl^- \rightarrow Cl_2$

$14H^+ + Cr_2O_7^{2-} \rightarrow 2Cr^{3+} + 7H_2O$

Step 3

$2Cl^- \rightarrow Cl_2 + 2e^-$

$2(-1) = 0 + 2(-1)$

$14H^+ + Cr_2O_7^{2-} \rightarrow 2Cr^{3+} + 7H_2O$

$14(+1) + (-2) = 2(+3)$

$+12 \neq +6$

Six electrons must be added to the left side.

$14H^+ + Cr_2O_7^{2-} + 6e^- \rightarrow 2Cr^{3+} + 7H_2O$

$+14 + (-2) \quad + (-6) = +6 = +6$

Step 4

$3[2Cl^- \rightarrow Cl_2 + 2e^-]$

$6Cl^- \rightarrow 3Cl_2 + 6e^-$

$14H^+ + Cr_2O_7^{2-} + 6e^- \rightarrow 2Cr^{3+} + 7H_2O$

Step 5

$$6Cl^- \rightarrow 3Cl_2 + 6e^-$$

$$14H^+ + Cr_2O_7^{2-} + 6e^- \rightarrow 2Cr^{3+} + 7H_2O$$

$$6Cl^- + 14H^+ + Cr_2O_7^{2-} \rightarrow 3Cl_2 + 2Cr^{3+} + 7H_2O$$

Step 6

6Cl	3 × 2 Cl
14 H	7 × 2 H
2 Cr	2 × 1 Cr
7 O	7 O

$$6Cl^- + 14H^+ + Cr_2O_7^{2-} \rightarrow 3Cl_2 + 2Cr^{3+} + 7H_2O$$

$$-6 + (14) + (-2) = +6 = 0 \qquad +2(+3) + 0 = +6$$

B. REACTIONS IN BASIC SOLUTIONS

Reactions that occur in basic solution may not contain H^+. H_2O molecules and OH^- ions must be added.

Example 14-6

Balance the equation for the oxidation of AsO_3^{3-} to AsO_4^{3-} by MnO_4^-. In basic solution MnO_4^- is reduced to MnO_2.

Step 1

$$AsO_3^{3-} \rightarrow AsO_4^{3-}$$

$$MnO_4^- \rightarrow MnO_2$$

Step 2

$$H_2O + AsO_3^{3-} \rightarrow AsO_4^{3-}$$

to provide 1 O.

In acid solution we could add H^+ to the right side. Since we cannot do so here because the reaction occurs in basic solution, we must add OH^- in place of H_2O.

$$OH^- + AsO_3^{3-} \rightarrow AsO_4^{3-}$$

to provide 1 O.

To balance the H introduced with the O in OH^-, we must add H_2O to the right side.

$$OH^- + AsO_3^{3-} \rightarrow AsO_4^{3-} + H_2O$$

Balance by multiplying OH^- by 2

$$2OH^- + AsO_3^{3-} \rightarrow AsO_4^{3-} + H_2O$$

$$2H_2O + MnO_4^- \rightarrow MnO_2 + 4OH^-$$

Step 3

$$2OH^- + AsO_3^{3-} \rightarrow AsO_4^{3-} + H_2O$$

$$2(-1) + (-3) = -5 \neq -3$$

Add $2e^-$ to the *right* side.

$$2OH^- + AsO_3^{3-} \rightarrow AsO_4^{3-} + H_2O + 2e^-$$

$$-2 \quad + (-3) = -5 = -3 \qquad + (-2) = -5$$

$$2H_2O + MnO_4^- \rightarrow MnO_2 + 4OH^-$$

$$0 \quad + (-1) \quad \neq 0 \quad + 4(-1)$$

Add $3e^-$ to the *left* side.

$$2H_2O + MnO_4^- + 3e^- \rightarrow MnO_2 + 4OH^-$$

$$0 \quad + (-1) \quad + (-3) = -4 = 0 + 4(-1) = -4$$

Step 4

$$3(2OH^- + AsO_3^{3-} \rightarrow AsO_4^{3-} + H_2O)$$

$$2(2H_2O + MnO_4^- \rightarrow MnO_2 + 4OH^-$$

Step 5

$$6OH^- + 3AsO_3^{3-} + 4H_2O + 2MnO_4^- \rightarrow$$

$$3AsO_4^{3-} + 3H_2O + 2MnO_2 + 8OH^-$$

$$3AsO_3^{3-} + H_2O + 2MnO_4^- \rightarrow 3AsO_4^{3-} + 2MnO_2 + 2OH^-$$

Step 6

18 O 18 O

2 H 2 H

3 As 3 As

2 Mn 2 Mn

$$3AsO_3^{3-} + H_2O + 2MnO_4^- \rightarrow 3AsO_4^{3-} + 2MnO_2 + 2OH^-$$

$$3(-3) \quad + 0 \quad + 2(-1) \quad = 3(-3) \quad + 0 \quad + 2(-1)$$

$$-11 = -11$$

Important: Whenever possible when writing half-reactions, use the actual species that exist in solution.

Example 14-7

HNO_3	$H^+ + NO_3^-$	not N^{5+}
Na_2CrO_4	$2Na^+ + CrO_4^{2-}$	not Cr^{6+}
KIO_3	$K^+ + IO_3^-$	not I^{5+}

14-5 STOICHIOMETRY OF OXIDATION-REDUCTION REACTIONS

A balanced ionic equation is often all that is required when considering an oxidation-reduction reaction. When the quantities of substances required to prepare the solution must be known, a complete "molecular" equation is necessary. The computation of a yield based on a starting material may also require such an equation.

Example 14-8

Write a stoichiometric equation for the reaction of potassium permanganate with iron(II) sulfate in sulfuric acid solution.

The balanced ionic equation is

$$5Fe^{2+} + 8H^+ + MnO_4^- \rightarrow 5Fe^{3+} + Mn^{2+} + 4H_2O$$

Potassium ions K^+ and sulfate ions SO_4^{2-} were omitted, since no change involved them. K^+ ions needed to balance the MnO_4^- anions; SO_4^{2-} ions are required to balance the cations Fe^{2+}, H^+, Fe^{3+}, and Mn^{2+}.

$$5FeSO_4 + 4H_2SO_4 + KMnO_4 \rightarrow 5Fe_2(SO_4)_3 + MnSO_4 + 4H_2O$$

This equation is not complete, since potassium is present on the left but not on the right. If it is added, it must be as the ion K^+ balanced by an anion, SO_4^{2-} in this case.

$$5FeSO_4 + 4H_2SO_4 + KMnO_4 \rightarrow 5Fe_2(SO_4)_3 + MnSO_4 + K_2SO_4 + 4H_2O$$

Check

5Fe	10Fe
1K	2K

$$10FeSO_4 + 4H_2SO_4 + 2KMnO_4 \rightarrow 5Fe_2(SO_4)_3 + 2MnSO_4 + K_2SO_4 + 4H_2O$$

Check

10Fe	10Fe
$14SO_4$	$18SO_4$
8H	8H
2K	2K
2Mn	2Mn
8O	8O

$$10FeSO_4 + 8H_2SO_4 + 2KMnO_4 \rightarrow 5Fe_2(SO_4)_3 + 2MnSO_4 + K_2SO_4 + 8H_2O$$

Check

10Fe	10Fe
$18SO_4$	$18SO_4$
16H	16H
2K	2K
2Mn	2Mn
8O	8O

Example 14-9

Calculate the weight of cadmium sulfide required to make 22.0 g of $Cd(NO_3)_2$ by the action of nitric acid on cadmium sulfide.

$$CdS(s) + NO_3^- \rightarrow Cd^{2+} + NO + S$$

$$3[CdS \rightarrow Cd^{2+} + S + 2e^-]$$

$$2[4H^+ + NO_3^- + 3e^- \rightarrow NO + 2H_2O]$$

$$3CdS(s) + 8H^+ + 2NO_3^- \rightarrow 3Cd^{2+} + 3S + 2NO + 4H_2O$$

$$3CdS(s) + 8HNO_3 \rightarrow 3Cd(NO_3)_2 + 3S + 2NO + 4H_2O$$

3 mol → 3 mol

1 mol → 1 mol

CdS		$Cd(NO_3)_2$	
	112.4		112.4
	32.1	2 × 14.0 =	28.0
	144.5 g/mol	5 × 16.0 =	80.0
			220.4 g/mol

$$\frac{22.0}{220} = 0.1000 \text{ mol of } Cd(NO_3)_2 \text{ prepared}$$

0.100 mol CdS required

0.100 × 145 = 14.5 g CdS

PROBLEMS

14-1 Indicate which elements are oxidized and which are reduced in the following reactions.

a. $Fe_3O_4 + 4H_2 \rightarrow 3Fe + 4H_2O$

b. $MnO_2 + 4HCl \rightarrow MnCl_2 + Cl_2 + 2H_2O$

c. $2K_2Cr_2O_7 + 2H_2O + 3S \rightarrow 3SO_2 + 4KOH + 2Cr_2O_3$

d. $2Ca_3(PO_4)_2 + 5C + 6SiO_2 \rightarrow 6CaSiO_3 + P_4 + 5CO_2$

e. $CaCO_3 \rightarrow CaO + CO_2$

f. $2KClO_3 \rightarrow 2KCl + 3O_2$

g. $H_2SO_4 + 2KOH \rightarrow K_2SO_4 + 2H_2O$

h. $NH_4NO_2 \rightarrow N_2 + 2H_2O$

i. $Mg_2Si + 4HCl \rightarrow 2MgCl_2 + SiH_4$

j. $6KOH + 3Br_2 \rightarrow 5KBr + KBrO_3 + 3H_2O$

14-2 Divide the following reactions into half-reactions. Designate whether oxidation or reduction has occurred in each case.

a. $NO_2 + HClO \rightarrow H^+ + NO_3^- + 2Cl^-$

b. $Mg + H^+ + NO_3^- \rightarrow Mg^{2+} + N_2 + H_2O$

c. $IO_3^- + H_2S \rightarrow I_2 + SO_2 + H_2O$

d. $H^+ + CrO_4^{2-} + HNO_2 \rightarrow Cr^{3+} + NO_3^- + H_2O$

e. $H^+ + MnO_2 + Cl^- \rightarrow Mn^{2+} + Cl_2 + H_2O$

14-3 Write balanced ionic equations for the following reactions that occur in neutral or acid solution.

a. $I_2(s) + S_2O_3^{2-} \rightarrow I^- + S_4O_6^{2-}$

b. $Br^- + BrO_3^- \rightarrow Br_2(\ell)$

c. $CuS(s) + NO_3^- \rightarrow Cu^{2+} + S_8(s) + NO(g)$

d. $Cl^- + MnO_4^- \rightarrow Cl_2(g) + Mn^{2+}$

e. $Zn(s) + NO_3^- \rightarrow Zn^{2+} + NH_4^+$

f. $Cu(s) + NO_3^- \rightarrow Cu^{2+} + NO(g)$

g. $Cu(s) + HSO_4^- \rightarrow Cu^{2+} + SO_2(g)$

h. $H_2S(aq) + Cr_2O_7^{2-} \rightarrow Cr^{3+} + S_8(s)$

i. $Cu^{2+} + 2I^- \rightarrow CuI(s) + I_2(s)$

j. $ClO_3^- + SO_2(g) \rightarrow ClO_2(g) + HSO_4^-$

k. $ICl(aq) \rightarrow Cl^- + IO_3^- + I_2$

l. $H_2SO_3(aq) + Cr_2O_7^{2-} \rightarrow HSO_4^- + Cr^{3+}$

m. $Mn^{2+} + S_2O_8^{2-} \rightarrow MnO_4^- + HSO_4^-$

n. $MnO_4^- + H_2O_2 \rightarrow Mn^{2+} + H_2O$

o. $Ag(s) + NO_3^- \rightarrow Ag^+ + NO_2(g)$

14-4 Write balanced ionic equations for the reactions that occur in basic solution.

a. $ClO_2(g) + SbO_2^- \rightarrow ClO_2^- + Sb(OH)_6^-$

b. $Cl_2(g) + IO_3^- \rightarrow Cl^- + IO_4^-$

c. $I_2 \rightarrow I^- + IO_3^-$

d. $Fe_3O_4(s) + MnO_4^- \rightarrow Fe_2O_3(s) + MnO_2(s)$

e. $MnO_2(s) + H_2O_2 \rightarrow MnO_4^-$

14-5 The following problems are a little more difficult than average to balance. Balance them.

a. $C_2H_5OH + K_2Cr_2O_7 + H_2SO_4 \rightarrow$

$HC_2H_3O_2 + Cr_2(SO_4)_3 + K_2SO_4 + H_2O$

b. $As_2S_5 + HNO_3 \rightarrow NO_2 + H_2O + H_3AsO_4 + H_2SO_4$

c. $CrI_3 + Cl_2 + KOH \rightarrow KIO_4 + K_2CrO_4 + KCl + H_2O$

d. $As_2S_3 + HClO_3 + H_2O \rightarrow HCl + H_3AsO_4 + H_2SO_4$

14-6 Calculate the mass of iodine that can form by the reaction of sulfuric acid with 1.66 g of potassium iodide. The HSO_4^- ions are reduced to HSO_3^-.

14-7 Determine the volume, in milliliters, of the NO gas formed by the reaction of 12.7 g of Cu with nitric acid at 27°C and 570 mm.

14-8 Sodium iodate is made by the reaction of iodine with an excess of sodium chlorate. The chlorate ion is reduced to chloride ion. How many grams of sodium chlorate would be required to react with 100 g of iodine? How many grams would be required to provide a 20% excess?

[H. H. Willard, *Inorg. Syn.*, *1*, 169 (1939)]

14-9 How many grams of sodium iodate would be required to prepare 225 g of $Na_3H_2IO_6$ by oxidation with chlorine gas in basic solution? The chlorine is reduced to chloride ion in the reaction.

[H. H. Willard, *Inorg. Syn.*, *1*, 169-170 (1939)]

14-10 K_2ReCl_6 has been made by reduction of ReO_4^- in hydrochloric acid solution by KI. How many grams of K_2ReCl_6 can be made from 8.00 g $KReO_4$ if the yield is 85%?

[L. C. Hurd and V. A. Reinders, *Inorg. Syn.*, *1*, 178-179 (1939)]

14-11 A vanadium sulfate containing 26.13% vanadium and 74.16% sulfate ion has been prepared by the action of sulfur on V_2O_5 in sulfuric acid solution. The sulfur was oxidized to SO_2 gas. The

weight of vanadium(V) was 36.4 g, and the weight of sulfur was 9.6 g. Was either of these substances in excess? The vanadium compound collected weighed 75 g. Calculate the percent yield.
[R. T. Claunch and M. M. Jones, *Inorg. Syn.*, *7*, 92-93 (1963)]

14-12 Cyanogen $(CN)_2$ can be made by the oxidation of aqueous sodium cyanide by aqueous copper(II) sulfate. Insoluble copper(I) cyanide is also formed. Sodium cyanide solution is added to the copper(II) sulfate solution until the copper(II) ion is completely reduced. How many grams of sodium cyanide will be required to reduce 500 g of copper(II) sulfate? The yield of cyanogen was 21 g. Calculate the percent yield. What volume would the cyanogen produced occupy at 27° and 760 mm?
[G. J. Janz, *Inorg. Syn.*, *5*, 43 (1957)]

14-13 Telluric acid, H_6TeO_6, is formed by the oxidation of TeO_2 with permanganate ions in nitric acid solution. A sample of H_6TeO_6 weighing 12.5 g was obtained from 100 g of TeO_2. Calculate the percent yield.
[F. C. Mathers, C. M. Rice, H. Broderick, and R. Forney, *Inorg. Syn.*, *3*, 145 (1950)]

14-14 Copper(I) iodide can be quantitatively prepared according to the following unbalanced reaction:

$$CuSO_4 + KI + Na_2S_2O_3 \rightarrow CuI + K_2SO_4 + Na_2S_4O_6 + Na_2SO_4$$

How many grams of CuI can be made starting with 25.0 g of $CuSO_4 \cdot 5H_2O$?
[G. B. Kauffman and R. P. Pinnell, *Inorg. Syn.*, *6*, 3 (1960)]
[G. B. Kauffman, *Inorg. Syn.*, *11*, 215 (1968)]

14-15 Potassium dithioferrate(III) can be prepared by the direct reaction of iron, sulfur, and potassium carbonate at 900°C. Potassium sulfate and carbon dioxide are the other products. Eight grams of Fe powder gave 18.5 g of potassium dithioferrate(III). What is the percent yield?
[J. L. Deutsch and H. B. Jonassen, *Inorg. Syn.*, *6*, 170 (1960)]

14-16 What mass of iodine can be obtained from the reaction of 0.4528 g of copper(II) nitrate in reaction with an excess of potassium iodide?

14-17 What volume of 0.1250 M sodium thiosulfate ($Na_2S_2O_3$) solution would be required to reduce the iodine formed in Problem 14-16? The thiosulfate ion would be oxidized to $S_4O_6^{2-}$.

14-18 Elemental mercury and $Hg(NH_2)Cl$ are formed by the reaction of aqueous ammonia with mercury(I) chloride (Hg_2Cl_2). Write a balanced equation for the reaction.

14-19 When oxygen gas is passed through a solution of cobalt(III) chloride, ammonia, and ammonium chloride, the coordination compound $[CO(NH_3)_6]Cl_3$ is formed. Write a balanced equation for the reaction.

14-20 The final reaction in the manufacture of nitric acid is the reaction of nitrogen dioxide with water. Nitrogen monoxide is also a product.

a. Write a balanced equation for the reaction.
b. Which element is oxidized?
c. Which element is reduced?

15
ELECTROCHEMISTRY

15-1 ELECTROLYSIS

When an electric current passes through a molten salt, such as sodium chloride, cations are reduced at the cathode and anions are oxidized at the anode. The two reactions may be written as follows.

Cathode reaction:

$$Na^+ + e^- \rightarrow Na(s) \quad \text{(reduction)}$$

Anode reaction:

$$Cl^- \rightarrow \frac{1}{2}Cl_2(g) + e^- \quad \text{(oxidation)}$$

Solutions of electrolytes behave in a similar fashion. Since the solvent water may also react at the electrodes, the situation may be more complicated.

The quantitative relationships of the reactions at the electrodes are the same for molten salts and for solutions of electrolytes. They are described by Faraday's laws.

15-2 FARADAY'S LAWS OF ELECTROLYSIS

1. The mass of a substance produced or consumed at an electrode is proportional to the quantity of electricity that passes through the solution .

The quantity of electricity is the amount of charge passed and is measured in coulombs (SI symbol is C). Charge is equal to current × time. Coulombs = amperes × seconds.

2. One *gram equivalent* of a substance is produced or consumed at an electrode by the passage of 96,485 C (1 faraday, F) of charge.

If 1 F of electric charge is passed through molten sodium chloride, the quantity of sodium deposited is 23.0 g, 1 gram equivalent of sodium.

$$Na^+ + e^- \rightarrow Na$$

A gram equivalent of a substance is its atomic or molecular mass divided by the number of electrons involved in the electrode process in question. The mass of a gram equivalent of copper is 63.54/2 = 31.77 g.

$$Cu^{2+} + 2e^- \rightarrow Cu$$

The coulomb is a measure of electric charge and is equal to 1 ampere flowing for 1 sec; coulombs = amperes × seconds. One faraday, 96,485 C, is the charge of 1 mole (6.022×10^{23}) of electrons. One mole of sodium ions (23.0 g) reacts with 1 F (6.022×10^{23} electrons) to form 1 mole (6.022×10^{23} atoms) of sodium. One mole of Cu^{2+} ions requires 2 F to produce 1 mole of Cu atoms. One gram equivalent of copper ions ($1/2 \times 6.022 \times 10^{23}$ ions) requires 1 F (6.022×10^{23} electrons).

15-3 DETERMINATION OF THE FARADAY

The recommended value of the faraday is 96,485.309 C/equivalent.

[*Science*, *235*(#6), 633 (1987)]

For calculations involving three significant figures, the value 96,500 C may be used.

Example 15-1

Calculate a value of the faraday from the observation that the passage of 3671.3 C resulted in the deposition of 4828.4 mg of iodine from a solution containing iodide ions.

[S. J. Bates and G. W. Vinal, *J. Am. Chem. Soc.*, *36*, 916 (1914)]

$I^- \rightarrow I + e^-$

1 gram equivalent of I weighs $\frac{126.90}{1}$ g

$$1\ F = \frac{3671.3\ C}{4.8284\ \not{g}} \times 126.90\ \frac{\not{g}}{\text{equiv.}} = 96,489\ \frac{C}{\text{equiv.}}$$

Example 15-2

Calculate the mass of copper deposited from a solution of copper(II) sulfate by the passage of a current of 2.21 A for 3 hr 15 min.

$Cu^{2+} + 2e^- \rightarrow Cu$

$$\text{Quantity of charge} = 2.21\ A \times 3.25\ \not{hr} \times \frac{60\ \not{min}}{1\ \not{hr}} \times \frac{60\ sec}{1\ \not{min}}$$

$$= 25,900\ C$$

$1\ F = (96,500\ C)$, will deposit $\frac{63.5}{2} = 31.8$ g

$$31.8\ g \times \frac{25,900\ \not{C}}{96,500\ \not{C}} = 8.54\ g\ Cu\ deposited$$

15-4 GALVANIC CELLS

In Chapter 14 it was shown that oxidation-reduction reactions could be divided into two half-reactions. The overall reaction is the combination of the two half-reactions, and electrons are transferred from the reducing agent to the oxidizing agent. It is possible to construct an apparatus in which the reactants are separated in such a way that the electrons are transferred through a wire. Such an apparatus is called a galvanic or voltaic cell. Practical galvanic cells, which provide a useful current for a sufficient length of time, are called batteries. Figure 15-1 represents one galvanic cell, often called the Daniell cell. In one compartment, the half-reaction $Zn \rightarrow Zn^{2+} + 2e^-$ occurs, and in the other, the reaction $Cu^{2+} + 2e^- \rightarrow Cu$ takes place. The metal

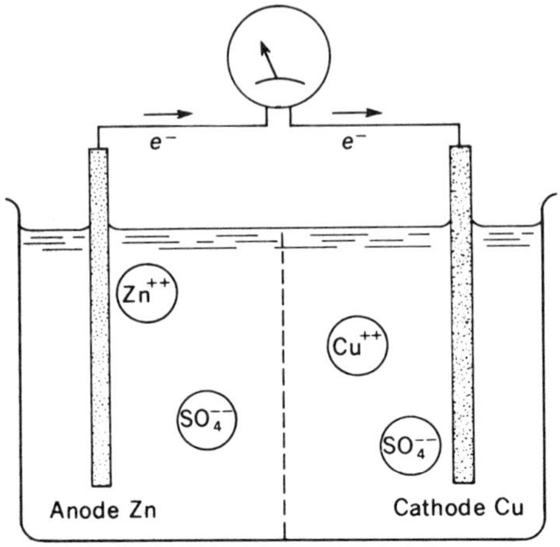

FIGURE 15-1 Schematic diagram of a Daniell cell.

electrodes are connected by a wire through which electric current flows. The two cell compartments are separated by a salt bridge or porous diaphragm, which enables ions to carry the current and complete the circuit. The electrode at which *oxidation* occurs is called the *anode*, and *reduction* occurs at the *cathode*. The two parts of the cell are called *half-cells*.

In the *half-cells* of the Daniell cell the chemical changes are as follows.

Anode reaction:

$$Zn \rightarrow Zn^{2+} + 2e^- \quad \text{(oxidation)}$$

Cathode reaction:

$$Cu^{2+} + 2e^- \rightarrow Cu \quad \text{(reduction)}$$

The total reaction is the sum of the two half-reactions:

$$Zn + Cu^{2+} \rightarrow Zn^{2+} + Cu$$

The electric current flows through the external circuit by virtue of a *potential* which is established between the *electrodes* when the circuit is completed. The value of the potential depends on the nature of the substances that comprise each *half-cell*, the activities of the substances involved, and the temperature. Many galvanic cells have been investigated; it is possible to measure the potential developed by a number of these cells.

15-5 STANDARD POTENTIALS: MEASUREMENT

Since each cell may be considered as the sum of two half-cells, it would be desirable to have a means of measuring the potential developed by each half-cell. Such a measurement is not possible, and it is necessary to adopt a table of relative potentials. The potentials of the half-cells are compared to the *normal hydrogen electrode*, to which the arbitrary value 0.000 volt (V) is assigned. The normal hydrogen electrode may be represented by

$$H^+(a = 1) + e^- \rightleftharpoons \frac{1}{2}H_2(1 \text{ atm}) \qquad E° = 0.000 \text{ V}$$

The pressure of $H_2(g)$ is 1 atm, the activity of H^+ is 1 m, and the temperature is 25°C. The standard potential of the Zn^{2+} + $2e^-$ = Zn electrode may be determined from a cell such as that illustrated in Fig. 15-1. The complete cell is represented by the notation $Zn|Zn^{2+}(1\ m)\|H^+(1\ m)|\ H_2(1\ atm)$, Pt. The single line denotes a junction between an electrode and a solution; the double line represents a junction between two solutions. By convention, the electrode at the left of the cell diagram is the anode, and that on the right is the cathode. The cell reaction is therefore found by writing an oxidation reaction for the left electrode and a reduction reaction for the right electrode, and adding the two reactions.

Example 15-3

What is the galvanic cell designation of the overall reaction Zn(s) + $CuSO_4$(aq) → Cu(s) + $ZnSO_4$(aq) if the activities are 1 m?

$$Zn|\ ZnSO_4(1\ m)\ \|\ CuSO_4(1\ m)|\ Cu$$

Example 15-4

Write equations for the half-cell reactions and the overall reaction for a cell consisting of a lead electrode immersed in a 1 m $Pb(NO_3)_2$

solution in one compartment and hydrogen gas at 1 atm bubbled over a platinum electrode immersed in 1 m HCl.

Oxidation	$Pb \rightleftharpoons Pb^{2+} + 2e^-$	(anode)
Reduction	$2H^+ + 2e^- \rightleftharpoons H_2$	(cathode)
	$Pb + 2H^+ \rightleftharpoons H_2 \quad Pb^{2+}$	

The cell designation is

$$Pb|Pb(NO_3)_2(1\ m)\|HCl(1\ m)|Pt,\ H_2(1\ atm)$$

15-6 STANDARD POTENTIALS: TABLE

Table 15-1 is a list of a few standard electrode potentials required for the solution of problems at the end of this chapter. The half-reactions are written as *reductions*. The standard potential E° is the potential of the electrode in a cell in which the other electrode is the normal hydrogen electrode. Similar tables may be found in many general chemistry textbooks. Since the reactions in Table 15-1 are written as reduction reactions, they are also known as standard reduction potentials.

In Table 15-1 the substances that are more easily reduced are at the top, and the substances that are more easily oxidized than hydrogen have E° values that are (-). Each of these substances is the negative electrode when placed in a battery in which the other electrode is the normal hydrogen electrode. Another widely used compilation of electrode potentials assigns a positive sign to these half-reactions and a negative sign to those below the normal hydrogen electrode. These potentials are known as oxidation potentials. The former assignment of signs has the approval of the International Union of Pure and Applied Chemistry and may be found in many physical chemistry and freshman textbooks. These are also referred to as reduction potentials.

15-7 USE OF STANDARD ELECTRODE POTENTIALS

If two half-reactions are combined and the potential is calculated, a positive value indicates that the reaction will proceed as written. If the electromotive force (emf) is negative, the reaction will not proceed spontaneously as written. When standard potentials are used, the species in solution are at unit activity and gases are at 1 atm. The emf is given as

$$E^\circ_{cell} = E^\circ_{right} - E^\circ_{left} = E^\circ_{cathode} - E^\circ_{anode}$$

where right and left represent the right and left electrodes as designated in the cell diagram. The right electrode is always the cathode; the left, the anode. Since the electrode potentials used in Table 15-1 represent reduction reactions, this convention will always reverse the reaction at the anode and make it an oxidation reaction. This is equivalent to saying

$$E^\circ_{cell} = E^\circ_{oxid} + E^\circ_{red}$$

Once a cell is written a given way in a problem, it must not be changed. Changing the cell is the same as changing the reaction in a chemical equation. In Example 15-3, the cell as diagrammed corresponds to the given reaction. If the electrodes are reversed, i.e.,

$$Cu|CuSO_4(1\ m)\|ZnSO_4(1\ m)|Zn$$

the cell reaction is reversed. It now becomes

$$Cu(s) + ZnSO_4(aq) \rightleftharpoons Zn(s) + CuSO_4(aq)$$

Its equilibrium constant and thermodynamic properties are different from the other reaction.

To determine the emf of a cell, the electrode reaction at the anode must be reversed from that given in the table. The table lists reduction reactions, whereas the anode reaction is an oxidation. This can be accomplished by writing both reactions as reduction reactions and then subtracting them in the same manner as was done for the electrode potentials (right electrode reaction − left electrode reaction)

Example 15-5

Will cadmium metal reduce Cu^{2+}?

The half-reactions are

$$Cu^{2+} + 2e^- \rightleftharpoons Cu \qquad E^\circ = 0.337\ V$$

$$Cd^{2+} + 2e^- \rightleftharpoons Cd \qquad E^\circ = -0.403\ V$$

TABLE 15-1 Standard Electrode Reduction Potentials

Electrode reactions	E° (V)
$F_2(g) + 2e^- = 2F^-$	+2.87
$Ag^{2+} + e^- = Ag^+$	+1.980
$H_2O_2 + 2H^+ + 2e^- = 2H_2O$	+1.776
$MnO_4^- + 8H^+ + 5e^- = Mn^{2+} + 4H_2O$	+1.51
$Cl_2 + 2e^- = 2Cl^-$	+1.3595
$Cr_2O_7^{2-} + 14H^+ + 6e^- = 2Cr^{3+} + 7H_2O$	+1.33
$Tl^{3+} + 2e^- = Tl^+$	+1.25
$O_2 + 4H^+ + 4e^- = 2H_2O(\ell)$	+1.229
$Br_2(aq) + 2e^- = 2Br^-$	+1.087
$Br_2(\ell) + 2e^- = 2Br^-$	+1.0652
$NO_3^- + 4H^+ + 3e^- = NO + 2H_2O$	+0.96
$Ag^+ + e^- = Ag$	+0.7991
$Fe^{3+} + e^- = Fe^{2+}$	+0.771
$Hg_2SO_4 + 2e^- = 2Hg + SO_4^{2-}$	+0.6151
$I_2 + 2e^- + 2I^-$	+0.5355
$Cu^+ + e^- = Cu$	+0.521
$Cu^{2+} + 2e^- = Cu$	+0.337
$Hg_2Cl_2 + 2e^- = 2Hg + 2Cl^-$	+0.2676
$AgCl + e^- = Ag + Cl^-$	+0.2222
$Cu^{2+} + e^- = Cu$	+0.153
$S + 2H^+ + 2e^- = H_2S(aq)$	+0.142
$UO_2^{2+} + e^- = UO_2^+$	+0.05
$2H^+ + 2e^- = H_2$	0.0000

(continued)

TABLE 15-1 (continued)

Electrode reactions	E° (V)
$Pb^{2+} + 2e^- = Pb$	-0.126
$Sn^{2+} + 2e^- = Sn$	-0.136
$V^{3+} + e^- = V^{2+}$	-0.256
$Co^{2+} + 2e^- = Co$	-0.277
$In^{3+} + 3e^- = In$	-0.343
$Cd^{2+} + Hg + 2e^- = Cd(Hg)$	-0.3516
$PbSO_4 + 2e^- = Pb + SO_4{}^{2-}$	-0.3588
$Cd^{2+} + 2e^- = Cd$	-0.4029
$Fe^{2+} + 2e^- = Fe$	-0.4402
$Zn^{2+} + 2e^- = Zn$	-0.7628
$V^{2+} + 2e^- = V$	-1.186
$Al^{3+} + 3e^- = Al$	-1.662
$Mg^{2+} + 2e^- = Mg$	-2.363
$Na^+ + e^- = Na$	-2.714
$Ca^{2+} + 2e^- = Ca$	-2.866
$K^+ + e^- = K$	-2.925
(Basic solutions)	
$NiO_2(s) + 2H_2O + 2e^- = Ni(OH)_2(s) + 2OH^-$	+0.490
$O_2(g) + 2H_2O(\ell) + 4e^- = 4OH^-(aq)$	+0.40
$HgO(s) + H_2O(\ell) + 2e^- = Hg(\ell) + 2OH^-(aq)$	+0.10
$Fe(OH)_3(s) + e^- = Fe(OH)_2(s) + OH^-(aq)$	-0.56
$Cd(OH)_2(s) + 2e^- = Cd(s) + 2OH^-$	-0.809
$ZnO(s) + H_2O + 2e^- = Zn(s) + 2OH^-$	-1.245

The electrode with the higher reduction potential will be the cathode in a cell made of these two electrodes.

$$E^\circ_{cell} = E^\circ_{cathode} - E^\circ_{anode}$$

$$E^\circ_{cell} = 0.337\ V - (-0.403\ V) = 0.740\ V$$

The actual reactions in the half-cells are

Cathode	$Cu^{2+} + 2e^- \longrightarrow Cu$	(reduction)
Anode	$Cd \rightleftharpoons Cd^{2+} + 2e^-$	(oxidation)
Cell reaction	$Cu^{2+} + Cd \rightleftharpoons Cd^{2+} + Cu$	

Since E°_{cell} is +, the reaction will occur as written.

Example 15-6

Will an acid solution of potassium permanganate oxidize bromide ions to bromine?

The half-reactions are

$$MnO_4^- + 8H^+ + 5e^- \rightleftharpoons Mn^{2+} + 4H_2O \qquad E^\circ = 1.51\ V$$

$$Br_2 + 2e^- \rightleftharpoons 2Br^- \qquad E^\circ = 1.06\ V$$

$$E^\circ_{cell} = 1.51\ V - 1.06\ V = +0.45\ V$$

The half-cell reactions are

$$2MnO_4^- + 16H^+ + 10e^- \rightleftharpoons 2Mn^{2+} + 8H_2O \qquad \text{(cathode)}$$

$$10Br^- \rightleftharpoons 5Br_2 + 10e^- \qquad \text{(anode)}$$

Cell reaction $\quad 10Br^- + 2MnO_4^- + 16H^+ \rightleftharpoons 2Mn^{2+} + 8H_2O + 5Br_2$

Note: The value of E° for each half-reaction is not altered when the equation is multiplied by an integer, since E° is an intensive property. The reaction proceeds as written, since E°_{cell} is positive.

Example 15-7

Calculate the emf of a cell in which an iron electrode is immersed in 1 m iron(II) in the left compartment and a zinc electrode is inserted in 1 m zinc nitrate in the right compartment.

$$Fe|Fe(NO_3)_2 \ (1\ m)\|Zn(NO_3)_2 \ (1\ m)|\ Zn$$

By definition, oxidation occurs in the left compartment.

	$Zn^{2+} + 2e^- \rightleftharpoons Zn$	$E° = -0.763$ V
	$Fe^{2+} + 2e^- \rightleftharpoons Fe$	$E° = -0.440$ V
Cell reaction	$Fe + Zn^{2+} \rightleftharpoons Fe^{2+} + Zn$	$E°_{cell} = -0.323$ V

$$E°_{cell} = E°_{right} - E°_{left} = -0.323 \text{ V}$$

The reaction will not occur spontaneously as written, since $E°_{cell}$ is negative.

Example 15-8

Calculate the emf of a cell comprising a $UO_2^+ \rightleftharpoons UO_2^{2+}$ and $V^{2+} \rightleftharpoons V^{3+}$ electrode.

The half-reactions are

$$V^{3+} + e^- \rightleftharpoons V^{2+} \qquad E° = -0.256 \text{ V}$$

$$UO_2^{2+} + e^- \rightleftharpoons UO_2^+ \qquad E° = -0.05 \text{ V}$$

Again, the electrode with the higher reduction potential will be the cathode in a cell. This means the second reaction is the cathode reaction, while the reverse of the first reaction is the anode reaction. (Remember, with negative numbers, the number with the smallest absolute value is the higher number.)

Cathode	$UO_2^{2+} + e^- \rightleftharpoons UO_2^+$
Anode	$V^{2+} \rightleftharpoons V^{3+} + e^-$
Cell reaction	$UO_2^{2+} + V^{2+} \rightleftharpoons UO_2^+ + V^{3+}$

$$E°_{cell} = -0.05 \text{ V} - (-0.256 \text{ V}) = 0.306 \text{ V}$$

The reaction as written will occur spontaneously.

15-8 THE EFFECT OF CONCENTRATION ON THE emf OF CELLS

The values of the standard electrode potentials in Table 15-1 are for reactions in which all gases are at 1 atm and all substances in solution are at unit activity. Pure solid substances are assigned an activity of 1. When the activities are at other values, the potential of an electrode is given by the Nernst equation:

$$E = E° - \frac{0.0592}{n} \log Q$$

where E° is the standard potential and n is the number of electrons involved in the electrode reaction. Q is the reaction quotient.

$$Q = \frac{[\text{right}]}{[\text{left}]}$$

Example 15-9

Calculate the potential of a zinc-zinc ion electrode in which the zinc ion activity is 0.00100 m.

$$E = E° - \frac{0.0592}{n} \log \frac{[\text{right}]}{[\text{left}]}$$

$$Zn^{2+} + 2e^- = Zn$$

$$n = 2$$

$$E° = -0.763 \text{ V}$$

The activity of the solid Zn is 1.

$$E = -0.763 - \frac{0.0592}{2} \log \frac{1}{[0.00100]}$$

$$= -0.763 - 0.0296 \times (+3.000)$$

$$= -0.763 - 0.0888 = -0.852 \text{ V}$$

The potential of a cell is given by an equivalent form of the Nernst equation:

$$E_{cell} = E°_{cell} - \frac{0.0592}{n} \log Q$$

Example 15-10

Calculate the potential of the cell

$$Cd\,|\,Cd^{2+}(0.100)\,\|\,H^+(0.200)\,|\,Pt,\ H_2(0.5\text{ atm})$$

$$2H^+ + 2e^- \rightleftharpoons H_2 \qquad E^\circ = 0.000\text{ V}$$

$$Cd^{2+} + 2e^- \rightleftharpoons Cd \qquad E^\circ = -0.403\text{ V}$$

$$Cd + 2H^+ \rightarrow Cd^{2+} + H_2 \qquad E^\circ_{cell} = 0.403\text{ V}$$

$$E_{cell} = E^\circ_{cell} - \frac{0.0592}{n}\log\frac{[\text{right}]}{[\text{left}]}$$

$$= E^\circ_{cell} - \frac{0.0592}{n}\log\frac{[Cd^{2+}]P_{H_2}}{[H+]^2}$$

$$= 0.403 - 0.0296\log\frac{0.1 \times 0.5}{(0.2)^2}$$

$$E_{cell} = 0.403 - 0.0296\log 1.25$$

$$= 0.403 - 0.0296(0.0969) = 0.403 - 0.003$$

$$= 0.400\text{ V}$$

Example 15-11

Calculate the emf of the following cell:

$$Pt,\ H_2(1\text{ atm})\,|\,HCl(1\text{ m})\,\|\,AgCl\,|\,Ag$$

First calculate the standard potential of the cell.

$$AgCl + e^- \rightleftharpoons Ag + Cl^- \qquad E^\circ = +0.2222\text{ V}$$

$$H^+ + e^- \rightleftharpoons \frac{1}{2}H_2 \qquad E^\circ = 0.0000\text{ V}$$

$$\frac{1}{2}H_2 + AgCl \rightleftharpoons H^+ + Cl^- + Ag \qquad E^\circ_{cell} = +0.2222\text{ V}$$

$$E = E^\circ_{cell} - \frac{0.0592}{1}\log\frac{[H^+][Cl^-]}{P_{H_2}^{1/2}}$$

$[H^+] = 1$ m

The chloride ion concentration must be calculated from the K_s of AgCl.

$$K_s = [Ag^+][Cl^-] = 1.8 \times 10^{-10}$$

$$[Cl^-] = [Ag^+] = x$$

$$x^2 = 1.8 \times 10^{-10}$$

$$x = \sqrt{1.8} \times 10^{-5} \text{ mol/L}$$

(At such low concentration, the molar concentration is equal to the molal concentration.)

$$E_{cell} = 0.222 - \frac{0.0592}{1} \log \frac{[1][1.34 \times 10^{-5}]}{1}$$

$$E_{cell} = 0.222 - 0.0592 \times (-4.89) = 0.222 + 0.289 = 0.511 \text{ V}$$

15-9 CALCULATION OF EQUILIBRIUM CONSTANTS

If current is drawn from a cell, the cell reaction will come to equilibrium and the emf will be zero.

For a cell at equilibrium:

$$E_{cell} = 0$$

$$E = E° - \frac{0.0592}{n} \log Q$$

at equilibrium:

$$E = 0 \text{ and } Q = K$$

$$E° - \frac{0.0592}{n} \log K = 0$$

$$\log K = \frac{nE°}{0.0592}$$

Example 5-12

Calculate the equilibrium constant for the reaction

$Zn + Cu^{2+} \rightleftharpoons Cu + Zn^{2+}$

Cathode (reduction)	$Cu^{2+} + 2e^- \rightleftharpoons Cu$	$E° = +0.337$ V
Anode (oxidation)	$Zn^{2+} + 2e^- \rightleftharpoons Zn$	$E° = -0.763$ V
$E°_{cell} = E°_{right} - E°_{left}$	$Zn + Cu^{2+} \rightleftharpoons Zn^{2+} + Cu$	$E°_{cell} = 1.100$ V

$$\log K = \frac{2 \times 1.10}{0.0592} = 37.2$$

$K = 1.6 \times 10^{37}$

PROBLEMS

15-1 In a very carefully executed experiment of S. J. Bates and W. Vinal, 4104.69 mg of silver was deposited.
[*J. Am. Chem. Soc.*, *36*, 916 (1914)]

a. What was the quantity of electricity passed through the solution?
b. Calculate the mass of iodine that could be deposited by the passage of the same current through a solution containing iodide ions.

(Bates and Vinal obtained 4828.62 mg.)

15-2 Calculate the mass of copper to be expected from an electrolysis in which 2.8197 g of silver was deposited in a coulometer in series with the electrolysis cell.
[T. W. Richards, E. Collins, and G. W. Heimrad, *Proc. Am. Acad. Arts Sci.*, *35*, 123 (1899–1900)]

15-3 A solution containing 0.0407 g of ruthenium was prepared by dissolving the appropriate mass of K_2RuCl_6. A current of 0.050 A was used to remove all of the ruthenium by electrolysis. What time was required?
[E. F. Smith and H. B. Harris, *J. Am. Chem. Soc.*, *17*, 652 (1895)]

15-4 Calculate the quantity of cadmium that may be deposited from a solution of cadmium sulfate by the passage of a 1.02 A current for 2 hr 20 min.

15-5 A quantity of electricity deposited 0.200 mole of copper from an aqueous solution. What mass of silver could be deposited from the same current?

15-6 Calculate the value of the electronic charge from the value of the faraday and Avogadro's number.

15-7 The Daniell cell is given in Example 15-3. What would be the potential of this cell with the following zinc and copper concentrations?

a. Zn^{2+} = (0.1 m)
   Cu^{2+} = (1.0 m)

b. Zn^{2+} = (1.0 m)
   Cu^{2+} = (0.1 m)

c. Zn^{2+} = (0.01 m)
   Cu^{2+} = (1.0 m)

15-8 What is the volume of hydrogen (at STP) that could be formed by the passage of 7.00 A for 6 hr through a solution of potassium azide in anhydrous hydrazoic acid?
[A. W. Browne and G. E. F. Lundell, *J. Am. Chem. Soc.*, *31*, 446 (1909)]

15-9 Water was electrolyzed by the passage of a current through a dilute acid solution. The oxygen and hydrogen were collected. What volume of each was collected at STP if a copper coulometer in series with the electrolysis apparatus had 0.636 g of copper deposited at the cathode?

15-10 A current of 300 mA was passed for 120 min through a solution in which a copper anode and a tellurium cathode were suspended. The loss in mass of the copper anode was 0.669 g, and of the tellurium cathode 1.120 g. What was the current efficiency at each electrode?
[A. J. Paulson, *Inorg. Chem.*, *3*, 941 (1964)]

15-11 Determine the value of the faraday from the weight of iodine, 4828.5 mg, released by the passage of 3671.3 C of electricity.

[S. J. Bates and G. W. Vinal, *J. Am. Chem. Soc.*, *36*, 916 (1914)]

15-12 Calculate the mass of silver that should be deposited by the current in Example 15-1; use the value of the faraday calculated from the data in that problem. Bates and Vinal obtained 4105.23 mg of silver in the experiment.

[S. J. Bates and G. W. Vinal, *J. Am. Chem. Soc.*, *36*, 916 (1914)]

15-13 The electrochemical equivalent of an element is the quantity of the element deposited by the passage of 1 C of electricity. Calculate the electrochemical equivalent of hydrogen, oxygen, and nickel in milligrams per coulomb.

15-14

a. Calculate the emf of each of the following cells.
b. Write equations for the cell reactions that occur.

$Zn|Zn^{2+}(1\ m)\|Cl^-(1\ m)|Pt,\ Cl_2$

$Cd|Cd^{2+}(1\ m)\|Cl^-(1\ m),\ AgCl(s)|Ag$

$Cu|Cu^{2+}(1\ m)\|V^{2+}(1\ m)|V$

$Pb|PbSO_4(s),\ SO_4{}^{2-}(1\ m)\|H_3O^+(1\ m)|Pt,\ H_2\ (1\ atm)$

$Cl_2,\ Pt|Cl^-(1\ m)\|AgCl(s),\ Cl^-(1\ m)|Ag$

15-15 Determine the standard potential of the cell

$In|In_2(SO_4)_3(1\ m)\|Hg_2SO_4(s),\ SO_4{}^{2-}(1\ m)|Hg$

E. M. Hattox and T. DeVries reported a value of 0.954 V.

[*J. Am. Chem. Soc.*, *58*, 2126 (1936)]

15-16 Find the standard potential of the cell

$Cd(Hg)|Cd^{2+}(1\ m)\|Cl^-(1\ m)AgCl|Ag.$

H. S. Harned and M. E. Fitzgerald reported 0.57300 V.

[*J. Am. Chem. Soc.*, *58*, 2624 (1936)]

15-17 Draw the cell diagram and write the cell reaction and the standard potential for each of the following reactions. (Balance each reaction.)

a. $Al + Fe^{2+} \rightarrow Al^{3+} + Fe$

b. $Al + Ag^{+} \rightarrow Ag + Al^{3+}$

15-18 It is claimed that silverware can be cleaned by putting it in an aluminum pan, adding salt and water, and heating the mixture. Write the possible reactions involved here. The tarnish on silverware is Ag_2S. Is this a spontaneous reaction?

15-19 Determine whether each of the following reactions will occur spontaneously to an appreciable extent. The equations are not balanced.

a. $Fe^{2+} + Cr_2O_7^{2-} \rightarrow Fe^{3+} + Cr^{3+}$

b. $MnO_4^{-} + Ag + H^{+} \rightarrow Mn^{2+} + Ag^{+} + H_2O$

c. $H_2S + NO_3^{-} \rightarrow S + NO$

d. $Co + Zn^{2+} \rightarrow Co^{2+} + Zn$

e. $Ag + Sn^{2+} \rightarrow Ag^{+} + Sn$

15-20 Calculate the emf of the cell

$Cd|Cd^{2+}(0.200\ m)\|Cu^{2+}(0.100\ m)|Cu$

15-21 Calculate the emf of the cell

$Pt,\ H_2(1\ atm)|H^{+}(1 \times 10^{-7}\ m)\|H^{+}(1\ m)|H_2(1\ atm),\ Pt$

15-22 Calculate the values of the equilibrium constants of the reactions in the cells in Problem 15-14 as written.

15-23 Draw the cell diagram and determine the cell potential for the following reaction:

$PbSO_4 + 2Ag \rightarrow Pb + Ag_2SO_4$

15-24 A sulfur vapor electrode has been developed that can be used in the temperature range 490–860°C. The cell diagram is

$Ag|Ag_2S,\ AgCl|S(vapor),\ graphite$

Here AgCl is the solvent and the sulfur vapor exists mainly as S_2.

a. Write the cathode reaction.
b. Write the anode reaction.
c. Write the cell reaction.
d. Write the Nernst equation for the emf of the cell.
e. What should be the slope of the line formed when E (cell) is plotted versus $\log(P_{S_2})$ when the cell is saturated with Ag_2S? (This information was used by the authors to verify that sulfur is S_2 in the vapor state.)

[W. T. Thompson and S. W. Flengas, *Can. J. Chem.*, *46*, 1611 (1968)]

15-25 Fredericks and Temple have measured the emf and determined the E° of the following cell at various temperatures.

$$Pt, O_2|O^{2-}\|AgNO_3(1.0\ m)|Ag$$

At 300°C with the O^{2-} concentration fixed at 10^{-5} m, the following data were obtained as a function of the oxygen pressures

$\log P_{O_2}$	E° (V)
-0.20	0.338
-0.40	0.344
-0.60	0.349
-0.80	0.356

a. Write the cell reaction.
b. Write the Nernst equation for the emf of the cell (remember Ag^+ = 1 m)
c. Does this cell obey the Nernst equation at these pressures?
d. What is E° for the cell at 300°C?

[*Inorg. Chem.*, *11*, 968 (1972)]

Supplementary Problems

15-26 Determine the quantity of iodine that could be deposited from a solution of iodide ions by a passage of 3671.06 C of electricity.
[S. J. Bates and W. Vinal, *J. Am. Chem. Soc.*, *36*, 916 (1914)]

15-27 What is the quantity of electricity required to deposit 0.636 g of copper by electrolysis?
[T. W. Richards, E. Collins, and G. W. Heimrad, *Proc. Am. Acad. Arts Sci.*, *35*, 123 (1899–1900)]

15-28 Compare the value of the faraday calculated from Avogadro's number and the electronic charge with that calculated from the data obtained by S. J. Bates and G. W. Vinal: the passage of 3666 C deposited 4099 mg of silver.
[*J. Am. Chem. Soc.*, *36*, 916 (1914)]

15-29 What time would be required to deposit 0.1174 g of nickel from a solution of nickel nitrate by the passage of a current of 193 mA?

15-30 What volume of hydrogen, measured at 25°C and 735 torr, could be collected by the passage of a current of 0.321 A for 4.00 hr through a solution of barium hydroxide?

15-31 The preparation of lithium hydroxide by the electrolysis of a 35% solution of lithium chloride using a platinum anode led to a current efficiency of 95%. What mass of lithium hydroxide was formed by the passage of 2.68 A for 1 hr?

15-32 Calculate the number of coulombs passed through a solution containing silver ions from the mass of silver deposited, 4099.03 mg. The quantity of electricity passed was determined from emf and resistance measurements to be 3666.0 C.
[S. J. Bates and G. W. Vinal, *J. Am. Chem. Soc.*, *36*, 916 (1914)]

15-33 An electrolysis was conducted with a silver anode, whose mass loss was 4.18685 g. How many coulombs of electricity passed through the cell?
[W. M. Bovard and G. A. Hulett, *J. Am. Chem. Soc.*, *39*, 1077 (1917)]

15-34 The electrolysis of a solution containing copper(I), nickel, and zinc as complex cyanides produced a deposit weighing 0.175 g. The deposit contained 72.8% by mass of copper, 4.3% nickel, and 22.9% zinc. Assume that no other element was released and calculate the number of coulombs passed through the solution.
[J. Faust and R. Montillon, *Trans. Electrochem. Soc.*, *65*, 361 (1934)]

15-35 In the experiment described in Problem 15-10, a precipitate weighing 1.746 g was collected. What was the precipitate and what was the current efficiency of its formation?
[A. J. Paulson, *Inorg. Chem.*, 3, 941 (1964)]

15-36 Calculate the standard potential of the cell

$$Ag|AgCl|HCl(1\ m)|Hg_2Cl_2|Hg$$

and compare the result with the emf (+45.5 mV) observed by M. H. Lietzke and J. V. Vaughan.
[*J. Am. Chem. Soc.*, 77, 876 (1955)]

15-37 Calculate the standard potential of the cell

$$Ag|AgCl(s),\ Cl^-(1\ m)\|Cu^{2+}(1\ m)|Cu$$

15-38 When aluminum gutters were first installed on houses, iron nails were used. After a period of time, the holes around the nails grew so large that the gutters fell off the nails. Write the cell diagram for this reaction and the cell reaction.

15-39 Determine whether it is worthwhile to attempt to prepare each of the following compounds:

a. CuF b. FeI_3

c. TlI_3 d. AgF_2

15-40 What is the emf of the cell

$$Ag|AgCl,\ Cl^-(1\ m)\|H^+(0.00100\ m)|H_2(0.5\ atm),\ Pt$$

15-41 Determine the emf of the cell

$$H_2(740\ mm),\ Pt|H^+,\ Cl^-(1\ m),\ AgCl|Ag$$

(Correct the pressure of hydrogen for the water content.)

15-42 The standard potential of the $Ag|AgSCN(s)$ electrode was found to be 0.08951 V. Calculate the potential of the cell

$$Pt,\ H_2(1\ atm)|HClO_4\ (m_1),\ KSCN\ (m_2)|AgSCN(s),\ Ag$$

when $m_1 = 1.4138 \times 10^{-3}$ m and $m_2 = 1.5270$ m.
[C. E. Vanderzee and W. E. Smith, *J. Am. Chem. Soc.*, 78, 723 (1956)]

15-43 What are the values of the equilibrium constants of the reactions in the cells in Problem 15-19?

15-44 The emf of the cell

$Pt, O_2|O^{2-}\|AgNO_3(1.0\ m)|Ag$

has been measured at various O^{2-} concentrations with the oxygen gas pressure fixed at 1 atm. The following data were obtained at 320°C.

log $[O^{2-}]$	E° (V)
-6.50	0.2403
-6.00	0.2694
-5.50	0.2986
-5.00	0.3277

a. Write the cell reaction.
b. Write the Nernst equation for the emf of the cell.
c. Does this cell obey the Nernst equation? (Plot E versus log $[O^{2-}]$.)
d. How does the slope of the line compare to the theoretical value expected at this temperature?
e. What is E° for this cell at 320°C?

[Fredericks and Temple, *Inorg Chem.*, *11*, 968 (1972)]

15-45 The standard reduction potential for the lead dioxide electrode used in the car battery is +1.455 V. Its reaction is

$PbO_2 + 4H^+ + 2e^- \rightarrow Pb^{2+} + 2H_2O$

a. What is the standard cell potential when this electrode is combined with a Pb electrode?
b. Write the cell reaction.
c. Write the reaction in a sulfuric acid solution.

15-46 The cell reaction for the nickel-cadmium (Nicad) battery used in many electrical pieces of equipment (some hand calculators) can be represented as

$NiO_2(s) + Cd(s) + 2H_2O(\ell) \rightarrow Ni(OH)_2(s) + Cd(OH)_2(s)$

The electrolyte is LiOH.

a. Write the anode reaction.
b. Write the cathode reaction.
c. What is the standard potential of the cell?

15-47 The cell reaction for the mercury battery used in pacemakers is as follows:

$$Zn(s) + HgO(s) \rightarrow ZnO(s) + Hg(\ell)$$

The electrolyte is KOH.

a. What is the standard potential for this cell?
b. Explain why this produces a constant voltage over the life of a battery.

16
ENTROPY AND FREE ENERGY

16-1 ENTROPY

A fundamental property of a system is the *entropy* of the system. Entropy is a measure of the disorder of a system.

Like enthalpy, entropy changes are a function of the initial and final states of a system and are independent of the path. For this reason entropy is called a state function. The change in entropy, for a system in going from state 1 to state 2, is equal to the *reversible* heat absorbed by the system divided by the absolute temperature of the system.

$$\Delta S = S_2 - S_1 = \frac{q_{rev}}{T}$$

The entropy of a system is an extensive property and, in chemistry, is usually measured in units of joules per Kelvin ($J\ K^{-1}$). In the past the usual unit of entropy was calories per degree, and this was often written as entropy units (eu); 1 eu = 1 cal K^{-1}. Now when the entropy unit is used, it is defined as 1 J K^{-1}. The entropy per gram, or per mole, of material is an intensive property.

16-2 ENTROPY CHANGES ACCOMPANYING CHANGES IN STATE

In Section 9-4 it was shown that the enthalpy content of a system changes with a change of state. Since these changes occur reversibly, the entropy changes of these systems can be calculated.

Example 16-1

Calculate the entropy change associated with the boiling of 1 mol of water at 100°C and 1 atm pressure. The heat of vaporization of water at the boiling point is 2230 J/g.

$$q_{rev} = 18.0\ \frac{g}{mol} \times 2230\ \frac{J}{g} = 44100\ \frac{J}{mol}$$

$$T = 100°C + 273 = 373\ K$$

$$\Delta S = \frac{44\ 100\ J/mol}{373\ K} = 118\ J\ mol^{-1}\ K^{-1}$$

or

$$\Delta S = 118\ eu\ mol^{-1}$$

The entropy of the gaseous phase is greater than that of the liquid phase at the normal boiling point of the liquid, since there is more disorder in the gaseous phase than the liquid phase.

Example 16-2

Calculate the entropy change associated with the melting of 1.12 mol of ice at 0°C to water at 0°C. The heat of fusion of ice is 6.01 kJ mol^{-1}.

$$\Delta S = \frac{1.12\ mol \times 6010\ J/mol}{273\ K} = 24.6\ \frac{J}{K}$$

The entropy of the liquid phase is greater than that of the solid phase at the melting point.

16-3 ENTROPY CHANGES AT CONSTANT TEMPERATURE

If an ideal gas undergoes a reversible compression or expansion at constant temperature, the entropy change of the gas is given as

$$\Delta S = nR\ \ln\left(\frac{V_2}{V_1}\right) = nR\ \ln\left(\frac{P_1}{P_2}\right)$$

here ln = natural logarithm

R = gas constant = 8.314 J K^{-1} mol^{-1}
V_2 = final volume of gas
V_1 = initial volume of gas
P_2 = final pressure of gas
P_1 = initial pressure of gas

Example 16-3

Calculate the entropy change for the reversible expansion of 1 mol of an ideal gas from 20 L to 40 L at constant temperature.

$$\Delta S = 1.00 \text{ mol} \times 8.31 \text{ J K}^{-1} \text{ mol}^{-1} \times \ln \frac{40.0 \text{ L}}{20.0 \text{ L}} = 5.76 \text{ J K}^{-1}$$

or: 5.76 eu

Example 16-4

Calculate the entropy change for the reversible compression of 1.25 mol of an ideal gas from 1 atm to 5 atm at contant temperature.

$$\Delta S = 1.25 \text{ mol} \times 8.31 \text{ J K}^{-1} \text{ mol}^{-1} \times \ln \frac{1.00 \text{ atm}}{5.00 \text{ atm}} = -16.7 \text{ J K}^{-1}$$

16-4 IRREVERSIBLE CHANGES

Whenever a system undergoes an irreversible change, the heat absorbed or given off by the system is always less than the heat change of a similar process under reversible conditions. Here the entropy change is greater than q_{irr}/T, or

$$\Delta S_{system} > \frac{q_{irr}}{T}$$

where q_{irr} is the heat change under irreversible conditions. To calculate the entropy change of such a system, a method in going from the same initial to the same final states (under reversible conditions) is assumed. Since entropy is a state function, the entropy change of the reversible process is the same as the entropy change of the irreversible process.

Example 16-5

Calculate the entropy change of 1 mol of an ideal gas that expands, isothermally (at constant temperature), against a vacuum from 20 L to 40 L. Here no heat is given off to or absorbed from the surroundings.

Examples 16-4 and 16-5 both measure the entropy change in going from the same initial state to the same final state. The entropy change is the same in both examples.

$$\Delta S = 1.00 \text{ mol} \times 8.31 \text{ J K}^{-1} \text{ mol}^{-1} \times \ln \frac{40.0 \text{ L}}{20.0 \text{ L}} = 5.76 \text{ J K}^{-1}$$

or: 5.76 eu

16-5 ENTROPY OF THE UNIVERSE

Whenever a system absorbs heat from its surroundings or gives off heat to the surroundings, both the system and the surroundings undergo a change in entropy. The heat change of the surroundings is the negative of the heat change of the system.

$$q_{system} = -q_{surroundings}$$

The sum of the entropy change of the system plus that of the surroundings is called the entropy change of the universe.

$$\Delta S_{universe} = \Delta S_{system} + \Delta S_{surroundings}$$

for reversible processes

$$\Delta S_{universe} = \frac{q_{system}}{T} + \frac{q_{surroundings}}{T} = 0$$

The entropy change of the universe for a reversible process is zero.

For an irreversible process, the entropy change of the *surroundings* is given as

$$\Delta S_{surroundings} = \frac{-q_{irr}}{T}$$

The entropy change of the universe is

$$\Delta S_{universe} = \Delta S_{system} - \frac{q_{irr}}{T} > 0$$

For any irreversible process there is always a net increase in the entropy of the universe.

Example 16-6

Calculate the entropy change of the universe in Example 16-5.

Since there was no heat absorbed by the surroundings, the entropy change of the surroundings was zero.

$$\Delta S_{\text{universe}} = 5.76 \text{ J K}^{-1} + 0 = 5.76 \text{ J K}^{-1}$$

Example 16-7

If 1 mol of an ideal gas is allowed to expand, isothermally at 215°C, from 20 L to 40 L against a constant pressure of 1 atm, approximately 488 cal will be absorbed from the surroundings. What is the entropy change of the universe in this expansion?

$$q_{\text{irr}} = 2040 \text{ J}; \quad T = 215 + 273 = 488 \text{ K}$$

Since this is Example 16-4 again, the entropy change of the system is 5.76 J K^{-1}. The entropy change of the universe is

$$\Delta S_{\text{universe}} = 5.76 \text{ J K}^{-1} - \frac{2040 \text{ J}}{488 \text{ K}} = 5.76 - 4.18 = 1.58 \text{ J K}^{-1}$$

16-6 BOLTZMANN EQUATION

The entropy of a system is related to the probability of that particular state of the system to exist. From statistical theory, the entropy per molecule can be shown to be given by

$$S = k \ln W$$

where k = Boltzmann constant = 1.38×10^{-16} erg/deg (this is the gas constant per molecule), and W = number of distinct ways in which the molecules can be distributed throughout the system and still end up with the same state of the system. This equation was first given by L. Boltzmann in 1896.

Example 16-8

When CO crystallizes in the solid state, there are two possible orientations of the molecule: one with the O atoms of one molecule opposite a C atom of another molecule, and the other with the O atom of a molecule opposite another O atom of another molecule. Calculate the entropy of CO if both these orientations have the same probability of occurrence.

The entropy per molecule of the system is

$$S = 1.38 \times 10^{-23} \times \ln 2 = 9.56 \times 10^{-24}\ \mathrm{J\ K^{-1}}$$

The entropy per mole is

$$S = 9.56 \times 10^{-24} \frac{\mathrm{J}}{\mathrm{molecule}} \times 6.02 \times 10^{-23} \frac{\mathrm{molecules}}{\mathrm{mole}}$$

$$S = 5.76\ \mathrm{J\ mol^{-1}}$$

[J. Clayton and W. F. Giauque, *J. Am. Chem. Soc.*, *54*, 2620 (1932)]

16-7 ABSOLUTE ENTROPIES

The Boltzmann equation is the theoretical basis for what is called the third law of thermodynamics. There is only one distinct way in which the molecules of a perfect pure substance can be arranged in the solid state, at absolute zero. W is therefore equal to 1 and S is equal to 0 for this case.

The third law of thermodynamics states that at *absolute zero* (0 K) *the entropy of a pure perfect crystalline substance is zero.* By using this as a reference point, it is possible to obtain entropy values of a substance at any temperature. These entropies are called absolute entropies (see Table 16-1).

Absolute entropies can be used to calculate the entropy changes that accompany chemical reactions, in the same manner that enthalpy changes were calculated:

$$\Delta S^\circ = S^\circ(\text{products}) - S^\circ(\text{reactants})$$

Example 16-9

Commercially, NH_3 is obtained by the fixation of N_2 from the atmosphere by the Haber process. The reaction is

$$N_2(g) + 3H_2(g) = 2NH_3(g)$$

What is the molar entropy change of this reaction?

TABLE 16-1 Absolute Entropies (S°)

Substance	Entropy cal K^{-1} mol^{-1}	Entropy J K^{-1} mol^{-1}
B(s)	1.56	6.52
BF_3(g)	60.70	254.0
C(graphite)	1.36	5.68
CO(g)	47.30	197.9
CO_2(g)	51.06	213.6
$CaCO_3$(c)	22.2	92.8
CaO(c)	9.5	40
CH_4(g)	44.50	186.2
Cl_2(g)	53.29	223.0
F_2(g)	48.6	203
HBr(g)	47.44	198.5
HCl(g)	44.62	186.7
H_2(g)	31.21	130.6
H_2O(g)	45.11	188.7
H_2O(ℓ)	16.72	69.96
NH_3(g)	46.01	192.5
N_2(g)	45.77	191.5
O_2(g)	49.00	205.0
P(s)	10.6	44.3
PH_3(g)	50.2	210
S(c)	7.62	31.8
SO_2(g)	59.4	248
SO_3(g)	61.24	256

Source: Roy A. Keller, *Basic Tables in Chemistry*, McGraw-Hill, New York, 1967, pp. 297–368.

$$\Delta S^\circ = 2S^\circ(NH_3) - S^\circ(N_2) - 3S^\circ(H_2)$$

$$\Delta S^\circ = 2(192.5) - [191.5 + 3(130.6)] = -199 \text{ J K}^{-1} \text{ (eu)}$$

There is a decrease in entropy in the above example, since four gas molecules are combining to form two gas molecules.

Example 16-10

$CaCO_3$ when heated decomposes into CaO and CO_2. What is the entropy change of this reaction?

$$CaCO_3(s) = CaO(s) + CO_2(g)$$

$$\Delta S^\circ = S^\circ(CaO) + S^\circ(CO_2) - S^\circ(CaCO_3)$$

$$\Delta S^\circ = 40 + 214 - 93 = 161 \text{ J K}^{-1}$$

There is an increase in entropy here, due to the formation of the gaseous product. Gases normally have a much higher entropy than solids or liquids.

16-8 GIBBS FREE ENERGY

Both the enthalpy change and the entropy change of a reaction are measures of the tendency of a reaction to occur spontaneously. Neither of these functions, by itself, can be used to determine whether a reaction will occur spontaneously. If the functions are properly combined, a new state function is formed that tells in which direction a chemical reaction will proceed. This new function is called the Gibbs free energy and is defined as

$$G = H - TS$$

Absolute values of the free energy G cannot be determined. The difference in free energy between products and reactants can be determined from standard free energies.

16-9 STANDARD FREE ENERGY OF FORMATION

The free energy change for 1 mol of compound formed from the elements at 25°C, each substance being in its standard state, is the *standard free energy of formation* of the compound (see Table 16-2). As with enthalpy, the standard free energy of

TABLE 16-2 Standard Free Energy of Formation ($\Delta G°_f$)

Substance	Free energy kcal mol^{-1}	Free energy kJ mol^{-1}
$BF_3(g)$	-261.3	-1093
$B_2O_3(s)$	-283.0	-1184
$CaCO_3(s)$	-269.78	-1128.8
CaO(s)	-406.5	-1701
$CH_4(g)$	-12.14	-50.79
$C_2H_4(g)$	+16.28	+68.12
$C_2H_6(g)$	-7.86	-32.9
CO(g)	-32.81	-137.3
$CO_2(g)$	-94.26	-394.4
HBr(g)	-12.72	-53.22
HCl(g)	-22.77	-95.27
HF(g)	-64.7	-270
$H_2O(g)$	-54.64	-228.6
$H_2O(\ell)$	-56.69	-237.2
$H_2S(g)$	-7.89	-33.0
$PH_3(g)$	+4.36	+18.2
$SO_2(g)$	-71.79	-300.4
$SO_3(g)$	-88.52	-370.4

Source: Roy A. Keller, *Basic Tables in Chemistry*, McGraw-Hill, New York, 1967, pp. 297–368.

formation of all elements in their standard state at 25°C has been assigned a value of zero. The standard free energy change of a reaction $\Delta G°$ can be obtained from the standard enthalpy of formation and from the entropy change of the formation reaction at 25°C.

$$\Delta G°_f = \Delta H°_f - T\Delta S°_f$$

Example 16–11

What is the Gibbs free energy change for the formation reaction of $H_2O(\ell)$ at 25°C?

This reaction is

$$H_2(g) + \frac{1}{2}O_2(g) \rightarrow H_2O(\ell)$$

From Table 9-3, $\Delta H^\circ_f = -285.8$ J mol^{-1}

$$\Delta S^\circ_f = 69.96 - \left[130.6 + \frac{1}{2}(205.0)\right] = -163.1 \text{ J mol}^{-1}$$

$$\Delta S^\circ_f = -0.163 \text{ kJ mol}^{-1}$$

$$\Delta G^\circ_f = -285.8 - (298)(-0.163) = -237.2 \text{ kJ mol}^{-1}$$

From tables of ΔG°_f of compounds, the standard free energy of any reaction can be determined.

Example 16–12

What is the standard free energy change for the combustion of 1 mol of methane?

$$CH_4(g) + 2O_2(g) \rightarrow CO_2(g) + 2H_2O(\ell)$$

$$\Delta G^\circ = \Delta G^\circ_f(CO_2)_g + 2\Delta G^\circ_f(H_2O)_\ell - \left[\Delta G^\circ_f(CH_4)_g + \Delta G^\circ_f(O_2)_g\right]$$

$$\Delta G^\circ = -394.4 + 2(-237.2) - [-50.79 + 0] = -818.0 \text{ kJ}$$

Note that this same result could have been obtained from tables of ΔH° and S° and the definition of ΔG°.

Example 16–13

Repeat the calculation in Example 16-12 using the values of ΔH° and S° from the appropriate tables.

$$\Delta H^\circ_{react} = \Delta H^\circ_f(CO_2)_g + \Delta H^\circ_f(H_2O)_\ell - \left[\Delta H^\circ_f(CH_4)_g + \Delta H^\circ_f(O_2)_g\right]$$

$$\Delta S^\circ_{react} = S^\circ(CO_2)_g + S^\circ(H_2O)_\ell - S^\circ(CH_4)_g - S^\circ(O_2)_g$$

$\Delta H^\circ_{react} = -393.2 - 571.7 - [-66.9 + 0] = -898.0$ kJ

$\Delta S_{react}^\circ = 213.6 + 139.9 - [186.2 + 410.0] = -242.7$ J K^{-1}

$\Delta G^\circ = -898.0 - (298)(-0.242) = -898.0 + 72.1 = -825.9$ kJ

$\Delta G^\circ = \Delta H^\circ_{react} - T\Delta S^\circ_{react}$

16-10 FREE ENERGY AND THE EQUILIBRIUM CONSTANT

The change in the standard free energy of a reaction is related to the equilibrium constant of the reaction by the equation

$\Delta G^\circ = -RT \ln K$

If your calculator does not have a natural logarithm function, you may use the relationship ln K = 2.303 log K either from the calculator or from the log table, page 462. This relationship can be used either to calculate ΔG° or to determine equilibrium constants from ΔG°.

Example 16-14

K_p for the dissociation of HBr at 1108 K was found to be 1.47×10^{-5}. What is the free energy change for this reaction?

$\Delta G^\circ_{1108} = -(8.314 \text{ J K}^{-1} \text{ mol}^{-1})(1108 \text{ K}) \ln 1.47 \times 10^{-5}$

$\Delta G^\circ_{1108} = 1.03 \times 10^5 \text{ J mol}^{-1}$

Example 16-15

What is the equilibrium constant for the reaction of sulfur dioxide with oxygen to form sulfur trioxide?

$SO_2(g) + \frac{1}{2}O_2(g) \rightarrow SO_3(g)$

$\Delta G^\circ = \Delta G^\circ_f(SO_3)_g - \frac{1}{2}\Delta G^\circ_f(O_2)_g - \Delta G^\circ_f(SO_2)_g$

$\Delta G^\circ = -370.4 \text{ kJ} - [0 + 300.4 \text{ kJ}] = -70.4 \text{ kJ} = -70{,}400 \text{ J}$

$\Delta G^\circ = -RT \ln K_p$

$$\ln K_p = -\frac{\Delta G^\circ}{RT} = -\frac{-70,400\ J}{(8.31\ J\ K^{-1})(298\ K)} = 28.4$$

$$K_p = 2.16 \times 10^{12}$$

16-11 FREE ENERGY AND SPONTANEOUS REACTIONS

For a reaction that is not at equilibrium, ΔG is given as

$$\Delta G = \Delta G^\circ + RT \ln Q$$

Here Q is the reaction quotient (see Section 15-8). The value of ΔG tells whether a reaction can occur spontaneously from a thermodynamic point of view. If ΔG is negative, the reaction, as written, can be spontaneous. The rate of such a reaction, though, may be too slow to measure. If ΔG is positive, the reaction will be nonspontaneous. If ΔG is zero, the system is at equilibrium.

Example 16-16

The equilibrium constant for the reaction of sulfur dioxide with oxygen to form sulfur trioxide was found in Example 16-15 to be 2.16×10^{12}. In which direction will this reaction tend to go, if these substances are mixed in a container at the following partial pressures: SO_3, 1 atm; SO_2, 1/4 atm; O_2, 1/4 atm?

$$SO_2 + \frac{1}{2}O_2 \rightleftharpoons SO_3$$

$$\Delta G = \Delta G^\circ + RT \ln Q$$

$$\Delta G = \Delta G^\circ + RT \ln \frac{(P_{SO_3})}{(P_{SO_2})(P_{O_2})^{1/2}}$$

$$\Delta G = -70,000\ J + (8.31\ J\ K^{-1})(298\ K) \ln \frac{(1)}{(1/4)(1/4)^{1/2}}$$

$$\Delta G = -70,000 + (8.31)(298)(2.08) = 64,850\ J$$

Since ΔG is negative, the reaction will go to the right, i.e., SO_2 will react with O_2 to form more SO_3.

16-12 CALCULATION OF FREE ENERGY CHANGES FROM CELL emf's

In Section 15-9 the equilibrium constant for a reaction was calculated from E° values by use of the equation

$$nFE^\circ = +RT \ln K$$

since

$$\Delta G^\circ = -RT \ln K$$

$$\Delta G^\circ = -nFE^\circ$$

Note that n is the number of electrons transferred in the reaction, F is the Faraday constant (96,480 coulombs), and E° is the standard reduction potential in volts. Fortunately, the unit of the product FE° is the volt-coulomb (VC) and 1 VC = 1 J.

Example 16-17

Johnson et al. determined E° for the cell

$$Li(\ell)|LiX \text{ (saturated with LiH(s))}|H_2(g), \text{ 1 atm, Fe}$$

to be

$$E^\circ = 0.9081 - 7.699 \times 10^{-4}\ T$$

Here LiX was any halide and T may range from 673 to 853 K. The cell reaction is

$$Li(\ell) + \frac{1}{2}H_2(g) \rightarrow LiH(s)$$

What is the free energy change for this reaction at 800 K? Here n = 1 and F = 96,480 C.

$$\Delta G^\circ = -(1)(96,480\ C)[0.9081 - (7.699 \times 10^{-4}\ T)]\ V$$

$$\Delta G^\circ = -28,190\ J$$

[C. E. Johnson, R. R. Heinrich, and C. T. Crouthamel, *J. Phys. Chem.*, *70*, 242 (1966)]

The free energy change of a reaction that is not at equilibrium can be calculated from

$$\Delta G = -nFE$$

Example 16-18

What is the free energy change for the cell reaction in Example 15-10?

The cell reaction is

$$Cd + 2H^+ \rightleftharpoons Cd^{2+} + H_2$$

and E = 0.400 V

$$\Delta G = -(2 \text{ mol})(96{,}500 \text{ C mol}^{-1})(0.400 \text{ V}) = -77{,}200 \text{ VC} = -77{,}200 \text{ J}$$

PROBLEMS

16-1 The heat of fusion of hydrogen fluoride is 196.4 J/g at 189.79 K. What is the entropy change of fusion?
[J. Hu, D. White, and H. Johnston, *J. Am. Chem. Soc.*, *75*, 1232 (1953)]

16-2 The heat of vaporization of diborane is 14,276 J mol^{-1} at 180.32 K. What is the entropy change of vaporization?
[J. Clarke, E. Rifkin, and H. Johnston, *J. Am. Chem. Soc.*, *75*, 781 (1953)]

16-3 The heat of sublimation of iodine at 298.1 K is 62,245 J mol^{-1}. The absolute entropy of iodine at 298.1 K is 117 J K^{-1} mol^{-1} for the solid state. What is the entropy of the gaseous state of iodine at this temperature?
[W. F. Giauque, *J. Am. Chem. Soc.*, *53*, 511 (1931)]

16-4 The absolute entropy of mercury was measured at the melting point for the solid and the liquid state of mercury. The absolute entropy of the solid state was 14.217 eu/mol and for the liquid state, 16.559 eu/mol. The melting point is 234.29 K. What is the heat of fusion of mercury? Here eu is calories per degree.
[R. Busey and W. Giauque, *J. Am. Chem. Soc.*, *75*, 806 (1953)]

16-5 What is the entropy change of the system associated with the compression of 1.33 mol of an ideal gas from 10.0 L to 5.00 L at constant temperature?

16-6 What is the entropy change of the system associated with the expansion of 0.750 mol of an ideal gas from 1.00 to 0.750 atm at constant temperature?

16-7 If in Problem 16-5, 2,930 J was absorbed by the surroundings at 25°C, what was the entropy change of the universe?

16-8 If in Problem 16-6, 104 J was absorbed by the system from the surroundings at 25°C, what was the entropy change of the universe?

16-9 Nitrous oxide is a linear molecule that can orientate itself in two possible ways at absolute zero. What is the entropy of this molecule at absolute zero, according to the Boltzmann equation?
[R. W. Blue and W. F. Giaque, *J. Am. Chem. Soc.*, *57*, 991 (1935)]

16-10 Linus Pauling developed a structure for the crystalline state of water which leads to 3/2 distinct ways in which the ice molecules may be orientated at absolute zero. How does the entropy of this structure compare to an observed value of 10.87 cal/deg/mol for ice?
[*J. Am. Chem. Soc.*, *57*, 2680 (1935)]

16-11 The absolute entropy of H_2S at its boiling point, 212.77 K, was found to be 194.0 J from heat capacity measurements and 194.3 J from spectroscopic measurements. Heat capacity measurements assume the entropy of a substance to be zero at absolute zero, while spectroscopic measurements do not. How many orientations of the molecule are predicted at absolute zero from the above data?

[W. Giauque and R. Blue, *J. Am. Chem. Soc.*, *58*, 831 (1936)]

16-12 Calculate the entropy change for each of the following reactions. Balance the reactions where necessary.

a. C(graphite) + O_2(g) = CO_2(g)

b. H_2(g) + Cl_2(g) = HCl(g)

c. CO(g) + O_2(g) = CO_2 (g)

d. S(s) + O_2(g) = SO_3(g)

e. SO_2(g) + O_2(g) = SO_3(g)

f. H_2(g) + O_2(g) = H_2O(g)

16-13 The absolute entropy of LiOH was found to be 42.806 J K^{-1} mol^{-1} at 298.16 K. The entropy change for the reaction

$$LiOH \cdot H_2O(s) \rightarrow LiOH(s) + H_2O(g)$$

was 160.2 J K^{-1} mol^{-1} at 298.16 K. What is the absolute entropy of LiOH·H_2O(s) at this temperature? (Use the value for the entropy of water given in Table 16-1.)

[T. Bauer, H. Johnston, and E. Keir, *J. Am. Chem. Soc.*, *72*, 5174 (1950)]

16-14 Calculate the free energy change for each of the following reactions. Balance the reactions where necessary. Tell whether the reactions are spontaneous or not spontaneous.

a. B(s) + O_2(g) $\rightleftharpoons$ B_2O_3(s)

b. C_2H_4(g) + O_2(g) $\rightleftharpoons$ CO_2(g) + H_2O(g)

c. C_2H_4(g) + O_2(g) $\rightleftharpoons$ CO_2(g) + H_2O(ℓ)

d. C_2H_6(g) + O_2(g) $\rightleftharpoons$ CO(g) + H_2O(ℓ)

e. $CaCO_3$(s) $\rightleftharpoons$ CaO(s) + O_2(g)

f. Br_2(ℓ) + HF(g) $\rightleftharpoons$ HBr(g) + F_2(g)

16-15 Using the enthalpy of formation of BF_3 from Table 9-3 and the absolute entropies of B(s), F_2(g), and BF_3(g) from Table 16-1, verify the value of the free energy of formation given in Table 16-2.

16-16 Using the value of the enthalpy of formation for HCl(g) from Table 9-3 and the absolute entropies of HCl(g), $H_2(g)$, and $Cl_2(g)$ from Table 16-1, verify the value for the free energy of formation of HCl(g) listed in Table 16-2.

16-17 The free energy change for the reaction

$$GdCl_3(s) + H_2O(g) = GdOCl(s) + 2HCl(g)$$

was determined by measuring the equilibrium pressures of H_2O and HCl. In one experiment at 895 K the pressures obtained for duplicate determinations were: HCl, 610, 636 (mm) and H_2O, 2.46, 2.87 (mm). What is $\Delta G°$ for the reaction?

[C. Koch and R. Cunningham, *J. Am. Chem. Soc.*, *75*, 796 (1953)]

16-18 The free energy of formation of Ag_2O was measured at 302°C by determining the equilibrium pressure of O_2. A value of 20.5 atm was obtained at this temperature. What is $\Delta G°_f$ of Ag_2O at this temperature?

[G. N. Lewis, *J. Am. Chem. Soc.*, *28*, 154 (1906)]

16-19 The E° of the cell

$$Ag(s)|AgCl(\ell)|Cl_2(g,\ 1\ atm),\ C\ rod$$

was found to be 0.9001 V at 500°C. What is the cell reaction? What is $\Delta G°$ for this reaction?

[E. J. Salstrom, *J. Am. Chem. Soc.*, *56*, 1274 (1934)]

16-20 Thompson and Flengas have determined the free energy of formation of $Ag_2S(s)$ from 500 to 800°C by use of the cell

$$Ag|Ag_2S\ (saturated)\ AgCl,\ KCl|S(g,\ 1\ atm),\ C$$

where AgCl and KCl are the solvents.

E° in millivolts was found to be given by E° = 455.3 - 0.1919 T in this temperature range, where T is in degrees Kelvin. What is $\Delta G°$ for the formation reaction at 500°C?

[*J. Electrochem. Soc.*, *118*, 419 (1971)]

16-21 E. J. Salstrom determined the free energy of formation of $PbBr_2(\ell)$ by measuring E° of the cell

$$Pb(\ell)|PbBr_2(\ell)|Br_2(g,\ 1\ atm)$$

A tungsten wire was used to make contact with the molten lead and a graphite rod in contact with the molten $PbBr_2$. Br_2 was passed over the carbon rod. At 500°C the researcher got a value of -199.16 kJ for the formation of 1 mol of $PbBr_2(\ell)$. What was E° for the cell?

[*J. Am. Chem. Soc.*, *52*, 4648 (1930)]

Supplementary Problems

16-22 The heat of vaporization of hydrogen fluoride at 741.4 mm is 374.2 J/g. The boiling point of HF at this pressure is 292.61 K. What is the entropy change of vaporization?

[J. Hue, D. White, and H. Johnston, *J. Am. Chem. Soc.*, *75*, 1232 (1953)]

16-23 Diborane has a heat of fusion of 4473.1 J mol^{-1} at 108.30 K. What is the entropy change of fusion?

[J. Clarke, E. Rifkin, and H. Johnston, *J. Am. Chem. Soc.*, *75*, 781 (1953)]

16-24 The entropy of the liquid state of chlorine trifluoride at 284.91 K, the boiling point of the substance, was found to be 182.4 J/deg-mol^{-1}. The heat of vaporization at this temperature is 27,531 J mol^{-1}. What is the entropy of the gaseous state at the boiling point?

[J. Grissard, H. Bernhardt, and G. Oliver, *J. Am. Chem. Soc.*, *73*, 5725 (1951)]

16-25 The absolute entropy of rhombic sulfur and monoclinic sulfur was measured at 368.6 K, the transition temperature. A value of 36.93 J K^{-1} was obtained for S(r) and 37.83 J K^{-1} for S(m). What is the entropy of transition from rhombic to monoclinic sulfur at this temperature? What is the enthalpy of transition for this same change?

[E. D. Eastman and W. C. McGavock, *J. Am. Chem. Soc.*, *59*, 150 (1937)]

16-26 The heat of fusion of molybdenum trioxide is 52,467 J mol^{-1} at 1068.36 K, the melting point of MoO_3. The absolute entropy of MoO_3(s) at 1068.36 K was found to be 195.91 J/deg-mol^{-1}. What is the absolute entropy of the liquid state of MoO_3 at this melting point?

[L. Cosgrove and P. Snyder, *J. Am. Chem. Soc.*, *75*, 1227 (1953)]

16-27 What is the entropy change of an ideal gas for the expansion of 2.20 mol of the gas from 100 atm to 1.00 atm at constant temperature?

16-28 If the process described in Problem 16-27 is carried out in such a way that no work is done on or by the gas, what is the entropy change of the system, the surroundings, and the universe?

16-29 Westrum and Pitzer have determined from heat capacity measurements that $F\text{-}H\text{-}F^-$ anion in KHF_2 is completely symmetrical. What would be the expected value of the absolute entropy of KHF_2 at absolute zero?

[E. Westrum, Jr., and K. Pitzer, *J. Am. Chem. Soc.*, *71*, 1740 (1949)]

16-30 The absolute entropy of nitrous oxide was determined at 184.59 K, the boiling point, in two different ways. Using spectroscopic data, a value of 48.50 cal/deg/mol was obtained. From specific heat data, a value of 47.36 cal/deg/mol was obtained. Specific heat measurements assume the entropy of nitrous oxide to be zero at absolute zero, while spectroscopic measurements do not. The difference in these results is due to the different orientations of the molecule at absolute zero. From the Boltzmann equation, how many different orientations are necessary to account for these data?

[R. W. Blue and W. F. Giauque, *J. Am. Chem. Soc.*, *57*, 991 (1935)]

16-31 The absolute entropy of $ReCl_3(s)$ was measured and found to be 123.9 J K/mol^{-1}, and the absolute entropy for Re(s) was 36.5 J K/mol^{-1}. The heat of formation of $ReCl_3(s)$ was -263 kJ mol^{-1}. What is the free energy of formation of $ReCl_3(s)$? Use the entropy value for Cl_2 in Table 16-1.

[R. B. Bevan, Jr., R. A. Gilbert, and R. H. Busey, *J. Phys. Chem.*, *70*, 147 (1966)]

16-32 Using the value of the enthalpy of formation for $PH_3(g)$ from Table 9-3 and the absolute entropies of P(s) and $H_2(g)$ from Table 16-1, verify the value for the free energy of formation of $PH_3(g)$ listed in Table 16-2.

16-33 The free energy change for the reaction

$$Hg(\ell) + H_2S(g) \rightleftharpoons HgS(s) + H_2(g)$$

at 25°C, was determined to be -9.75 kJ. Using the values from Table 16-2, calculate the free energy of formation of HgS.
[J. Goates, A. Cole, and E. Gray, *J. Am. Chem. Soc.*, *73*, 3596 (1951)]

16-34 The free energy change for the reaction

$$SmCl_3(s) + H_2O(g) = SmOCl(s) + 2HCl(g)$$

was determined by measuring the equilibrium pressures of H_2O and HCl. In one experiment at 893 K, the pressures obtained for duplicate determinations were HCl: 418.8, 406.3 mm; and H_2O: 1.60, 1.60 mm. What is $\Delta G°$ for this reaction?
[C. Koch and B. Cunningham, *J. Am. Chem. Soc.*, *75*, 797 (1953)]

16-35 The free energy of formation of Ag_2O was determined by measuring the pressure of O_2 in equilibrium with Ag and Ag_2O at various temperatures. At 173°C a pressure of 422 mm was found. What is $\Delta G°_f$ of Ag_2O at this temperature?
[A. Benton and L. Drake, *J. Am. Chem. Soc.*, *54*, 2188 (1932)]

16-36 The emf of the cell

$$Hg, HgS|H_2S(1 \text{ atm})|HCl\ (0.1\text{ m})|H_2(1 \text{ atm})\ Pt$$

was determined at 25°C. E° was 0.0504 V. The cell reaction is

$$Hg(\ell) + H_2S(g) \rightleftharpoons HgS(s) + H_2(g)$$

Calculate $\Delta G°$ for this reaction.
[J. Goates, A. Cole, and E. Gray, *J. Am. Chem. Soc.*, *73*, 3596 (1951)]

16-37 The free energy of formation of PbS(s) was determined from 500 to 800°C using a proper electrochemical cell. The E°, in millivolts, of the cell was determined to be E° = 867.4 - 0.4884 T. Here T is degrees Kelvin. What is $\Delta G°$ for this reaction at 500°C?
[W. T. Thompson and S. N. Flengas, *J. Electrochem. Soc.*, *118*, 419 (1971)]

16-38 The E° of the cell

$$Ag(s)|AgBr(\ell)|Br_2(g,\ 1 \text{ atm}),\ C\ (\text{rod})$$

was found to be 0.7865 V at 500°C. Write the cell reaction. What is $\Delta G°_f$ for this reaction?

[E. J. Salstrom and J. H. Hildebrand, *J. Am. Chem. Soc.*, *52*, 4653 (1930)]

16-39 Wachter and Hildebrand have measured E° for the cell

$$Pb(\ell)|PbCl_2(\ell)|Cl_2(g, 1\ atm)$$

A tungsten wire was used to contact the molten lead and a graphite rod for the chlorine electrode. E° at 500°C was 1.2730 V. What is the cell reaction? What is $\Delta G°$ for this reaction?

[*J. Am. Chem. Soc.*, *52*, 4655 (1930)]

17
CHEMICAL KINETICS

17-1 INTRODUCTION

Chemical kinetics includes the study of the rates at which reactions occur, the factors that influence the rates, and detailed mechanisms by which the reactions occur. The factors that influence the rate of a chemical reaction are concentration, temperature, and catalysts.

17-2 RATE OF REACTION

The rate of a chemical reaction is usually defined as the change in concentration of a *reactant* per unit of time. The units are the units of concentration divided by units of time, for example, mole per liter per second.

17-3 RATE EQUATION

At constant temperature, the rate of a chemical reaction is proportional to the concentration of the reacting substances (law of mass action). An equation describing this relationship for a particular reaction is called the *rate equation* of the reaction.

In a generalized reaction represented by $A \rightarrow C + D$, the rate of the reaction can be represented by the rate equation

$$\text{Rate} = k[A]^n$$

where [A] represents the concentration of the species in *moles per liter*. The value of the exponent n in the rate equation is called the *order of the reaction.* The constant k is the *specific rate constant* of the reaction.

The nature of the proportionality between the rate and the concentration of a reactant in a particular reaction is given by the value of n, the order of the reaction, and the value of k. The numerical values of n and k are calculated from *experimental data.* The usual data are the concentration of a reactant observed at time intervals during the course of the reaction.

Example 17-1

The decomposition of N_2O_5 has been investigated several times, both in the gas phase and in solution. The following data have been calculated from one such investigation by H. C. Ramsperger, M. E. Nordberg, and R. C. Tolman.

Time (sec)	Concentration of N_2O_5 (mol/L)
110	1.92×10^{-3}
980	1.65×10^{-3}
2310	1.31×10^{-3}
3670	1.04×10^{-3}
4770	0.865×10^{-3}
6580	0.644×10^{-3}
8988	0.428×10^{-3}

Source: Proc. Natl. Acad. Sci., USA, 15, 450 (1929).

These data are graphed in Fig. 17-1. The upper curve is the log of the concentration, and the lower curve is the concentration. The *average rate* of the decomposition during any time interval may be calculated from the data.

If the concentration of N_2O_5 is represented by $[N_2O_5]$, the average rate of decomposition will be

$$r_{av} = \frac{[N_2O_5]_2 - [N_2O_5]_1}{t_2 - t_1} = \frac{\Delta[N_2O_5]}{\Delta t}$$

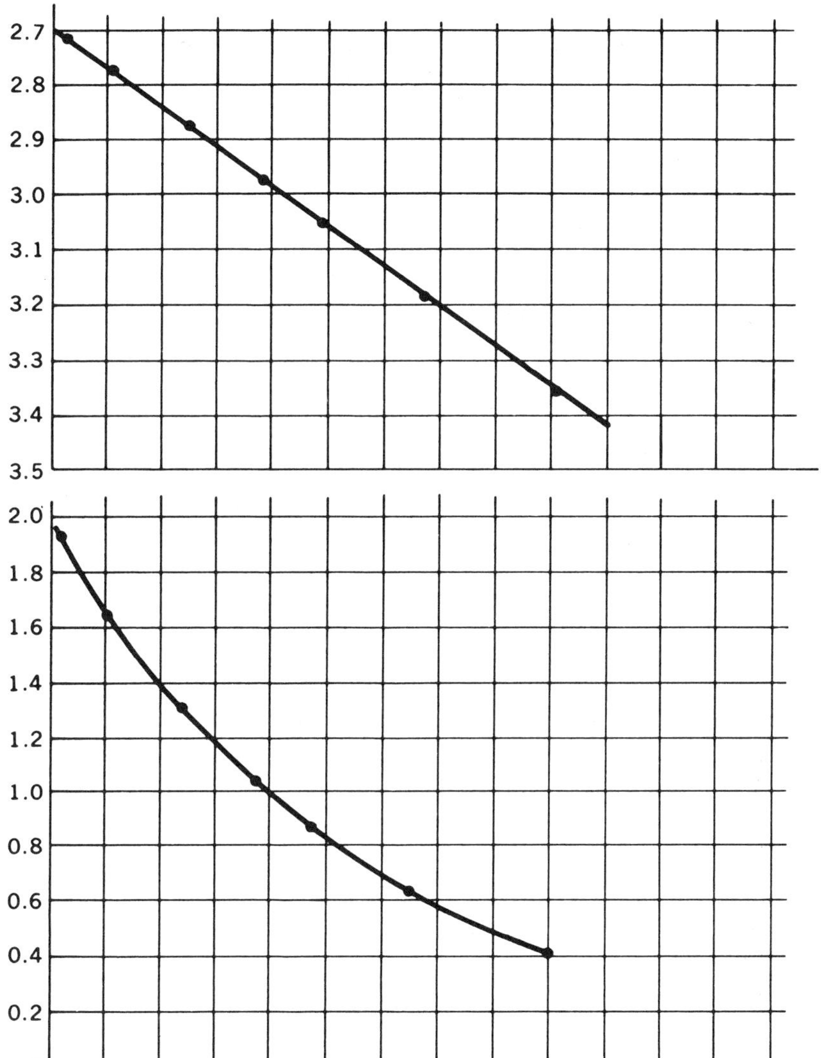

FIGURE 17-1 Log of concentration of N_2O_5 versus time (upper curve); concentration of $N_2O_5 \times 10^3$ versus time (lower curve).

Since $[N_2O_5]_2 < [N_2O_5]_1$, $\Delta[N_2O_5]$ will be negative and r_{av} will be negative, signifying that it is the rate of disappearance of N_2O_5.

Since the concentration of N_2O_5 is continually decreasing, the rate of the reaction will continually diminish. The *average rate* of reaction is not a useful concept, but the *instantaneous rate* is useful. If the time interval Δt over which the rate is observed is made smaller and smaller, it approaches zero as a limit. By using calculus notation

$$r = \frac{-d[N_2O_5]}{dt} = \lim_{\Delta t \to 0} \frac{-\Delta[N_2O_5]}{\Delta t}$$

From experimental observation, the rate of this reaction is proportional to the concentration of N_2O_5:

$$\frac{-d[N_2O_5]}{dt} = k[N_2O_5]$$

This is the rate equation for the decomposition of dinitrogen pentoxide. In the general case of a reaction $A \rightarrow C + D$, the rate equation takes the form

$$\frac{-d[A]}{dt} = k[A]^n$$

As noted at the beginning of this section, the value of n, the exponent in the rate equation, is the *order*. The rate of decomposition of N_2O_5 is *first-order* with respect to the N_2O_5 concentration.

17-4 DETERMINATION OF n AND k

The order of a reaction and its specific reaction rate constant may be determined by means of the integrated form of the rate equation

$$\frac{-d[A]}{[A]} = k\ dt$$

by integration

$$-\int_{[A_o]}^{[A]} \frac{d[A]}{[A]} = k\int_o^t dt$$

$$-(\ln[A] - \ln[A_o]) = k(t - 0)$$

$$\ln[A_o] - \ln[A] = kt$$

$$2.303\,(\log[A_o] - \log[A]) = kt$$

$$\log[A_o] - \log[A] = \frac{k}{2.303}\,t$$

$[A_o]$ is the initial concentration of the reactant A. The equation may be written in the form

$$\log[A] = -\frac{k}{2.303}\,t + \log[A_o]$$

where $-\frac{k}{2.303}$ is the slope and $\log[A_o]$ is the intercept.

Example 17-2

The rate of decomposition of azomethane was determined by H. C. Ramsperger. The reaction that occurred was

$$CH_3NNCH_3 \rightarrow C_2H_6 + N_2$$

Time (min)	$[CH_3NNCH_3]$ (mol/L)
0	1.21×10^{-2}
10	1.04×10^{-2}
20	0.880×10^{-2}
33	0.706×10^{-2}
46	0.575×10^{-2}
65	0.434×10^{-2}

Source: *J. Am. Chem. Soc.*, *49*, 912 (1927)].

Calculate the specific reaction rate constant.

A plot of log $[CH_3NNCH_3]$ against time gives a straight line, indicating that the reaction is first-order (see Fig. 17-2). (The use of semilogarithmic graph paper minimizes the work involved.)

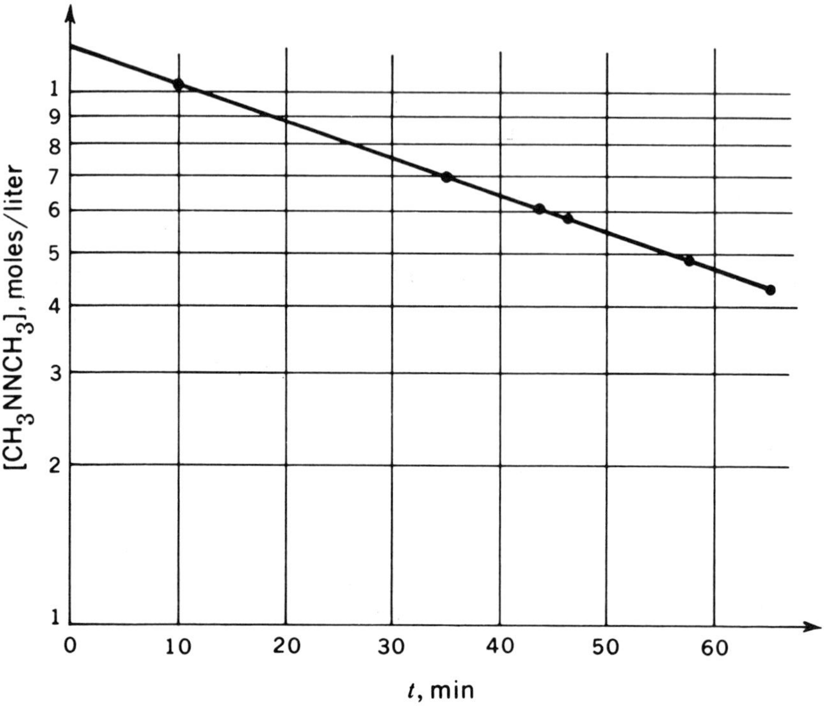

FIGURE 17-2 Log of $[CH_3NNCH_3]$ versus time.

From the slope of the curve, using the logarithms of the concentrations,

$$k = 1.57 \times 10^{-2}$$

The specific reaction rate constant k may be calculated from the integrated form of the first-order rate equation

$$\log[A] = -\frac{k}{2.303}t + \log[A_o]$$

$$k = \frac{2.303}{t} \times (\log[A_o] - \log[A])$$

$$= \frac{2.303}{10} \times (\log 1.21 \times 10^{-2} - \log 1.04 \times 10^{-2})$$

$$= 0.230 \times 0.0658 = 1.51 \times 10^{-2}\ \text{min}^{-1}$$

In a similar way, other values of k may be calculated from the data.

t (min)	k (min^{-1})
20	1.55×10^{-2}
33	1.60×10^{-2}
46	1.60×10^{-2}
65	1.51×10^{-2}

The average of these calculated values is 1.54×10^{-2} min^{-1}.

Example 17-3

F. O. Rice and D. Getz found that at 64.0°C, 0.00091 mol/L of N_2O_5 had decomposed after 2.474 min from an initial concentration of 0.001943 mol/L. Calculate k for this first-order decomposition.

$$\log 1.943 \times 10^{-3} - \log 1.032 \times 10^{-3} = \frac{k}{2.303} 2.474$$

$$\frac{2.303 \times 0.2749}{2.474} = k$$

$$k = 0.256 \text{ min}^{-1}$$

[*J. Phys. Chem.*, *31*, 1572 (1927)]

17-5 HALF-LIFE

The time required for one-half of the initial quantity of a reactant to be consumed in a reaction is the half-life of that substance. For a first-order reaction, the half-life $t_{1/2}$ is given by

$$t_{1/2} = \frac{0.693}{k}$$

Example 17-4

In the reaction described in Example 17-2, calculate the half-life by using the average value of k, 1.54×10^{-2}.

$$t_{1/2} = \frac{0.693}{1.54 \times 10^{-2}} = 45.0 \text{ min}$$

($t_{1/2}$ estimated from the curve is 43.5 min.)

17-6 OTHER INTEGRATED EQUATIONS FOR FIRST-ORDER REACTIONS

Another, practical form of the first-order rate equation is

$$-\frac{d(a - x)}{dt} = k(a - x)$$

where a is the initial concentration of the reacting species and x is the quantity that has reacted in time t.

After integration and rearrangement, the equation may be expressed as

$$k = \frac{2.303}{t} \log \frac{a}{a - x}$$

Example 17-5

Calculate the value of k for the decomposition of N_2O_5 at 64.4° from the data of F. O. Rice and D. Getz. After 2.123 min, 0.001018 mol of N_2O_5 had decomposed when the initial quantity of N_2O_5 was 0.002347 mol. (Assume that the volume was constant.)

[*J. Phys. Chem.*, *31*, 1572 (1927)]

$$k = \frac{2.303}{t} \log \frac{a}{a - x}$$

$a = 0.002347$

$x = 0.001018$

$t = 2.123$ min

$a - x = 0.001329$

$$k = \frac{2.303}{2.123} \log \frac{0.002347}{0.001329}$$

$$k = \frac{2.303 \times 0.2470}{2.123} = 0.268 \text{ min}^{-1}$$

If the rate equation is integrated between limits, another useful integrated form results:

$$2.303 \log \frac{a - x_1}{a - x_2} = k(t_2 - t_1)$$

where x_1 and x_2 are the concentrations of the substance that have reacted at times t_1 and t_2.

Example 17-6

Calculate k for the decomposition of N_2O_5 at 45°C from the data of F. Daniels. The pressure of N_2O_5 after 1800 sec was 140 mm, and after 3600 sec it was 58 mm.

[*Chemical Kinetics*, Cornell University Press, Ithaca, New York, 1938, p. 9]

$$2.303 \log \frac{a - x_1}{a - x_2} = k(t_2 - t_1)$$

The pressures 140 and 58 mm are measures of the concentrations of N_2O_5 at times t_1 and t_2, $a - x_1$, and $a - x_2$.

$$2.303 \log \frac{140}{58} = k \times (3600 - 1800)$$

$$2.30 \times 0.382 = k \times 1800$$

$$k = \frac{2.30 \times 0.382}{1.80 \times 10^{+3}} = 4.87 \times 10^{-4} \text{ sec}^{-1}$$

17-7 HIGHER-ORDER REACTIONS

In the general reaction

$$aA + bB \rightarrow \text{products}$$

the rate of the reaction may be described by the rate equation

$$\frac{-d[A]}{dt} = k[A]^m[B]^n$$

The exponent m is the "order of the reaction with respect to reactant A," and n is the "order with respect to the reactant B." The *overall order of the reaction* is the sum of n and m. As in the case of first-order reactions, the order must be determined by experiment.

Example 17-7

An investigation of the rate of reaction of $2NO_2 + F_2 \rightarrow 2NO_2F$ has shown that it is first-order with respect to NO_2 and first-order with respect to F_2.

$$\frac{-d[F_2]}{dt} = k'[NO_2]$$

$$\frac{-d[F_2]}{dt} = k''[F_2]$$

The *overall order* of the reaction is second-order, the sum of the two exponents.

$$\frac{-d[F_2]}{dt} = k[NO_2][F_2]$$

[R. L. Perrine and H. S. Johnston, *J. Chem. Phys.*, *21*, 2202 (1953)]

Rate equations for higher reactions are complicated by the fact that the number of moles of each reactant involved may be different. In the generalized equation

$$aA + bB \rightarrow cC + dD$$

the rate of the reaction is defined as:

$$-\frac{1}{a}\frac{d[A]}{dt} = -\frac{1}{b}\frac{d[B]}{dt} = \frac{1}{c}\frac{d[C]}{dt} = \frac{1}{d}\frac{d[D]}{dt}$$

Example 17-8

In the reaction described in Example 17-7, the rate may be expressed as

$$\frac{-d[F_2]}{dt} = -\frac{1}{2}\frac{d[NO_2]}{dt} = k[NO_2][F_2]$$

17-8 SECOND-ORDER REACTIONS

Just as the use of calculus led to a differential equation for a first-order reaction, so a rate equation for a second-order reaction may be found:

$$\frac{-d[A]}{dt} = k[A]^2$$

Integration leads to

$$\frac{1}{[A]} = \frac{1}{[A_O]} = kt$$

This equation holds for *second-order reactions* in which *the reactants* are present in *equivalent concentrations*. More involved equations result from more complex situations. The problems here are limited to this simple type.

A plot of 1/[A] versus t will show that a reaction is second-order if a straight line results.

Example 17-9

What is the order of the reaction

$CH_3COOCH_3 + OH^- \rightarrow CH_3CO_2^- + CH_3OH$?

$[CH_3COOCH_3] = [OH^-] = 0.0100$ M at the start of the reaction.

t (min)	$[CH_3COOCH_3]$	$1/[CH_3COOCH_3]$
0	0.0100	100
3	0.00740	135
4	0.00683	147
5	0.00634	158
6	0.00589	170
7	0.00550	182
8	0.00519	193
10	0.00464	216
12	0.00416	241
15	0.00363	276

Source: J. Walker, *Proc. Roy. Soc. (London), Ser. A, 78*, 157 (1906).

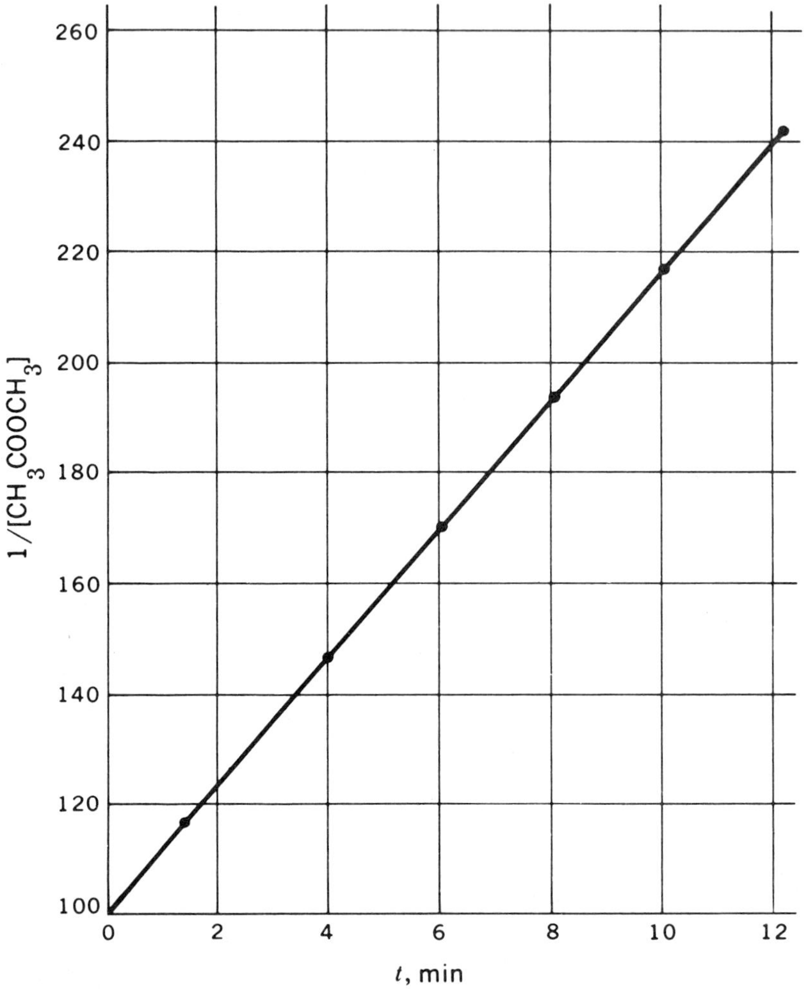

FIGURE 17-3 $1/[CH_3COOCH_3]$ versus time.

The plot of $1/[CH_3COOCH_3]$ versus t indicates that the reaction is second-order with respect to $[CH_3COOCH_3]$ (see Fig. 17-3). k may be calculated from the equation

$$\frac{1}{[A]} - \frac{1}{[A_0]} = kt$$

$$135 - 100 = 3k$$

$$\frac{35}{3} = k = 11.7\ (\text{mol/L})^{-1}\ \text{min}^{-1}$$

$$147 - 100 = 4k$$

$$\frac{47}{4} = k = 11.7\ (\text{mol/L})^{-1}\ \text{min}^{-1}, \text{ etc.}$$

Unlike k for first-order reactions, the units of concentration must be included in k for second-order reactions.

17-9 HALF-LIFE OF SECOND-ORDER REACTIONS

The time required for the initial concentration of the reactants to decrease to one-half of their initial values for second-order reactions is given by

$$t_{1/2} = \frac{1}{ka}$$

where a is the initial concentration of one of the reactants.

Example 17-10

What is the half-life of the reaction in Example 17-9?

$$t_{1/2} = \frac{1}{ka}$$

$$t_{1/2} = \frac{1}{11.7 \times 0.0100} = 8.55\ \text{min}$$

$$t_{1/2} = \frac{1}{11.7\ \dfrac{\not{L}}{\not{\text{mol}}\ \text{min}} \times 0.0100\ \dfrac{\not{\text{mol}}}{\not{L}}} = 8.55\ \text{min}$$

17-10 ZERO-ORDER REACTIONS

Occasionally you may encounter a reaction in which the rate of the reaction is independent of the concentration of any chemical. This can be expressed as

$$\text{Rate} = -\frac{d[A]}{dt} = k$$

The power of the concentration of any chemical on the right side of the equation is zero. This type of reaction is therefore called a *zero order* reaction. The integrated equation for this reaction is

$$[A] = [A_o] - kt$$

A plot of the concentration of a reactant versus time is a straight line. This type of kinetics is often encountered in enzyme kinetics when the substrate (reactant) concentration is high compared to the enzyme concentration.

The half-life of this type of reaction is given by

$$t_{1/2} = \frac{A_o}{2k}$$

Note that it is directly proportional to the original concentration of reactant.

17-11 INITIAL RATE METHOD

The method used previously in this chapter to find the rate of a reaction is known as the method of integration. In each case, the final equation solved was obtained by integration of a differential equation. Another common method is the differential method. Here the rates of reaction are used directly to determine the order of a reaction. A common differential method is the *initial rate method*.

To use this method, the initial rate of a reaction is measured over various initial concentrations of reactants. The order of the reaction with respect to each reactant is found by substituting into the rate expression and solving the expression algebraically. With this method it is relatively easy to measure orders of reaction that are fractional integers (e.g., $n = 1/2$).

Example 17-11

The rate of reaction for the reaction

$$A + B + C \rightarrow \text{products}$$

was studied using the initial rate method. The results are summarized in the following table:

	Rate (mol/dm^{3}s)	[A] (mol/dm^3)	[B] (mol/dm^3)	[C] (mol/dm^3)
1	5.0×10^{-5}	3.0×10^{-2}	1.7×10^{-2}	1×10^{-7}
2	10.0×10^{-5}	6.0×10^{-2}	1.7×10^{-2}	1×10^{-7}
3	10.0×10^{-5}	3.0×10^{-2}	3.4×10^{-2}	1×10^{-7}
4	10.0×10^{-5}	3.0×10^{-2}	1.7×10^{-2}	2×10^{-7}

What is the order of the reaction with respect to each reactant? What is the overall order?

$$\text{Rate} = k[A]^m[B]^n[C]^p$$

To solve this algebraically, substitute the data from trial 2 into the rate expression, and divide this by the same expression containing the data from trial 1.

$$\frac{10.0 \times 10^{-5}}{5.0 \times 10^{-5}} = \frac{\cancel{k}[6.0 \times 10^{-2}]^m[1.7 \cancel{\times} 10^{-2}]^n[1 \cancel{\times} 10^{-7}]^p}{\cancel{k}[3.0 \times 10^{-2}]^m[1.7 \cancel{\times} 10^{-2}]^n[1 \cancel{\times} 10^{-7}]^p}$$

$$2 = \frac{[6.0 \times 10^{-2}]^m}{[3.0 \times 10^{-2}]^m} = 2^m$$

$$m = 1$$

Using the data from trials 1 and 3,

$$\frac{10.0 \times 10^{-5}}{5.0 \times 10^{-5}} = \frac{\cancel{k}[3.0 \times 10^{-2}]^m[3.4 \times 10^{-2}]^n[1 \cancel{\times} 10^{-7}]^p}{\cancel{k}[3.0 \times 10^{-2}]^m[1.7 \times 10^{-2}]^n[1 \cancel{\times} 10^{-7}]^p}$$

$$2 = \frac{[3.4 \times 10^{-2}]^n}{[1.7 \times 10^{-2}]^n} = 2^n$$

$$n = 1$$

Finally, using the data from trials 1 and 4,

$$\frac{10.0 \times 10^{-5}}{5.0 \times 10^{-5}} = \frac{k[3.0 \times 10^{-2}]^m[1.7 \times 10^{-2}]^n[2 \times 10^{-7}]^p}{k[3.0 \times 10^{-2}]^m[1.7 \times 10^{-2}]^n[1 \times 10^{-7}]^p}$$

$$2 = \frac{[2 \times 10^{-7}]^p}{[1 \times 10^{-7}]^p} = 2^p$$

$$p = 1$$

The reaction is first-order with respect to A, first-order with respect to B, and first-order with respect to C. It is third-order overall.

Example 17-12

A reaction follows the rate expression

$$-\frac{d[A]}{dt} = k[A]^m[B]^n$$

If the concentration of A is doubled and that of B held constant, the reaction rate is doubled. If the concentration of A is held constant and the concentration of B is doubled, the rate decreases in half. What are the values of m and n for this reaction?

To solve this problem, let X = the rate of the reaction when [A] = A and [B] = B. Therefore, 2X is the reaction rate when [A] = 2A and [B] is B, whereas X/2 is the rate when [A] is A and [B] is 2B.

Set up the algebraic expressions to be solved.

$$\text{Rate} = X = k[A]^m[B]^n$$

$$2X = k[2A]^m[B]^n$$

$$\frac{X}{2}\ k[A]^m[2B]^n$$

Dividing the second expression by the first gives

$$\frac{2X}{X} = \frac{k[2A]^m[B]^n}{k[A]^m[B]^n}$$

$$2 = 2^m$$

$$m = 1$$

Now divide the third equation by the first equation.

$$\frac{X/2}{X} = \frac{\not{k}[\not{A}]^m[2B]^n}{\not{k}[\not{A}]^m[B]^n}$$

$$\frac{1}{2} = 2^n$$

To find n here it is best to take the log of both sides and solve for n:

$$\log \frac{1}{2} = n \log(2)$$

$$-0.301 = 0.301\, n$$

$$n = -1$$

The order of the reaction with respect to B is -1. This type of material is called an inhibitor.

17-12 ARRHENIUS ACTIVATION ENERGY

Before a chemical reaction can take place, the reacting molecules must approach one another with sufficient energy to start the reaction. The minimum energy that must be overcome before a reaction will take place is called the Arrhenius activation energy (E_a). As the temperature increases, more and more molecules have an energy greater than E_a. A greater number of molecules will react and thus the reaction will speed up. The rate of the reaction thus increases with increasing temperature. The equation used to describe this is known as the Arrhenius equation and can be represented as

$$\log k = -\frac{E_a}{2.303\, R}\frac{1}{T} + b$$

This is an equation of a straight line in which log k (the rate constant) is the Y-axis and 1/T (in degrees kelvin) is the X-axis. The slope of the equation is $-E_a/(2.303\, R)$ and b is the y intercept. R is the gas constant. The activation energy can be determined by plotting log k versus 1/T and measuring the slope of the line.

If the rate is measured at two different temperatures, then this equation can be rewritten as

$$\log\left(\frac{k_2}{k_1}\right) = \left(\frac{E_a}{2.303\ R}\right)\left(\frac{(T_2 - T_1)}{T_2 T_1}\right)$$

where the subscript 2 represents one temperature and the subscript 1 the other. From this equation we can obtain the rule used in laboratories that, for every 10° rise in temperature of a system, the reaction rate doubles.

Example 17-13

For many organic reactions the activation energy is around 45 kJ/mol. Show that a rise in temperature from 273°K to 283°K will approximately double the rate of the reaction.

$$\log \frac{k_2}{k_1} = \frac{45 \times 10^3\ \text{J/mol}\ (283-273)\ \text{K}}{(2.303)(8.314)\ \text{J/K mol}\ (283\ \text{K})(273\ \text{K})}$$

$$\log \frac{k_2}{k_1} = 0.304$$

$$\frac{k_2}{k_1} = 2.01$$

Thus the rate of the reaction has doubled for the 10° rise in temperature.

PROBLEMS

17-1 In an experiment in which the thermal decomposition of N_2O_5 was investigated, an initial concentration of 0.002347 mol/L had decreased to 0.001329 mol/L in 2.123 min at 64.4°C. Calculate k for the first-order reaction.

[F. O. Rice and D. Getz, *J. Phys. Chem.*, *31*, 1572 (1927)]

17-2 In a second Rice and Getz experiment, 0.000902 mol/L of N_2O_5 decomposed from an initial concentration of 0.001993 mol/L in 2.226 min. Determine k for the reaction.

[F. O. Rice and D. Getz, *J. Phys. Chem.*, *31*, 1572 (1927)]

17-3 In a kinetics study the initial concentration of one of the reactants was 0.188 mM in 4.00 ml of solution. After 25 min, 68.4% of the substance remained. Assuming that the reaction was first-order with respect to the substance, calculate k.

[W. A. Pryor, *J. Am. Chem. Soc.*, *80*, 6483 (1958)]

17-4 Calculate the specific reaction-rate constant for the first-order decomposition of azomethane, $(CH_3)_2N_2$, at 320.4 K. The initial pressure was 21.23 cm in a 210-cc container; the pressure after 9 min was 10.20 cm.

[H. C. Ramsperger, *J. Am. Chem. Soc.*, *49*, 912 (1927)]

17-5 The vapor phase thermal decomposition of nitroglycerin was studied at 160°C. It was found to follow first-order kinetics. The following data were obtained from a series of experiments:

Initial concentration g cm^{-3} (M/V) nitroglycerin	Time (sec)	% Decomposition
1.11	300	52.0
0.80	300	52.9
0.52	300	53.2
0.41	300	53.9
1.21	180	34.6
0.92	180	35.9
0.67	180	36.0
0.39	180	35.4

What is the average value of the rate constant determined from the above data?

[C. E. Waring and G. Krastins, *J. Phys. Chem.*, *74*, 999 (1970)]

17-6 Determine the order of the reaction and the value of k from the following data for the decomposition of N_2O_5:

t (sec)	$P_{N_2O_5}$ (mm)
110	0.826
980	0.708
2310	0.562
3670	0.449
4770	0.372
6580	0.277
8988	0.184

Source: H. C. Ramsperger, M. E. Nordberg, and R. C. Tolman, *Proc. Natl. Acad. Sci. U.S.*, *15*, 456 (1929).

17-7 In an investigation of the decomposition of N_2O_5 in carbon tetrachloride solution, R. H. Lueck found that the specific reaction-rate constant was 3.21×10^{-4} sec^{-1}. What was the half-life of the reaction?

[*J. Am. Chem. Soc.*, *44*, 757 (1922)]

17-8 The rate of decomposition of azomethane was studied by H. C. Ramsperger. In one experiment, conducted in a 210-cc chamber at 296.8°C, the following data were obtained:

t (min)	Pressure of azomethane (cm)
0	3.62
15	3.00
30	2.49
48	1.93
75	1.31

Source: *J. Am. Chem. Soc.*, *49*, 912 (1927).

Calculate the specific reaction-rate constant and the half-life of the reaction.

17-9 The rate of hydrolysis of a complex boron cation was followed by measuring the volume of hydrogen given off by the reaction. The concentration of the unreacted boron complex is proportional to $V_\infty - V_t$ (volume of hydrogen after the reaction is completed -volume of hydrogen at time t). Determine the order and the rate constant for this reaction.

Time (sec)	100 × $(V_\infty - V_t)$ ml
152	90
248	87
344	83
401	81
540	75
589	73
660	70
775	65
955	58
1024	56
1075	54
1164	51
1254	48
1392	44
1561	39
1799	34
2992	16
3160	14
3715	11.5
4010	8.0
4676	6.0
5330	3.0

Source: N. E. Miller, *J. Am. Chem. Soc.*, *92*, 4564 (1970).

17-10 The thermal isomerization of cyclopropane occurs according to the equation

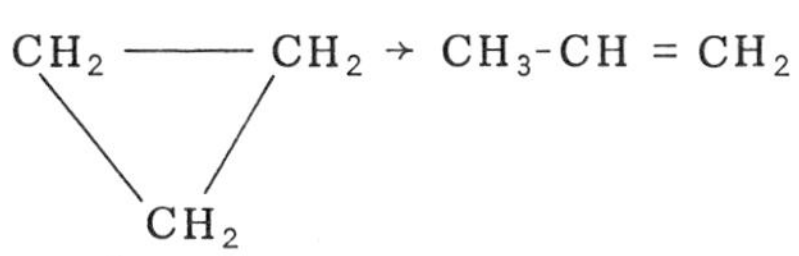

cyclopropane propene

T. S. Chambers and G. B. Kistiakowsky found that the specific reaction-rate constant for this reaction at 499°C was 5.95×10^{-4} sec^{-1}. What is the half-life of the reaction? What fraction of the cyclopropane will remain after 1.00 hr at 499.5°C?

[*J. Am. Chem. Soc.*, *56*, 399 (1934)]

17-11 The thermal isomerization of cyclobutane, described in Problem 17-34, was carried out in a 0.53-L vessel at 150°C. The initial pressure was 50 mm. Calculate the concentration of cyclobutane, in moles per liter, after 30 min. $k = 2.0 \times 10^{-4}$ sec^{-1}.

[W. Cooper and W. D. Walters, *J. Am. Chem. Soc.*, *80*, 4220 (1958)]

17-12 The initial pressure of cyclopropane, C_3H_6, in a 500-cc vessel was 702.6 mm at 499.5 K. Calculate the concentration of cyclopropane that remained after 720 sec if the specific reaction-rate constant for its isomerization was 5.95×10^{-4} sec^{-1}.

[T. S. Chambers and G. B. Kistiakowsky, *J. Am. Chem. Soc.*, *56*, 399 (1934)]

17-13 Chlorine dioxide reacts with iodide ions in aqueous solutions at 25°C according to the following equation:

$$2ClO_2 + 2I^- \rightarrow 2ClO_2^- + I_2$$

When a large excess of $[I^-]$ is present, the reaction is pseudo first-order. A $ClO_2 \cdot I^-$ complex is formed that decomposes following first-order kinetics. The concentration of this complex can be measured spectroscopically.

Time (sec)	$[ClO_2\ I^-]M \times 10^4$
0.00	4.77
1	4.31
2	3.91
3	3.53
4	3.19
5	2.89
10	1.76
15	1.06
20	0.64
30	0.24
40	0.087
50	0.032

What is k for this pseudo first-order reaction?

[H. Fukutomi and G. Gordon, *J. Am. Chem. Soc.*, *89*, 1362 (1967)]

17-14 What is the half-life of a second-order reaction where k = 1.8×10^{-4} $(\mathrm{mmol/L})^{-1}$ min^{-1}?

17-15 Calculate k for the second-order reaction in which the initial concentration of the reacting substance was 123 mmol/L and the concentration after 10 min was 90.0 mmol/L.

17-16 The decomposition of NO_2 was studied by M. Bodenstein and I. Ramstetter, from whose paper the data below have been taken. The decomposition proceeded according to the equation

$$2NO_2 \rightleftharpoons 2NO + O_2$$

t (min)	$[NO_2]$ (mmol/L)	$1/[NO_2]$
0	123	8.13×10^{-3}
1	107.6	9.26×10^{-3}
2	105.1	9.54×10^{-3}
3	102.8	9.74×10^{-3}
4	100.8	9.90×10^{-3}
5	99.1	10.1×10^{-3}
6	97.3	10.3×10^{-3}

Source: I. Ramstetter, *Z. Physik. Chem. (Leipzig)*, *100*, 106 (1922).

Determine the order of the reaction.

17-17 The rate of reduction of a series of oxonides with triphenyl phosphine at 25°C was found to follow second-order kinetics. In a series of experiments, the half-life of the reaction was measured at various initial concentrations of both reactants. At all times the initial concentration of each reactant was the same.

Initial concentration of reactants (mole liter^{-1})	$t_{1/2}$ (sec)
0.05	302
0.10	152
0.15	101
0.20	75

Source: J. Charles and S. Flisgár, *Can. J. Chem.*, *48*, 1309 (1970).

Determine k for this reaction.

17-18 The reduction of an ozonide of styrene with triphenylphosphine was followed spectroscopically. The initial concentration of each reactant was 0.1 M. The fraction of each substance reacted is given by D/D_∞ from the following data:

t (sec)	$D \times 10^2$
0	0
355	22
638	29
949	32
1265	34
2130	37.5
2480	38.0
2870	38.5
∞	43.0

Source: J. Charles and S. Flisgár, *Can. J. Chem.*, *48*, 1309 (1970).

What is the second-order rate constant k for this reaction?

17-19 A study of the acid hydrolysis of organic acetals and ketals showed they followed pseudo-first-order kinetics. The reactions were first-order with respect to the concentration of acid and first-order with respect to the acetal concentration. Since the acid acted only as a catalyst, its concentration did not change with time, making the observed reaction pseudo first-order. In a dioxane-water mixture at 34.3°C the following data were obtained:

Time (min)	$10^3 \times$ [acetal] mmol/mL
0	23.6
9	22.9
21	22.0
48	20.8
81	18.7
128	16.7
168	14.0
226	12.1
296	10.3
348	8.8
416	7.6

Source: M. S. Newman and R. E. Dickson, *J. Am. Chem. Soc.*, *92*, 6880 (1970).

a. What is k for the pseudo-first-order reaction?
b. If the $[H^+]$ was 0.02 M, what is the second-order rate constant?

17-20 The rate of the reaction $NH_4NO_2 \rightarrow N_2 + H_2O$ was investigated by S. K. Ghosh. The following data were obtained:

t (hr)	$[NH_4^+]$	$[NO_2^-]$
0	0.500	0.500
3	0.432	0.459
6	0.401	0.430
9	0.373	0.410
24	0.340	0.361
30	0.330	0.351

Source: *Z. Physik. Chem. (Leipzig)*, *206*, 321 (1957).

a. What is the order of the reaction with respect to NH_4^+?
b. What is the order of the reaction with respect to NO_2^-?
c. What is the overall order of the reaction?

17-21 Wilson and Kelly recently determined the order of the reaction of the hypochlorite ion with morpholine-borane (Mbn) using the initial rate method. From the following data, determine the order of the reaction with respect to NaOCl and to Mbn.

$-10^4 d[NaOCl]/dt$ MS^{-1}	$10^3[NaOCl]$ M	$10^3[Mbn]$ M
4.36	0.160	10
8.75	0.160	20
29.4	0.535	20
270.0	1.97	50

[*Inorg. Chem.*, *21*, 1622 (1982)]

17-22 An experiment was carried out using the initial rate method. The rate expression for the reaction was assumed to be

$$\text{Rate} = k[A]^m[B]^n$$

Using the following data, determine m and n and the overall order of the reaction.

[A] mol/dm^3	[B] mol/dm^3	Rate
0.0125	0.0125	R
0.0250	0.0125	2R
0.0250	0.0250	5.66R

17-23 A reaction was carried out using a new substance. The half-life of the reaction for different amounts of starting material was as follows:

A (mol/dm^3)	0.0125	0.0150	0.0175	0.0200
$t_{1/2}$ (min)	2.20	2.64	3.08	3.52

What is the order of the reaction?

17-24 The rate of reaction of chromium(II) with a complex of cobalt(III) was recently measured as a function of T. Determine the activation energy for the following data and compare your results to the authors' average value of 8.4 kcal/mol.

T°C	20	25.0
k S^{-1}	578	735

[R. Balahura and A. J. Johnston, *Inorg. Chem.*, *22*, 3309 (1983)]

Supplementary Problems

17-25 What is the first-order reaction-rate constant for the isomerization of cyclopropane from the observations of E. S. Comer and R. N. Pease? They observed an initial pressure of 910 mm, and the pressure due to the cyclopropane that had reacted in 1200 sec was 450 mm.

[*J. Am. Chem. Soc.*, *67*, 2067 (1945)]

17-26 Find k for the thermal decomposition of N_2O_5 at 64.6°C from the observation that, of an initial quantity of 0.002193 mol, 0.000977 mol had decomposed after 2.194 min.

[F. O. Rice and D. Getz, *J. Phys. Chem.*, *31*, 1572 (1927)]

17-27 Calculate k for the decomposition of N_2O_5 at 35°C from the data of F. Daniels and E. H. Johnston. The initial pressure of N_2O_5 was 308.2 mm, and the partial pressure due to the N_2O_5 that had decomposed after 2400 sec was 90.0 mm.
[*J. Am. Chem. Soc.*, *43*, 53 (1921)]

17-28 The rate of decomposition of N_2O_5 in CCl_4 solution may be followed by measuring the amount of decomposition X in terms of the volume of oxygen liberated after various times. Determine the value of k for the decomposition from the data of R. H. Lueck, measured at 40°C.

t (sec)	V_{O_2} (cc)
0	0.00
600	6.30
900	8.95
1200	11.40
1800	15.53
2400	18.90
3000	21.70
∞	34.75

Source: *J. Am. Chem. Soc.*, *44*, 757 (1922).

17-29 Calculate the value of k for the isomerization of cyclopropane from the observation that, of an initial pressure of 649 mm, 143 mm had reacted after 420 sec.
[E. S. Comer and R. N. Pease, *J. Am. Chem. Soc.*, *67*, 2067 (1945)]

17-30 Find k for the decomposition of N_2O_5 at 37.30°C from the pressures of N_2O_5 after 2310 and 6580 sec, which were 0.562 and 0.277 mm, respectively.
[H. C. Ramsperger, M. E. Nordberg, and R. C. Tolman, *Proc. Natl. Acad. Sci, USA*, *15*, 456 (1929)]

17-31 In an experiment of F. O. Rice and D. Getz, a CCl_4 solution containing 8.86×10^{-5} mol of N_2O_5 per gram of solution was observed while decomposition of N_2O_5 occurred at 34.2°C. After 77.2 min, 3.67×10^{-5} mol of N_2O_5 per gram of solution had decomposed; after 105 min, 4.44×10^{-5} mol/g of solution had decomposed. Find k.

[*J. Phys. Chem.*, *31*, 1572 (1927)]

17-32 The reaction of dimethylmagnesium with 4-methylmercaptoacetophenone follows pseudo-first-order kinetics of the $(CH_3)_2Mg$ is present in a large excess. In one experiment with the ketone concentration at 10^{-3} M and $(CH_3)_2Mg$ at 0.185 M, the following data were obtained:

Time (sec)	Rel. A
0.050	9.26
0.075	7.08
0.100	5.48
0.125	4.30
0.150	3.24
0.175	2.49
0.200	1.92
0.225	1.46
0.250	1.12
0.275	0.89
0.300	0.70

Rel. A is proportional to the concentration of the 4-methylmercaptoacetophenone. What is k for this reaction?

[S. Smith and J. Billet, *J. Am. Chem. Soc.*, *89*, 6948 (1967)]

17-33 The decomposition of N_2O_5 in carbon tetrachloride at 35° was found to be first-order, and the specific rate constant was 8.67×10^{-2} min^{-1}. In one experiment the initial concentration of N_2O_5 was 8.54×10^{-5} mol/g of solution. Calculate the concentration of N_2O_5 after 15.0 min.

17-34 The thermal isomerization of cyclobutene

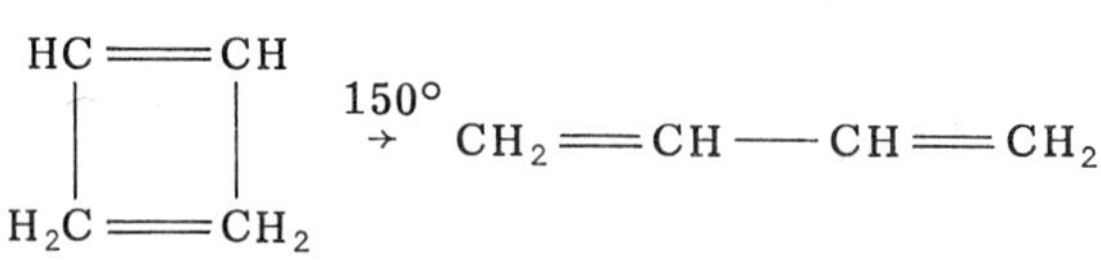

is first-order, and its specific reaction-rate constant was observed to be 2.0×10^{-4} sec^{-1}. Calculate the half-life of the reaction. What fraction of the cyclobutane will remain after 20.0 min?

[W. Cooper and W. D. Walters, *J. Am. Chem. Soc.*, *80*, 4220 (1958)]

17-35 The initial concentration of a carbon tetrachloride solution of N_2O_5 was 5.33 mol/L. H. Eyring and F. Daniels found that the specific reaction-rate constant was 6.74×10^{-4} min^{-1}. Calculate the concentration of N_2O_5 after 126 min.

[*J. Am. Chem. Soc.*, *52*, 1472 (1930)]

17-36 Determine the concentration of azomethane remaining after 33 min if the initial concentration was 1.21×10^{-2} mol/L and $k = 1.60 \times 10^{-2}$ min^{-1}.

[H. C. Ramsperger, *J. Am. Chem. Soc.*, *49*, 912 (1927)]

17-37 The reaction of chlorite ion with peroxymonosulfate ion was shown to be first-order in each of the reactant concentrations

$$ClO_2^- + HSO_5^- = ClO_3^- + SO_4^{2-} + H^+$$

What is the overall rate equation for the reaction?

[R. W. Johnson and J. O. Edwards, *Inorg. Chem.*, 5, 2073 (1966)]

17-38 The hydrolysis of C_6H_5COCl was studied by B. L. Archer and R. F. Hudson, who found that after 186 sec, 0.00084 mol/L had reacted, and that after 236 sec, 0.00102 mol/L had reacted. The initial concentration was 0.0100 mol/L. Calculate k for the reaction.

[*J. Chem. Soc.*, *1950*, 3259]

17-39 Determine the order of the reaction with respect to the methyl acetate concentration in the reaction

$$CH_3COOCH_3 + OH^- = CH_3CO_2^- + CH_3OH$$

The data are from the work of J. Walker. The initial concentration was 0.01000 M. Calculate k for this reaction.

t (min)	X (mol/L)
0	0
3	0.00260
4	0.00317
5	0.00366
6	0.00411
7	0.00450
8	0.00481
10	0.00536
12	0.00584
15	0.00637

Source: J. Walker, *Proc. Roy. Soc. (London), Ser. A, 78*, 157 (1906).

17-40 The rate of oxidation of I^- by ClO^- was studied by Y. T. Chia and R. E. Connick, at 25°C. Starting with 2.00×10^{-3} M I^- and ClO^-, the second-order rate constant was found to be 60.6 $(M/L)^{-1}$ sec^{-1}. Calculate the concentration of each of the ionic species after 5.00 sec.

$I^- + ClO^- = IO^- + Cl^-$

[*J. Phys. Chem., 63*, 1518 (1959)]

17-41 What is the half-life of the reaction in Problem 17-40?

17-42 Calculate the k for a reaction in which the initial concentration of the reactant was 0.500 M and the concentration after 360 min was 0.401 M.

[S. K. Ghosh, *Z. Physik. Chem. (Leipzig), 206*, 321 (1957)]

17-43 The initial rate of a reaction between A and B was found to be as follows:

Rate mol/dm^3S^{-1}	[A] mol/dm^3	[B] mol/dm^3
2.07×10^{-2}	1.54×10^{-2}	1.54×10^{-2}
8.28×10^{-2}	3.08×10^{-2}	1.54×10^{-2}
8.28×10^{-2}	3.08×10^{-2}	3.08×10^{-2}

What is the order of the reaction with respect to A, B, and the overall reaction?

17-44 In order to determine the order of a reaction, the concentration of a reactant (A) was followed as a function of time. In one trial the following results were obtained:

[A] mol/dm^3	0.0100	0.0085	0.0070	0.0055	0.0040
time (sec)	0	60	120	180	240

a. What is the order of the reaction?
b. What is the value of the rate constant?

17-45 The efficiency of a car battery decreases as the temperature drops. If we assume the power output of a brand-new battery is 100% at 80°F (300 K), then the power output drops to 60% at 32°F. What is the Arrhenius activation energy for a car battery? (Assume the power can be substituted for the rate constant.)

17-46 The rate at which a Co(III) complex was reduced by Fe(II) in water was recently measured. From the following data, determine the activation energy of the reaction.

Temp K	10^2 k S^{-1}
293	5.4
298	10.0

[I. Krack and R. van Eldik, *Inorg. Chem.*, *25*, 1743 (1986)]

18
COORDINATION COMPOUNDS

18-1 INTRODUCTION

A coordination compound is a compound containing one or more coordinate covalent bonds. These compounds consist of a central ion or atom bonded by coordinate covalent bonds to a definite number of atoms, molecules, or ions called ligands. The ligands are distributed about the central atom (or ion) in a definite spatial configuration.

The coordination number is the total number of coordinate covalent bonds formed between ligands and the central atom (or ion). A coordination compound, also called a complex, may be a neutral complex or a charged complex.

18-2 NAMING OF COORDINATION COMPOUNDS

A systematic method of naming complex compounds has been developed by the International Union of Pure and Applied Chemistry (IUPAC). A summary of the important rules is as follows:

1. If the complex is an ionic compound, the cation is named first, then the anion, with a space between the names.

2. Within the complex, first give the names of the ligands using the Greek prefixes *di*, *tri*, *tetra*, etc., followed by the metal ion with its oxidation number given in Roman numerals in parentheses.

Example 18-1

a. $[Ag(NH_3)_2]^+$ diamminesilver(I) ion

b. $[Cu(H_2O)_4]^{2+}$ tetraaquocopper(II) ion

c. $[Ni(CO)_4]$ tetracarbonylnickel(0)

d. $[Cr(NH_3)_6]Br_3$ hexaamminechromium(III) bromide

3. When more than one type of ligand is present in the complex, the ligands are named alphabetically according to the *names* of the ligand. The prefixes di, tri, etc., used to tell how many ligands of a given kind are present in a complex, are ignored in determining the alphabetic order. Negative ligands end in -o, neutral ligands are named as molecules except for H_2O, NH_3, CO, and NO (see Table 18-1), while positive ligands end in -ium. (Prior to 1971 the order of naming ligands was different from the rule listed above. Previously, negative ligands were named first, then neutral ligands, and then positive ligands.)

Example 18-2

a. $[Cu(NH_3)_2(H_2O)_2]^{2+}$ diamminediaquocopper(II) ion

b. $[Co(NH_3)_4Cl_2]^+$ tetraamminedichlorocobalt(III) ion

c. $[Pt(NH_3)_3CN]^+$ triamminecyanoplatinum(II) ion

4. If the ligands are complicated, enclose them in parentheses and use the prefixes *bis*, *tris*, *tetrakis*, etc.

Example 18-3

a. $[Cu(en)_2]^{2+}$ bis(ethylenediamine)copper(II) ion

b. $[Co(en)_3]^{3+}$ tris(ethylenediamine)cobalt(III) ion

5. If the complex is an anion, the metal name ends with -ate. For neutral complexes or cation complexes, the metal name is unchanged.

Example 18-4

a. $[Fe(CN)_6]^{4-}$ hexacyanoferrate(II) ion

b. $[Fe(CN)_6]^{3-}$ hexacyanoferrate(III) ion

c. $K_3[FeF_6]$ potassium hexafluoroferrate(III)

TABLE 18-1 Somd Common Ligands

Monodentate

F^-	fluoro	CN^-	cyano
Cl^-	chloro	SCN^-	S-thiocyanato or thiocyanato (S-bonded)
Br^-	bromo		
I^-	iodo	SCN^-	N-thiocyanato or isothiocyanato (N-bonded)
OH^-	hydroxo		
H_2O	aquo	NO_2^-	nitro (N-bonded)
CO	carbonyl	NO_2^-	nitrito (O-bonded)
NH_3	ammine	NO	Nitrosyl

Bidentate

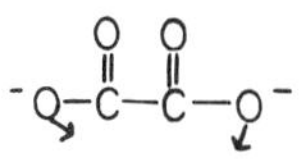

oxalato

$H_2NCH_2CH_2NH_2$

(en)

ethylenediamine

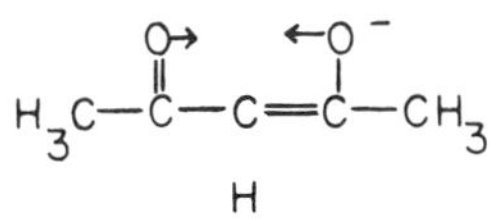

(acac)

acetylacetonato ion

Multidentate

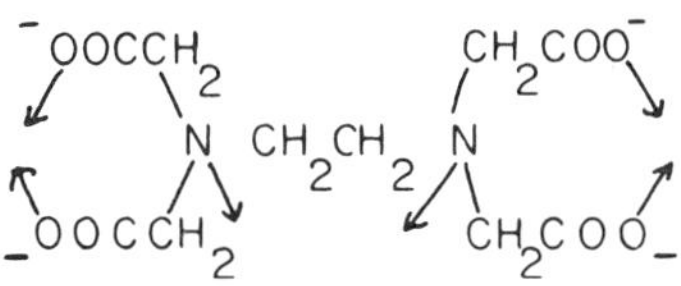

(EDTA)
anion
ethylenediaminetetraacetic acid
(hexadentate)

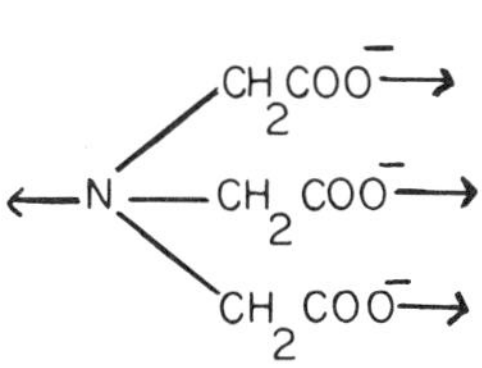

(NTA)
anion
nitrilotriacetic acid
(tetradentate)

18-3 CHELATES

The number of donor atoms that a ligand contains is called the denticity of the ligand. Ammonia (NH_3), which has one nitrogen to form a bond, is a monodentate ligand. Ethylenediammine has two nitrogen atoms to form coordinate bonds and is called a bidentate ligand, while ethylenediaminetetraacetic acid, which has four oxygen and two nitrogen atoms to form coordinate bonds, is called a hexadentate ligand. Ligands that form two or more coordinate bonds with a given metal ion are also called polydentate ligands. A polydentate ligand that is capable of forming two or more coordinate bonds with the same central metal ion is called a *chelating agent* and the resulting complex is known as a *chelate*. The complexes in Example 18-3 are chelate complexes. Table 18-1 is a list of some common ligands. The arrows in the bi- and multidentate ligands show the donor atoms for complex formation.

Example 18-5

Draw a schematic structure of the complex formed with EDTA chelates with Ca^{2+}.

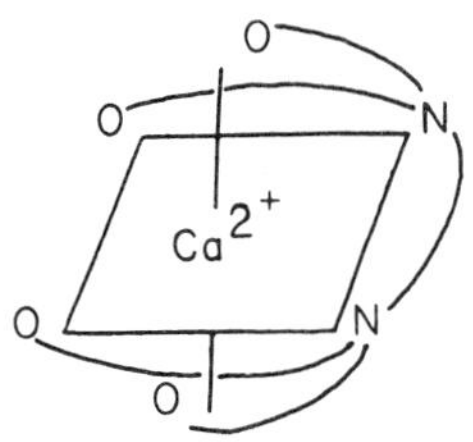

18-4 GEOMETRY AND COMPLEXES

In Chapter 6 the geometry of molecules was discussed. The shape of the complex for coordination compounds in which the neutral metal ion has no unpaired electrons is related to the number of ligands around the metal ion: e.g., diamminesilver(I) ion is linear corresponding to two bonding electron pairs and tetraaquoberyllium(II) is tetrahedral due to four electron pairs. For those complexes in which the central metal ion has unfilled d or f orbitals, the geometry is not directly related to the coordination number (Table 18-2).

TABLE 18-2 Coordination Number, Geometry, and Hybridization of Some Representative Complex Ions

Coordination number	Geometry	Hybridization	Example
2	Linear	sp	$Ag(NH_3)_2^+$
4	Tetrahedral	sp^3	$Ni(CO)_4$
4	Square planar	dsp^2	$Ni(DMG)_2$
5	Trigonal bipyramid	d^2sp^2	$Fe(CO)_5$
6	Octahedron	d^2sp^3	$Cr(NH_3)_6^{3+}$

18-5 GEOMETRICAL ISOMERISM

In square planar complexes, isomers are possible when two different types of ligands are present in a complex, owing to different spatial distributions of the ligands. These are called *cis* and *trans* isomers. In the cis-form, like ligands are adjacent to one another, while in the trans-form they are opposite one another.

Example 18-6

Draw the cis and trans isomers of square planar diamminedichloroplatinum(II).

cis form

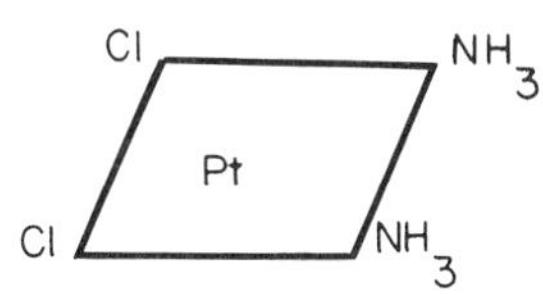

trans form

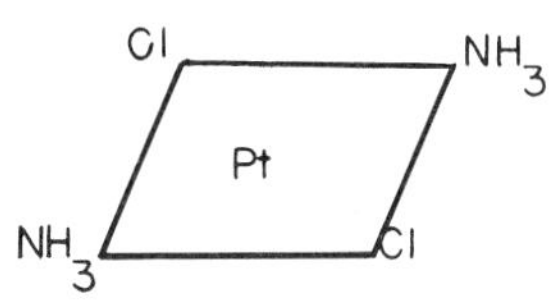

For octahedral complexes, cis and trans isomers can exist if two or more different ligands are present in the complex.

Example 17-8

B. M. Fung has recently determined the NMR spectra of cis- and trans-dichlorobis(ethylenediamine)cobalt(III) chloride. Draw the structure of each of these molecules.

[*J. Am. Chem. Soc.*, *89*, 5788 (1967)]

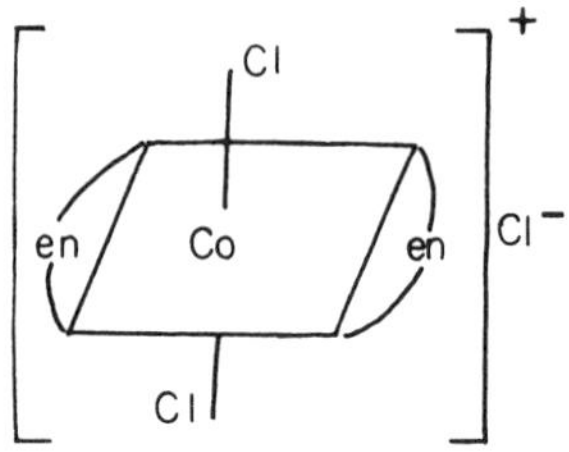

Trans

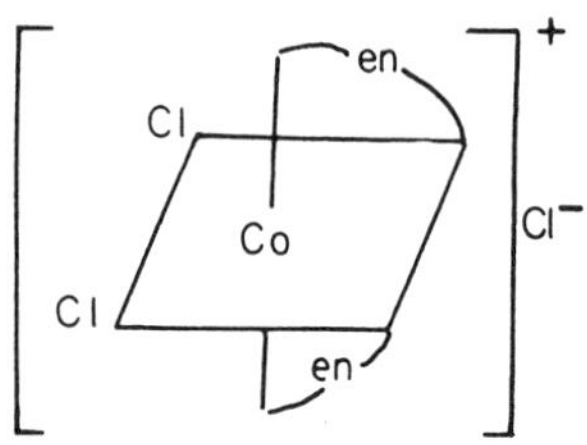

Cis

18-6 VALENCE BOND THEORY

Valence bond theory describes the behavior and properties of coordination complexes using hybridized orbitals from the central metal ion, as given in Table 18-2. Hund's rule is applied to the electrons in nonbonding orbitals. Those structures with unpaired electrons are paramagnetic, while those with no unpaired electrons are diamagnetic.

Example 18-8

Draw the valence bond diagram for the tetrachloronickelate(II) ion. This complex ion has a magnetic moment corresponding to two unpaired electrons. What is the structure of the ion? Is the ion paramagnetic or diamagnetic?

	3d	4s	4p	4d
Ni^{+2} ion	↑↓ ↑↓ ↑↓ ↑ ↑	_	_ _ _	_ _ _ _ _
$[Ni(Cl)_4]^{2-}$	↑↓ ↑↓ ↑↓ ↑ ↑	↑↓	↑↓ ↑↓ ↑↓	_ _ _ _ _

The $[Ni(Cl)_4]^{2-}$ ion is an sp^3 complex; hence the structure is tetrahedral. There are two unpaired electrons; so the ion is paramagnetic.

Example 18-9

The tetracyanonickelate(II) ion is diamagnetic. What would be its structure?

	3d	4s	4p	4d
Ni^{+2}	$\uparrow\downarrow$ $\uparrow\downarrow$ $\uparrow\downarrow$ $\uparrow$ $\uparrow$	__	__ __ __	__ __ __ __ __
$[Ni(CN)_4]^{2-}$	$\uparrow\downarrow$ $\uparrow\downarrow$ $\uparrow\downarrow$ $\uparrow\downarrow$ $\uparrow\downarrow$	$\uparrow\downarrow$	$\uparrow\downarrow$ $\uparrow\downarrow$ __	__ __ __ __ __

This ion is a dsp^2 complex and is square planar. There are no unpaired electrons in the structure.

The tetrachloronickelate(II) ion that has the same number of unpaired electrons as does the nickel(II) ion is called a *high-spin* complex. The tetracyanonickelate(II) ion that has fewer unpaired electrons than the nickel(II) ion is called a *low-spin* complex.

Example 18-10

The hexacyanoferrate(III) ion has a magnetic moment corresponding to one unpaired electron, while the hexafluoroferrate(III) ion has a magnetic moment corresponding to five unpaired electrons. Draw a valence bond diagram to account for the difference in these two structures.

	3d	4s	4p	4d
Fe^{3+}	$\uparrow$ $\uparrow$ $\uparrow$ $\uparrow$ $\uparrow$	__	__ __ __	__ __ __ __ __
$Fe(CN)_6{}^{3-}$	$\uparrow\downarrow$ $\uparrow\downarrow$ $\uparrow$ $\uparrow\downarrow$ $\uparrow\downarrow$	$\uparrow\downarrow$	$\uparrow\downarrow$ $\uparrow\downarrow$ $\uparrow\downarrow$	__ __ __ __ __
$Fe(F)_6{}^{3-}$	$\uparrow$ $\uparrow$ $\uparrow$ $\uparrow$ $\uparrow$	$\uparrow\downarrow$	$\uparrow\downarrow$ $\uparrow\downarrow$ $\uparrow\downarrow$	$\uparrow\downarrow$ $\uparrow\downarrow$ __ __ __

The hexacyanoferrate(III) ion is a low-spin complex while the hexafluoroferrate(III) ion is a high-spin complex. Both are octahedral.

18-7 CRYSTAL FIELD AND LIGAND FIELD THEORY

Crystal field theory, and an extension of this theory known as ligand field theory, describes the properties of coordination complexes in terms of the d-valence orbitals of the central atom.

In crystal field theory the ligands are treated as electrostatic charges. In ligand field theory refinements have been made to account for neutral molecules as ligands. Both give the same type of splitting of the d orbitals.

The d orbitals in the uncoordinated metal ion all have the same energy. Under the influence of ligands the d orbitals are no longer equivalent. The way the d orbitals split depends on the geometry of the ligands. In an octahedral field the d_{z^2} and $d_{x^2-y^2}$ orbitals (known as e_g orbitals) are equivalent to each other while the d_{xy}, d_{xz}, and d_{yz} orbitals (known as the t_{2g} orbitals) are equivalent to each other. Under the influence of ligands, the e_g orbitals are higher than the t_{2g} orbitals. The difference in energy between these two sets of orbitals is called Δ_o. The stronger the interaction of the ligand with the metal ion, the larger the value of Δ_o.

In a tetrahedral field the relative positions of the d orbitals are reversed. The e_g orbitals now have a lower energy than the t_{2g} orbitals. The difference in energy is called Δ_t.

Example 18-11

Explain the difference in the hexocyanoferrate(III) ion and the hexafluoroferrate(III) ion using ligand field theory.

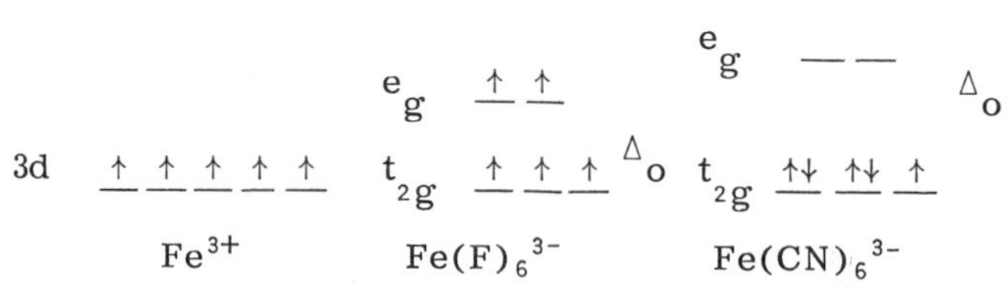

Since energy is needed to pair electrons, in the $Fe(F)_6^{3-}$ complex the splitting (Δ_o) is not large enough to have the electrons pair off, while in $Fe(CN)_6^{3-}$, Δ_o is large enough so that the electrons do force-pair.

For equilibrium constants involving complex ions see Chapter 13, Sections 5 and 6.

PROBLEMS

18-1 Name the following complexes:

a. $[Cu(NH_3)_4]^{2+}$
b. $K_4[NiF_6]$
c. $[Hg(SCN)_4]^{2-}$ (S-bonded)
d. $[Co(SCN)_4]^{2-}$ (N-bonded)
e. $Ni(CO)_4$
f. $[Co(en)_3]_2(SO_4)_3$
g. $K_2[PtCl_3F_3]$
h. $[Co(NH_3)_3Cl_3]^0$

18-2 Write the formula for each of the following complex compounds.

a. Tetraaminecadmium(II) nitrate
b. Hexamminenickel(II) chloride
c. Hexacarbonylchromium(0)
d. Tetraamminedichlorocobalt(III) nitrate
e. Sodium dichlorodifluoroplatinate(II)
f. Tris(ethylenediamine)rhodium(III) chloride
g. Hexamminechromium(III) pentachlorocuprate(II)

18-3 Name the following:

(a)

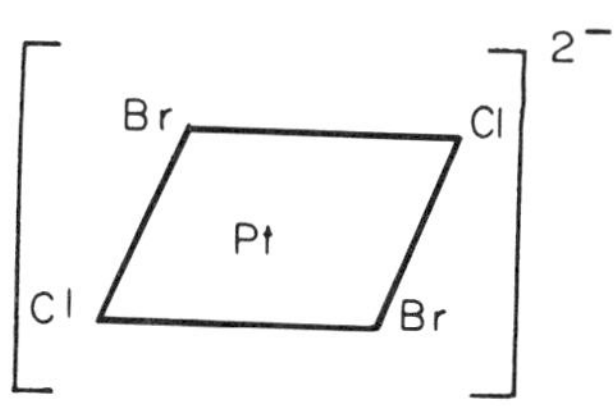

(b)

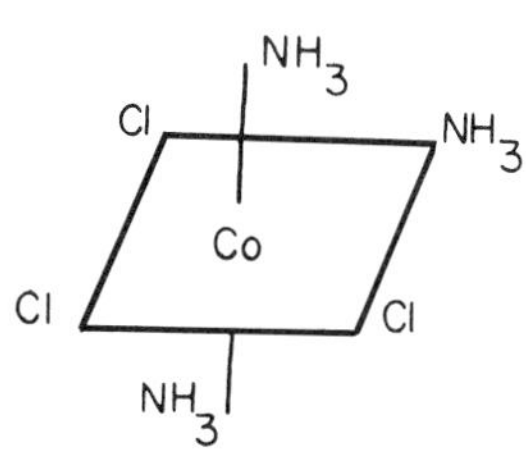

(c)

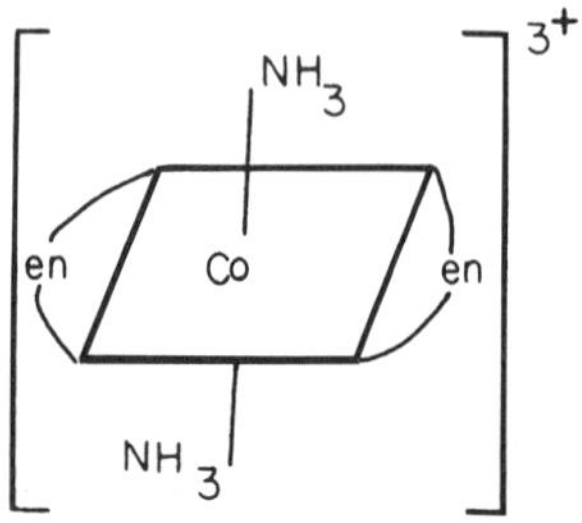

18-4 Draw the structure of each isomer of $[PtClFBrI]^{2-}$.

18-5 Basolo et al. gave a method of preparing each of the following compounds. Name them.

a. $[Cr(NH_3)_6][CuCl_5]$

b. $[Cr(NH_3)_6][Ni(CN)_5]$

c. $[Co(NH_3)_6][FeCl_6]$

d. $[Cr(en)_3][Ni(CN)_5]$

[*Inorg. Syn.*, *11*, 47 (1968)]

18-6 When $NbCl_5$ is dissolved in acetonitrile and KCNS is added, a deep-blue niobium salt is formed. The salt contained K, 8.4%; Nb, 19.1%, S, 39.8%, C, 15.1%; and N, 17.3%. An IR spectrum showed the ligands are N-bonded. What is the formula and name of this compound?

[T. M. Brown and G. F. Knox, *J. Am. Chem. Soc.*, *80*, 5296 (1967)]

18-7 When hexammine cobalt(II) dinitrate is heated under vacuum at 100°C for 24 hr, a compound having the following composition is obtained; Co, 26.93%, N(total), 25.76%; N(ammine), 12.86%; H, 2.79%; and the rest oxygen. What is the empirical formula of this complex compound? Name the compound.

[G. L. McPherson, J. A. Weil, and J. K. Kinnaird, *Inorg. Chem.*, *10*, 1574 (1971)]

18-8 When 1.5 g of dichloroethylenediamineplatinum(II) was treated with 30% H_2O_2 in water, 1.5 g of dichloroethylenediamine-dihydroxyplatinum(IV) was formed. What was the percent yield?

[F. Basolo, J. Bailar, and B. Tarr, *J. Am. Chem. Soc.*, *72*, 2437 (1950)]

18-9 A solution of hexachloroplatinic acid was made by dissolving 4.5 g of pure platinum in aqua regia. This solution was evaporated almost to dryness and concentrated HCl was added. This was evaporated to dryness; then 10 ml of distilled water was added. 8.0 g of NaCl was added to the solution and the solution was cooled. 4.5 g of solid yellow-orange disodium hexachloroplatinate(IV) was recovered. What was the percent yield?

[L. Cox and D. Peters, *Inorg. Syn.*, *13*, 174 (1972)]

18-10 Potassium thiocyanate (5.83 g) was added to niobium pentachloride (2.70 g) in acetonitrile. The solution was warmed (60–70°) and stirred. After purification 4.0 g of a deep-blue crystalline solid was obtained. The ligands were N-bonded. Analysis of the solid: K, 8.44%; Nb, 19.06%; N, 17.25%; C, 15.10%; and S, 39.79%.

a. What is the formula of the compound?
b. What is the name of the compound?
c. What was the percent yield?

[G. Knox and T. M. Brown, *Inorg. Syn.*, *13*, 226 (1972)]

18-11 $[V(CO)_6]^0$ has a magnetic moment corresponding to one unpaired electron. Draw a valence bond structure for this molecule. What is the geometrical structure for the complex?

18-12 Classify each of the following compounds as low-spin or high-spin complexes.

Compound	Number of unpaired electrons
$[Co(NH_3)_5Cl]Cl_2$	0
$K_4[Cr(CN)_6]$	2
$K_3[Fe(CN)_6]$	1
$[Pt(NH_3)_6]Cl_4$	0
$(NH_4)_2[FeF_5H_2O]$	5

[L. Pauling, *J. Am. Chem. Soc.*, *53*, 1367 (1931)] (This is the original paper on valence bond theory and is worth reading.)

18-13 $[Ti(H_2O)_6]^{3+}$ exists in solution as a reddish-violet complex. Δ_o is 20,300 cm^{-1} for the complex. Magnetic measurements show one unpaired electron.

a. Draw a valence bond diagram of this complex to explain the above information.
b. Draw a ligand field diagram.
c. Show that the reddish-violet color is reasonable from the Δ_o value.

18-14 The following Δ_t values were obtained for tetrahalocobaltate(II) ions in nitromethane.

Ion	$\Delta_t(cm^{-1})$
$CoCl_4^{2-}$	3214
$CoBr_4^{2-}$	2987
CoI_4^{2-}	2830

Arrange the ligands in order of decreasing ligand field strength. [F. A. Cotton, D. M. L.. Goodgame, and M. Goodgame, *J. Am. Chem. Soc.*, *83*, 4690 (1961)]

Supplementary Problems

18-15 Name the following complexes.

a. $[Cd(NH_3)_4]^{2+}$
b. $[Ag(CN)_2]^-$
c. $[Co(NH_3)_2Cl_4]^-$
d. $K_4[Mo(CN)_8]$
e. $Fe(CO)_5$
f. $[Co(NH_3)_5(H_2O)](ReO_4)_3$
g. $[Ir(NH_3)_5Cl]Cl_2$
h. $[Ru(en)_2Cl_2]Cl$
i. $[Cr(en)_3][Ni(CN)_5]$

18-16 Write the formula for each of the following complex compounds.

a. Tetraamminechloronitrocobalt(III) chloride
b. Tetraamminedichlorocobalt(III) nitrite
c. Tricarbonylnitrosylcobalt(0)
d. Potassium tetracyanozincate(II)
e. Pentaamminechlororuthenium(III) chloride
f. Tris(ethylenediamine)chromium(III) sulfate
g. Potassium octacyanomolybdate(IV)

18-17 Name the following.

(a)

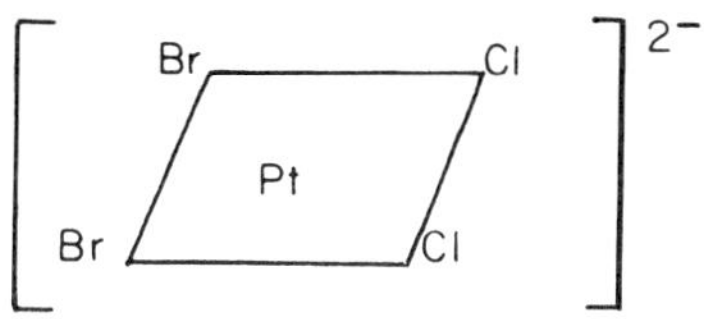

(b)

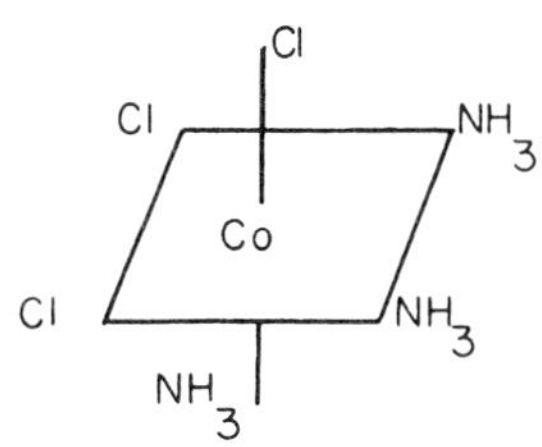

(c)

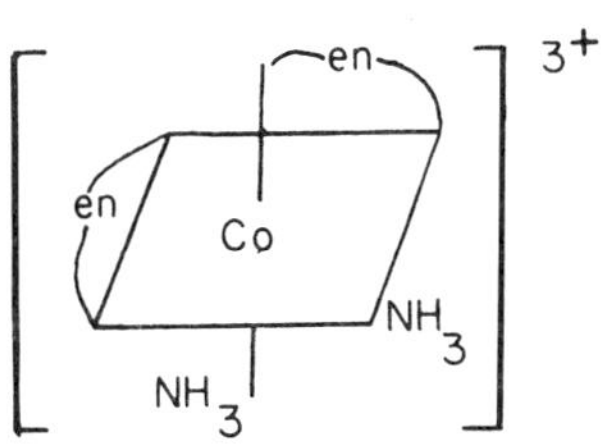

18-18 Name the following compounds prepared by McCutcheon and Schuele.

a. $[Co(H_2O)(NH_3)_5][Co(C_2O_4)_3]$

b. $[Co(H_2O)(NH_3)_5][Cr(SCN)_4(NH_3)_2]_3$ (S-bonded)

c. $[Co(H_2O)NH_3)_5][Co(NO_2)_4(NH_3)_2]_3$

[*J. Am. Chem. Soc.*, *75*, 1845 (1953)]

18-19 A complex cobalt compound was recently isolated having the following percent composition: Co, 16.2; C, 20.1; H, 5.60; N, 26.9; and the rest O. What is the empirical formula of the compound?

[D. Lambert and J. Mason, *J. Am. Chem. Soc.*, *88*, 1637 (1966)]

18-20 When $TaCl_5$ is dissolved in acetonitrile and KCNS is added, an orange compound is formed having the following percent composition; K, 7.2; Ta, 32.1; S, 33.6; C, 12.5; N, 14.6. Infrared analysis showed the ligands are N-bonded. What is the formula and name of this compound?

[T. M. Brown and G. F. Knox, *J. Am. Chem. Soc.*, *89*, 5296 (1967)]

18-21 A blue solid precipitated upon the addition of a solution of cobaltous thiocyanate (1.54 g) in hot absolute ethanol to a solution of tetramethylammonium thiocyanate (2.46 g) in hot absolute ethanol. The yield of the complex was 3.50 g. Analysis found: C, 32.62%; H, 5.52%; N, 19.05%; Co, 13.4%; and the rest S. X-ray analysis showed the ligands were N-bonded.

a. What is the formula of the compound?
b. What is the name of the compound?
c. What was the percent yield?

[F. A. Colton, D. M. L. Goodgame, M. Goodgame, and A. Sacco, *J. Am. Chem. Soc.*, *83*, 4157 (1961)]

18-22 When 2.5 g of ammonium aquopentachloromolybdate(III) and 0.5 g of ammonium chloride are added to 12 M HCl, and HCl is bubbled through the solution, 1.95 g of ammonium hexachloromolybdate(III) is recovered. What is the percent yield?

[S. V. Brencic and F. A. Cotton, *Inorg. Syn.*, *13*, 172 (1972)]

18-23 When chlorine is bubbled through 3.9 g of dichloroethylenediamineplatinum(II) for 45 min, 4.4 g of yellow tetrachloroethylenediamineplatinum(IV) is produced. What is the percent yield?

[F. Basolo, J. Bailor, and B. Tarr, *J. Am. Chem. Soc.*, *72*, 2437 (1950)]

18-24 According to magnetic susceptibility measurements, $[CoF_6]^{3-}$ is a high-spin complex, while $[Co(H_2O)_6]^{3+}$ is a low-spin complex. How many unpaired electrons are present in each complex?

[H. L. Friedman, J. P. Hunt, R. H. Plane, and H. Taube, *J. Am. Chem. Soc.*, *73*, 4028 (1951)]

18-25 Recently the visible and uv spectra of a series of ruthenium compounds were measured. Δ_o for the octahedral complexes was as follows:

Complex	$\Delta_o(cm^{-1})$
$[Ru(H_2O)_6]^{2+}$	19,800
$[Ru(en)_3]^{2+}$	28,100
$[Ru(pn)_3]^{2+}$	28,400
$[Ru(dien)_2]^{2+}$	28,800
$[Ru(NH_3)_6]^{2+}$	27,100

where en = ethylenediamine, pn = propylenediamine, and dien = diethylenediamine. Name each of these complexes and arrange the ligands in order of decreasing ligand field strength. Such an order is known as a spectrochemical series.

[H. Schmidtke and F. Garthoff, *Helv. Chem. Acta*, *49*, 2039 (1966)]

18-26 The directions for the preparation of hexammine cobalt(II) chloride call for the use of a freshly opened bottle of carbon black as a catalyst. A student followed the directions carefully and obtained a light-tan solid. A 0.2250-g sample of the product was dissolved in water and silver nitrate was added to the solution. The silver chloride that was formed weighed 0.3620 g. Does this result support the formula of the product expected?

18-27 Another student did not follow the directions described in Problem 18-26 and used a bottle of carbon black that had been in the storeroom for some time and had been used several times previously. He obtained a purple solid as the product. Treatment of a 0.2250-g sample of the product with silver nitrate gave 0.2413 g of silver chloride.

a. What is the formula of the compound he obtained?
b. Name the compound.

18-28 Draw a diagram representing the structure of the complex ion in Problem 18-27.

18-29 A general chemistry laboratory experiment called for the preparation of tetramminecopper(II) sulfate tetrahydrate.

a. What is the formula of the compound?
b. Draw a diagram representing its structure.

18-30 How many grams of lead sulfate could be obtained by adding an excess of lead acetate to 0.500 g of the compound in Problem 18-29?

18-31 Complex ions are common in the qualitative analysis scheme. To a mixture of copper(II), cadmium(II), and bismuth ions concentrated ammonia was added. A precipitate of bismuth hydroxide was formed, but the copper and cadmium remained in solution. What was the formula of the copper ion that remained in the solution?

18-32 Which of the two compounds in Problems 18-26 and 18-27 would have the greater molar conductivity in aqueous solution?

19
CHEMISTRY OF THE ELEMENTS

HYDROGEN

19-1 The average abundance of deuterium in natural hydrogen is 0.0145%. The masses of ^1H and ^2H are 1.007825 and 2.014102 amu, respectively. Calculate the atomic mass of hydrogen.
[A. E. Cameron and E. Wickers, *J. Am. Chem. Soc.*, *84*, 4181 (1962)]

19-2 Calculate the quantity of electricity required to produce 1.00 m^3 of hydrogen gas (at STP) by the electrolysis of water.

19-3 The equilibrium constant K_p for the reaction $CO_2(g) + H_2(g) \rightleftharpoons CO(g) + H_2O(g)$ was found to be 1.57 by W. Nernst and F. Haber. Calculate the equilibrium partial pressure of each gas if the initial pressure of CO_2 was 1.00 atm and that of H_2 was 1.00 atm.
[*Z. Anorg. Chem.*, *38*, 5 (1904)]

19-4 Calculate the pH of a 0.0316 M solution of formic acid, HCOOH.

19-5 What volume of hydrogen (at 27°C and 190 torr) would result from the hydrolysis of 2.10 g of calcium hydride?

19-6 Calculate the average bond-dissociation energy of the Si–H bonds in SiH_4.

19-7 E. W. Morley obtained 34.1559 g of water from the reaction of 3.8211 g of hydrogen with oxygen. What mass of oxygen was required for the reaction?
[*Am. Chem. J.*, *17*, 267 (1895)]

19-8 Determine the average O—H bond-dissociation energy in water. Compare the average bond-dissociation energy with the bond-dissociation energy $H_2O(g) \rightarrow H(g) + OH(g)$ and $OH(g) \rightarrow O(g) + H(g)$.

19-9 A bulb containing 0.0470 g of acetic acid vapor at 60.6°C had a pressure of 28.22 mm. The volume of the bulb was 359.8 mL. Calculate the molecular mass of the vapor and account for the result.
[M. D. Taylor, *J. Am. Chem. Soc.*, *73*, 315 (1951)]

19-10 What is the hydronium ion concentration and pH of a 0.125% solution of hydrochloric acid the density of which is 1.0623 g/mL?

19-11 The density of trifluoroacetic acid vapor at 81.2%C and 207.5 mm was 1.555 g/L. Calculate the molecular mass of the vapor and account for the result.
[R. E. Lundin, F. E. Harris, and L. K. Nash, *J. Am. Chem. Soc.*, *74*, 4654 (1952)]

19-12 Calculate the emf of each of the following cells:

a. H_2(1 atm), Pt|H_3O^+(1 m)‖Cl^-(1 m), AgCl(s)| Ag

b. H_2(0.100 atm), Pt|H_3O^+(0.0200 m)‖Cl^-(0.0200 m), AgCl(s)| Ag

19-13 Calculate the pH of a 0.212 M solution of nitrous acid.

19-14 Calculate the emf of a cell consisting of a normal hydrogen electrode and a silver-saturated solution of silver chloride electrode.

19-15 The equilibrium constant for the formation of ammonia from the elements $1/2N_2(g) + 3/2H_2(g) \rightleftharpoons NH_3(g)$ is 6.59×10^{-3} atm^{-1}. Find the partial pressure of ammonia if the pressure of $H_2 = N_2$ is 2.12 atm.

OXYGEN

19-16 Calculate the atomic mass of oxygen from the isotopic abundances of atmospheric oxygen: ^{17}O = 0.0374%, ^{18}O = 0.2039%. The nuclidic masses are ^{16}O = 15.994915, ^{17}O = 16.999134, and ^{18}O = 17.999160 amu.
[A. O. C. Nier, *Phys. Rev.*, *77*, 789 (1950)]

19-17 Oxygen gas was passed over copper metal. The volume of oxygen (at STP) that was consumed was 17.8 mL, and the weight of copper used was 0.1019 g. Was the reaction quantitative?
[M. G. Habashy, *Anal. Chem.*, *36*, 1782 (1964)]

19-18 When oxygen gas was passed over 0.1013 g of pulverized tin, 13.5 mL of oxygen was consumed. What percent of the tin reacted?
[M. G. Habashy, *Anal. Chem.*, *36*, 1782 (1964)]

19-19 For the reaction $2H_2O(g) \rightleftharpoons 2H_2(g) + O_2(g)$, $K_p = 4.5 \times 10^{-21}$ at 1000 K. Calculate the equilibrium partial pressures of hydrogen and oxygen if 1.00 mol of water is heated to 1000 K in a 1.00-L vessel.

19-20 A hydrate was analyzed and found to contain 18.05% CaO, 64.66% CrO_3, and 17.43% H_2O. What is the formula of the hydrate?
[W. H. Hartford, K. A. Lane, and W. A. Meyer, Jr., *J. Am. Chem. Soc.*, *72*, 3353 (1950)]

19-21 Calculate the standard enthalpy of the reaction $2H_2O_2(\ell) \rightarrow 2H_2O(\ell) + O_2(g)$.

19-22 Calculate the average bond-dissociation energy of the O–F bonds in OF_2.
[S. W. Benson, *J. Chem. Educ.*, *42*, 517 (1965)]

19-23 Calculate the standard enthalpy for the reaction $2F_2(g) + H_2O(\ell) \rightarrow 2HF(g) + OF_2(g)$.

19-24 Find E° at 25°C for the cell $Pt|Fe^{2+}, Fe^{3+}\|H_3O^+Pt, O_2$.

19-25 Calculate the volume of oxygen expected from the passage of a current of 1.25 amp for 4.00 hr through a solution of sodium fluoride.

19-26 An equimolar mixture of CO(g) and O_2(g) is contained in a bulb at a total pressure of 1.00 atm. A spark is passed through the gas and a reaction occurs to form CO_2(g). What is the final pressure in the bulb?

19-27 The standard oxidation potential of water is -1.23 V; $2H_2O = O_2(g) + 4H^+ + 4e^-$. Calculate the potential for the half-cell O_2(1 atm) | H_2O (pH = 7.00).

19-28 Calculate the mass of copper that can be obtained from the reduction of 1.25181 g of copper(II) oxide with hydrogen.
[R. Ruer and K. Bode, *Z. Anorg. Allgem. Chem.*, *137*, 101 (1924)]

19-29 What is the formula of the oxide of tungsten formed by the reaction of 2.44588 g with oxygen if the product weighed 3.08318 g?
[G. E. Thomas, *J. Am. Chem. Soc.*, *21*, 373 (1899)]

19-30 A sample of nitric oxide weighing 0.61862 g was catalytically decomposed with nickel. Calculate the mass of oxygen and nitrogen anticipated.
[R. W. Gray, *J. Chem. Soc.*, *87*, 1616 (1905)]

ALKALI METALS

19-31 Calculate the atomic mass of lithium from the ratio LiCl/Ag, 0.392992.
[T. W. Richards and H. H. Willard, *J. Am. Chem. Soc.*, *32*, 4 (1910)]

19-32 A 0.0779 m solution of C_2H_5Li in benzene had a freezing-point depression of 0.065°. What conclusion may be drawn from this experiment?
[T. L. Brown and M. T. Rogers, *J. Am. Chem. Soc.*, *79*, 1861 (1957)]

19-33 In a determination of atomic masses, 4.24171 g of lithium chloride was dissolved in perchloric acid. When the solution was evaporated to dryness, 10.64596 g of lithium perchlorate was collected. Compare the result of this experiment with the theoretical yield of lithium perchlorate expected.
[T. W. Richards and H. H. Willard, *J. Am. Chem. Soc.*, *32*, 47 (1910)]

19-34 The measured enthalpy of hydrolysis of sodium hydride was found to be -128.2 kJ mol^{-1}. The measured enthalpy of reaction of sodium metal with water was -185.1 kJ mol^{-1}. Calculate the enthalpy of formation of sodium hydride.

[C. E. Messer, L. G. Fasolino, and C. E. Thalmayer, *J. Am. Chem. Soc.*, *77*, 4525 (1955)]

19-35 The measured heat of hydrolysis of lithium hydride was -132.9 kJ mol^{-1}. The measured heat of the reaction of lithium metal with water was -222.2 kJ mol^{-1}. Calculate the heat of formation of lithium hydride.

[C. E. Messer, L. G. Fasolino, and C. E. Thalmayer, *J. Am. Chem. Soc.*, *77*, 4525 (1955)]

ALKALINE-EARTH METALS

19-36 Determine the atomic mass of calcium from an observation that 0.387200 g of calcium chloride yields 1.00000 g of silver chloride.

[O. Honigschmid and K. Kempter, *Z. Anorg. Allgem. Chem.*, *195*, 1 (1931)]

19-37 Calculate the volume of carbon dioxide, measured at 17°C and 280 torr, that can be obtained by heating 0.480 g of calcium carbonate.

19-38 Find the barium ion concentration of a solution that is also 0.100 M in sodium sulfate.

19-39 Calculate the standard enthalpy of the reaction

$$MgO(s) + CO_2(g) = MgCO_3(s) \qquad (\Delta H°_f = -144, -94.0, -266 \text{ kcal mol}^{-1})$$

19-40 What volume of hydrogen gas (at STP) can be formed by the reaction of 0.682 g of calcium hydride with water?

BORON AND ALUMINUM

19-41 The isotopic composition of boron in nature is variable. For a series of minerals the $^{11}B/^{10}B$ ratio varied from 3.95 to 4.10. Calculate the atomic mass corresponding to these ratios and compare them with the accepted value for the atomic mass of boron.

[A. E. Cameron and E. Wichers, *J. Am. Chem. Soc.*, *84*, 4182 (1962)]

19-42 A compound containing boron and chlorine was prepared by A. Stock and O. Priess. A sample weighing 0.1319 g gave 0.4850 g of silver chloride after reaction with silver nitrate. A sample weighing 0.3671 g occupied a volume of 149.8 cc at 379 mm and 19°C. What were the name and structure of the compound?

[*Chem. Ber.*, *47*, 3109 (1914)]

19-43 Boron was heated in a stream of bromine vapor at red heat. A sample of the product weighing 0.4398 g was dissolved in 15.99 g of benzene. The solution had a freezing-point depression of 0.542°C. Calculate the molecular mass of the compound.

[A. Stock and E. Kuss, *Chem. Ber.*, *47*, 3113 (1914)]

19-44 A sample of the compound prepared as in Problem 5-27 that weighed 67.0 mg exerted a pressure of 435 mm in a container whose volume was 93.2 cc at 0°C. What was the molecular mass of the vapor?

[A. Stock, E. Kuss, and O. Priess, *Chem. Ber.*, *47*, 3118 (1914)]

19-45 A boron hydride, mp 177 to 178.5°, that was stable in air was prepared by A. R. Pitochelli. The compound contained 89.11% B and 10.44% H. Its molecular mass from boiling-point-elevation experiments was 212.7. When it was decomposed to form the elements, 1.98×10^{-4} mol gave 2.21×10^{-3} mol of hydrogen. What is the formula of the hydride?

[*J. Am. Chem. Soc.*, *84*, 3218 (1962)]

19-46 A mixture of boric acid and urea was heated in an atmosphere of ammonia over the temperature range 500–950°C. The product formed contained 44.2% B and 56.0% N. What was it?

[J. Thomas, N. E. Weston, and T. E. O'Connor, *J. Am. Chem. Soc.*, *84*, 4619 (1962)]

19-47 Calculate the molecular mass of the fluoride of boron whose vapor density at standard conditions was 3.00 g/L, as determined by J. Dumas.

[*Traite Chim.*, *1*, 382 (1828)]

19-48 What is the shape of the molecules of the compound in Problem 19-47?

19-49 BF_3 and NH_3 gases were mixed in a glass bulb. A white solid formed and on analysis was found to contain 19.93% NH_3 and 12.72% B. A solution containing 8.15 g of the white solid in 1000 g of water had a freezing-point depression of 0.00179°C. What were the formula and structure of the white solid?

[A. W. Laubengayer and G. F. Condike, *J. Am. Chem. Soc.*, *70*, 2274 (1948)]

19-50 Boron trichloride was passed through an electric discharge, and a product that had a melting point of -92.6°C was formed. A sample of the compound was found to contain 0.0878 g of Cl and 0.0131 g of B. A 0.0810-g sample exerted a pressure of 48.9 mm at 26.5° in a volume of 191.6 cc. What was the formula of the compound?

[G. Urry, T. Wartik, R. E. Moore, and H. I. Schlesinger, *J. Am. Chem. Soc.*, *76*, 5296 (1954)]

19-51 A 0.0995-g sample of the compound described in Problem 19-50 produced 13.26 cc of hydrogen gas (at STP) when it reacted with sodium hydroxide. Sodium borate, sodium chloride, and water were the other products. Was the quantity of hydrogen produced consistent with the formula of the compound as found in Problem 19-50?

[G. Urry, T. Wartik, R. E. Moore, and H. I. Schlesinger, *J. Am. Chem. Soc.*, *76*, 5296 (1954)]

19-52 Gallium metal and iodine were heated at 350° for 24 hr; a lemon-yellow solid containing 15.56% gallium and 84.45% iodine was formed. The lemon-yellow compound was then heated in an atmosphere of oxygen for 2 hr at 300–350°; a white solid containing 32.34% gallium and 59.20% iodine was formed. When fluorine was passed over the white solid for 30 min at room temperature, a compound containing 66.42% gallium and 18.90% fluorine was formed. Write equations for the three reactions that occurred.

[F. M. Brewer, P. L. Goggin, and G. S. Reddy, *J. Inorg. Nucl. Chem.*, *28*, 361 (1966)]

19-53 The enthalpy of formation of aluminum fluoride (-1488 kJ mol^{-1}) was determined from the reaction of aluminum with lead(II) fluoride; calculate the enthalpy change of the reaction.

$\Delta H°_f$ (PbF_2) = -663.2 kJ mol^{-1}

[P. Gross, C. Hayman, and D. L. Levi, *Trans. Faraday Soc.*, *50*, 477 (1954)]

19-54 Assume ideal behavior and calculate the pressure exerted by 1.34 g of aluminum chloride vapor in a 200-mL bulb at 180°.

19-55 A. Stock and C. Massanex prepared a boron hydride that contained 81.7% B and 18.3% H. When 2.0 cc of the gas reacted with water, 21.6 cc of H_2 was formed. Do these observations support the opinion that the compound was B_4H_{10}, which reacted with water according to the equation

$$B_4H_{10} + 12H_2O = 4B(OH)_3 + 11H_2$$

[*Chem. Ber.*, *45*, 3559 (1912)]

CARBON AND SILICON

19-56 A sample of a compound containing 69.7% sulfur and weighing 0.1221 g was allowed to react with water. The white precipitate of silicon dioxide that was formed weighed 0.0794 g after drying. What was the compound?

[H. Gabriel and C. Alvarez-Tostado, *J. Am. Chem. Soc.*, *74*, 263 (1952)]

19-57 What are the formulas, Lewis structures, and shapes of the molecules of the two compounds whose analyses were 5.28% Si and 94.66% I, and 34.05% Ge and 66.0% Cl?

[W. C. Schumb and D. W. Breck, *J. Am. Chem. Soc.*, *74*, 1758 (1952)]

19-58 A solution of 350 g of phosgene ($COCl_2$) in acetonitrile as a solvent was added to a mixture of 380 g of sodium fluoride and acetonitrile, and the entire mixture was heated at 30–40°. The volatile product, COF_2, weighed 190 g, and the other product was sodium chloride.

a. Which of the reactants was in excess?
b. What was the percent yield of carbonyl fluoride?

c. Write a Lewis structure of carbonyl fluoride.
d. What is the shape of the molecules of carbonyl fluoride?

[F. S. Fawcett, C. W. Tullock, and D. D. Coffman, *J. Am. Chem. Soc.*, *84*, 4278 (1962)]

19-59 Write a Lewis structure for carbon suboxide, C_3O_2, that is linear. The C–C distance is 1.28 Å.
[R. L. Livingston and C. N. R. Rao, *J. Am. Chem. Soc.*, *81*, 285 (1959)]

19-60 Predict the shape of the molecules of formylfluoride, HCOF.
[M. E. Jones, K. Hedberg, and V. Schomaker, *J. Am. Chem. Soc.*, *77*, 5278 (1955)]

19-61 A hydride of silicon was prepared by A. Stock and C. Somieski. A sample weighing 0.0751 g exerted a pressure of 509 mm at 21° in a bulb whose volume was 43.3 cc. What was the molecular mass of the hydride?
[*Chem. Ber.*, *49*, 130 (1916)]

19-62 Calculate the volume of silane (at STP) expected from the reaction of 0.390 g of Mg_2Si with HCl(aq), assuming that no other hydride is formed.

19-63 Calculate the volume of carbon tetrafluoride, measured at 30°C and 742 torr, that can be obtained from the reaction of fluorine with 0.940 g of silicon carbide. What mass of sodium fluorosilicate would be formed if the mixture of gaseous products was passed over sodium hydroxide?

19-64 Write a Lewis structure and predict the shape of the molecules of thiocarbonyl fluoride, F_2CS.

19-65 Calculate the mass of silicon tetrachloride that can be formed by treatment of a mixture of 10.0 g of silica and excess carbon with chlorine gas at an elevated temperature assuming 100% yield. The other product of the reaction is carbon monoxide.

19-66 For the reaction C(graphite) + O_2(g) → CO_2(g) at 25°C, $\Delta H° = -393.5$ kJ mol^{-1}; for the reaction C(diamond) + O_2(g) → CO_2(g), $\Delta H° = -395.4$ kJ mol^{-1}. Calculate $\Delta H°_{trans}$ for the change C(graphite) → C(diamond).

19-67 Calculate E° for the cell $Sn|Sn^{2+}\|Cu^{2+}|Cu$.

19-68 Write a Lewis structure and predict the shape of the molecules of tin tetrachloride. In hydrochloric acid solution, tin tetrachloride forms $SnCl_6^{2-}$ ions. What is the expected shape of these ions?

19-69 The electrolytic reduction of 2.194802 g of $SnCl_4$ produced 1.000000 g of tin. Calculate a value for the atomic mass of tin from this observation.
[G. P. Baxter and H. W. Starkweather, *J. Am. Chem. Soc.*, *42*, 905 (1920)]

19-70 Determine the lead ion concentration in a solution that is 0.300 M in hydrochloric acid.

NITROGEN, PHOSPHORUS, ARSENIC, AND ANTIMONY

19-71 Calculate the atomic mass of nitrogen from the $^{14}N/^{15}N$ ratio in atmospheric nitrogen 272. The nuclidic masses are ^{14}N = 14.003074 and ^{15}N = 15.000108.
[A. E. Cameron and E. Wichers, *J. Am. Chem. Soc.*, *84*, 4182 (1962)]

19-72 A sample of phosphorus vapor weighing 10.46 mg in a bulb whose volume was 18.08 cc exerted a pressure of 246.5 torr at 575°C. What was the molecular mass of the vapor?
[A. Stock, G. E. Gibson, and E. Stamm, *Chem. Ber.*, *45*, 3529 (1912)]

19-73 One liter of a compound of arsenic and hydrogen weighed 3.50 g at 0°C and 1.00 atm. Calculate the molecular mass of the gas.
[J. Dumas, *Ann. Chim. Phys.*, *33*, 357 (1826)]

19-74 What is the shape of the molecules of the compound in Problem 19-73?

19-75 In an investigation of the decomposition of hydroxylamine in alkaline solution, 1.4848 g of hydroxylamine hydrochloride

produces 7.189 mmol of ammonia. Write an equation for the decomposition that is consistent with the results of this experiment.
[K. B. Morris, D. F. Johnson, and J. B. Morris, *J. Am. Chem. Soc.*, *72*, 2776 (1950)]

19-76 The reaction of hydrazine with hydrochloric acid led to the formation of a compound containing 26.9% N, 6.2% H, and 67.2% Cl. What is the formula of the compound?
[T. Curtius, *Chem. Ber.*, *20*, 1633 (1887)]

19-77 Nitrogen trifluoride gas was passed over copper metal at 375°, and copper(II) fluoride and a gas containing 71.92% fluorine were obtained. The molecular mass of the gas was found to be 107 by gas-density measurements.

a. Write an equation for the reaction.
b. What mass of NF_3 would be required to produce 1.00 L of the gaseous product if the yield was 62.0%?

[C. B. Colburn and A. Kennedy, *J. Am. Chem. Soc.*, *80*, 5004 (1958)]

19-78 A sample of gas, believed to be NF_3, was allowed to react with water. If 23.0 cc of the gas produced 11.4 cc of NO, is the result consistent with the equation $2NF_3 + 3H_2O = 6HF + NO_2 + NO$?
[O. Ruff, J. Fischer, and F. Luft, *Z. Anorg. Algem. Chem.*, *172*, 423 (1928)]

19-79 The electrolysis of molten ammonium hydrogen fluoride produced a gas at the anode. A sample of the gas (20.5 cc) was mixed with hydrogen gas until the total volume was 57.5 cc and a spark was passed through the gas mixture. One of the products was HF, which was removed by absorption in sodium hydroxide solution. The volume of gas remaining was 18.2 cc. What was the formula of the nitrogen-fluorine compound obtained in the electrolysis?
[O. Ruff, J. Fischer, and F. Luft, *Z. Anorg. Allgem. Chem.*, *172*, 421 (1928)]

19-80 Another sample of gas produced in the electrolysis in Problem 19-79, which weighed 0.5532 g, exerted a pressure of 730.6 torr at 0°C in a flask whose volume was 180.85 cc.

a. What is the molecular mass of the gas? Analysis indicated the presence of 80.2% fluorine in the compound.
b. What is the formula of the compound?

[O. Ruff, J. Fischer, and F. Luft, *Z. Anorg. Allgem. Chem.*, *172*, 421 (1928)]

19-81 Write a Lewis structure for the P_4 molecule. Electron and X-ray diffraction studies show that it is tetrahedral.

19-82 Dimethylaminodifluorophosphine, $(CH_3)_2NPF_2$, reacted with hydrogen iodide at room temperature to yield a compound containing 15.37% P, 18.09% F, and 64.34% I. Its observed molecular mass was 192.9.

a. What is the formula of the compound?
b. What is the shape of the molecules of the compound?

[R. W. Rudolph, J. G. Morse, and R. W. Parry, *Inorg. Chem.*, 5, 1464 (1966)]

19-83 The compound prepared in Problem 19-82 reacted with mercury at room temperature to produce a compound containing 44.1% phosphorus and 53.8% fluorine. Its molecular mass, calculated from vapor-density measurements, was 138.7.

a. What is the compound?
b. Write a Lewis structure for the molecules.
c. Describe the geometrical structure of the molecule.
d. Write a balanced equation for the preparation.

[M. Lustig, J. K. Ruff, and C. B. Colburn, *J. Am. Chem. Soc.*, *88*, 3875 (1966)]

19-84 Phosphorus pentachloride (98 g) reacted with calcium fluoride at 400° to give 41 g of phosphorus pentafluoride. Calculate the percent yield.

[E. L. Muetterties et al., *J. Inorg. Nucl. Chem.*, *16*, 54 (1961)]

19-85 In the same paper as Problem 19-84 it was reported that 87 g of phosphorus trifluoride was prepared by the reaction of 180 g of phosphorus trichloride with 234 g of calcium fluoride at 350° for 8 hr.

a. Which reactant was in excess?
b. What was the percent yield?

[E. L. Muetterties et al., *J. Inorg. Nucl. Chem.*, *16*, 54 (1961)]

19-86 The action of anhydrous hydrogen fluoride on KH_2AsO_4 gave a solid that contained 32.2% As, 48.8% F, and the remainder potassium. Predict the shape of the anion in the compound.
[H. M. Bess and R. W. Parry, *J. Am. Chem. Soc.*, *79*, 1591 (1957)]

19-87 Account for the fact that the three Sb-Cl basal distances in antimony pentachloride are 2.29 Å and the two Sb-Cl apical distances are 2.34 Å.
[S. M. Ohlberg, *J. Am. Chem. Soc.*, *81*, 811 (1959)]

19-88 Fluorine, diluted with nitrogen, was passed over a mixture of arsenic trichloride and arsenic(III) oxide. Among the products was a compound containing 49.91% As and 38.21% F; the remainder was oxygen. The vapor density of the compound was about 148 g per molar volume. What is the structure of the compound?
[G. Mitra, *J. Am. Chem. Soc.*, *80*, 5639 (1958)]

19-89 A. Stock and O. Guttermann prepared a hydride of antimony. A sample weighing 0.3333 g exerted a pressure of 754 mm in a 62.86-cc bulb at 15°C. What was the molecular mass of the hydride?
[*Chem. Ber.*, *37*, 887 (1904)]

19-90 In a determination of the atomic mass of nitrogen, R. W. Gray found that 0.35851 g of nitric oxide occupied 267.43 cc at 759.91 torr and 0°C. Calculate a value for the atomic mass of nitrogen by using the value 15.999 for oxygen.
[*J. Chem. Soc.*, *87*, 1608 (1905)]

SULFUR, SELENIUM, AND TELLURIUM

19-91 At 27°C and 420 mm, 0.170 g of H_2S was collected in a 223-mL vessel that also contained 0.622 g of tin. The tin was heated until a reaction occurred. What was the pressure after the reaction occurred (measured at 27°C), if tin(II) sulfide was formed?

19-92 The solubility of rhombic sulfur in carbon disulfide is 46.1 g/100 g of solvent at 22°C. What are the molar concentration and the mole fraction of sulfur?

19-93 Sulfur and silver fluoride were heated to 140°C, and a blue product from the reaction was collected at -70°C. A sample of the product contained 0.4327 g of sulfur and 0.5055 g of fluorine. What is the empirical formula of the compound?
[D. K. Padma and S. R. Satyanarayana, *J. Inorg. Nucl. Chem.*, *28*, 2432 (1966)]

19-94 A 1.290-g sample of the compound prepared in Problem 19-93 was treated with HI. Hydrogen fluoride, 9.378 g of iodine, and 0.6274 g of H_2S were obtained as products.

a. What is the formula of the compound?
b. Write a Lewis structure of the compound.
c. What is the shape of the molecules of the compound?

19-95 $SeOCl_2$ reacted with fluorine at 50° to give a compound whose analysis was Se, 37.23, 37.01; F, 54.68, 54.31. A molecular mass determination gave 208.5 g/mol.

a. What is the formula of the compound?
b. Propose a structure for the molecules of the compound.

[G. Mitra and G. H. Cady, *J. Am. Chem. Soc.*, *81*, 2646 (1959)]

19-96 A 0.1865-g sample of a compound containing 56.4% P and 43.6% S raised the boiling point of 17.40 g of carbon disulfide 0.110°. What was the compound?
[A. Heiff, *Z. Physik. Chem.*, *12*, 209 (1893)]

19-97 The conversion of 1.0000 g of sodium carbonate to sodium sulfate gave 1.3402 g of the latter. Calculate a value for the atomic mass of sulfur from this observation.
[T. W. Richards and C. R. Hoover, *J. Am. Chem. Soc.*, *37*, 108 (1915)]

19-98 Determine the sulfide ion concentration of a saturated solution of H_2S (0.1 M) that is also 0.3 M in HCl.

19-99 Calculate the Cu^{2+} ion concentration and the Zn^{2+} ion concentration in the solution in Problem 19-98.

19-100 An oxyfluoride of sulfur was prepared by F. B. Dudley, G. H. Cady, and D. F. Eggers. Analysis found: S, 26.1, 27.03, 26.6, 26.4; F, 31.5, 31.3, 32.0, 31.4. The molecular mass from vapor-density measurements was 117--119.5 in eight determinations. What is the formula of the compound? Propose a structure.
[*J. Am. Chem. Soc.*, *78*, 291 (1956)]

19-101 The compound CH_3SH was fluorinated, and the principle product contained 16.1% sulfur, 77.3% fluorine, and the remainder carbon. The molecular mass from gas-density measurements was 197. What is the formula of the compound? The carbon and sulfur atoms are joined by a single bond. Draw a diagram representing the structure of the molecules of the compound.
[G. A. Silvey and G. H. Cady, *J. Am. Chem. Soc.*, *72*, 3624 (1950)]

19-102 The reaction of a stream of HCl gas with S_4N_4 gave a compound containing 60.0% sulfur, 20.4% nitrogen, and 17.5% chlorine. What is the formula of the compound?
[A. G. MacDiarmid, *J. Am. Chem. Soc.*, *78*, 3872 (1956)]

19-103 In 1884 it was claimed that the reaction of P_4S_3 with S led to the formation of P_2S_4 and P_8S_{11}. To confirm this claim, A. R. Pitochelli and L. F. Audrieth dissolved 27.5 g of P_4S_3 and 4 g of S_8 in 100 mL of CS_2 and heated the solution under pressure at 170 to 200° for 3 hr. A mixture of two solids resulted. One contained 43.6% P and 57.9% S; the other contained 35.99% P and 64.4% S. What were the products? Write an equation for the reaction that occurred.
[*J. Am. Chem. Soc.*, *81*, 4458 (1959)]

19-104 A solution of 3.795 g of sulfur in 100 g of carbon disulfide boils at 46.66°C. The boiling point of carbon disulfide is 46.30°C; K_b is 2.41. What is the formula of the sulfur molecules in the solution?

19-105 What is the formula of the oxide of tellurium if 743.2 mg gave 148.8 mg of oxygen when decomposed to the elements?
[A. Metzner, *J. Am. Chem. Soc.*, *21*, 206 (1899)]

HALOGENS

19-106 Calculate a value of the atomic mass of bromine from the isotopic abundances and isotopic masses: ^{79}Br = 50.53%, 78.9434 amu; ^{81}Br = 49.463%, 80.9421 amu.

[A. E. Cameron and E. Wichers, *J. Am. Chem. Soc.*, *84*, 4187 (1962)]

19-107 Determine the mass of sodium bromide required to react with an acidified solution containing 0.755 g of sodium bromate.

19-108 Calculate the equilibrium concentration of HOCl from the equilibrium constant $K_c = 4.66 \times 10^{-4}$ for the reaction of chlorine with water. The initial concentration of chlorine was 0.0100 mol/L.

19-109 Find the standard electrode potential of the cell

$$Cl_2, Pt|Cl^- \| Mn^{2+}, MnO_4^-, H_3O^+|Pt$$

19-110 Calculate the potential of the cell

$$Cl_2(1\ atm), Pt|Cl^-(1\ m) \| Mn^{2+}(0.01\ m),$$
$$MnO_4^-(2.0\ m)H_3O^+(2.0\ m)|Pt$$

19-111 The equilibrium $BrF_3(g) + Br_2(g) \rightleftharpoons 3BrF(g)$ was studied by R. K. Steunenberg, R. C. Vogel, and J. Fischer. At 65.2°C, initial partial pressures of Br_2 and BrF_3 of 30.3 and 30.2 mm, respectively, led to a total pressure of 79.0 mm at equilibrium. What is K_p for the equilibrium?

[*J. Am. Chem. Soc.*, *79*, 1320 (1957)]

19-112 Calculate the value of K_c for the dissociation of I_3^-, where the triiodide ion concentration is 1.032×10^{-2} M, the iodide ion concentration is 1.019×10^{-2} M, and the iodine concentration is 1.320×10^{-3} M.

[L. I. Katzin and E. Gebert, *J. Am. Chem. Soc.*, *77*, 5814 (1955)]

19-113 Find the iodine concentration of a solution prepared by dissolving 0.100 mol of iodine in 1.00 L of 0.100 M KI solution. K_c for the formation of I_3^- is 0.752.

19-114 What is the shape of the molecules of each of the compounds in Problem 19-111?

19-115 The magnetic susceptibility of a sample of iodine dioxide was measured and the compound was found to be diagmagnetic. What is the structure of the compound?

[W. K. Wilmarth and S. S. Dharmatti, *J. Am. Chem. Soc.*, *72*, 5789 (1950)]

19-116 Predict the shapes of the molecules of IF_5 and IF_7 and compare your predictions with the shapes proposed to account for their infrared and raman spectra.

[R. C. Lord, M. A. Lynch, W. C. Shumb, and E. J. Slowinski, Jr., *J. Am. Chem. Soc.*, *72*, 522 (1950)]

19-117 The equilibrium $BrF_3 + Br_2 \rightleftharpoons 3BrF$ was studied at 74.5°C. The initial pressure of BrF_3 and Br_2 was 49.9 mm for each. At equilibrium the total pressure was 129.1 mm. Calculate K_p for the equilibrium at 74.5°C.

[R. K. Steunenberg, R. C. Vogel, and J. Fischer, *J. Am. Chem. Soc.*, *79*, 1320 (1957)]

19-118 At 84.8°C, K_p for the equilibrium $BrF_3 + Br_2 \rightleftharpoons 3BrF$ was found to be 2430. Calculate the partial pressure of BrF at equilibrium if the partial pressures of BrF_3 and Br_2 were each equal to 18.7 mm.

[R. K. Steunenberg, R. C. Vogel, and J. Fischer, *J. Am. Chem. Soc.*, *79*, 1320 (1957)]

19-119

a. Predict the shape of the molecules of bromine pentafluoride.
b. Why is the bromine atom below the basal plane of the molecule?

[R. S. McDowell and L. B. Asprey, *J. Chem. Phys.*, *37*, 165 (1962)]

19-120 A gaseous mixture of IF_5 and F_2 in which the initial pressure of fluorine was 562.1 mm and that of IF_5 was 99.4 mm was prepared. After reaction was complete, what was the total pressure?

[J. Fischer and R. K. Steunenberg, *J. Am. Chem. Soc.*, *79*, 1876 (1957)]

19-121 The reaction of iodine with fluorine was studied by E. B. R. Prideaux. He obtained a sample of a compound that contained 0.0770 g of iodine and 0.0564 g of fluorine. What was the empirical formula?

[*J. Chem. Soc.*, *89*, 317 (1906)]

19-122 A saturated solution of chlorine in water at 25°C under a pressure of 1 atm is about 0.06 M in total chlorine. Calculate the concentration of free chlorine if the equilibrium constant for the reaction of chlorine with water is 4.66×10^{-4}.

19-123 What are the shapes of the chlorite ions and the hypochlorous acid molecules?

19-124 When O_2F_2 liquid (5.468 g) decomposed to oxygen and fluorine at 190 K in a calorimeter, 88.0 cal of heat was evolved. Calculate ΔE for the decomposition $O_2F_2(\ell) \rightarrow O_2(g) + F_2(g)$ at 190 K.

[A. D. Kirshenbaum, A. V. Grosse, and J. G. Aston, *J. Am. Chem. Soc.*, *81*, 6400 (1959)]

19-125 Use ΔE from Problem 19-124 to calculate ΔH for the reaction $O_2(g) + F_2(g) \rightarrow O_2F_2(\ell)$ at 190 K.

[A. D. Kirshenbaum, A. V. Grosse, and J. G. Aston, *J. Am. Chem. Soc.*, *81*, 6400 (1959)]

NOBLE GASES

19-126 A mixture of fluorine and xenon was heated to 400° in a nickel reactor for 1 hr. A colorless solid, thermally stable at room temperature, was formed. In one experiment 0.2952 g of xenon reacted with 0.1733 g of fluorine. Calculate the simplest formula of the compound.

[H. H. Claasen, H. Selig, and J. G. Malin, *J. Am. Chem. Soc.*, *84*, 3593 (1962)]

19-127 A sample of the compound described in Problem 19-126 and weighing 0.4006 g reacted with an excess of hydrogen at

400° to form hydrogen fluoride and 0.2507 g of xenon. The hydrogen fluoride was absorbed in a solution of sodium hydroxide of known concentration. Analysis of the solution showed that 0.1435 g of fluoride ion was present. Are these observations consistent with the formula proposed in Problem 19-126?
[H. H. Claasen, S. Selig, and J. G. Malin, *J. Am. Chem. Soc.*, *84*, 3595 (1962)]

19-128 Write a Lewis structure and draw a diagram representing the geometrical structure of the compound formed in Problem 19-126.

19-129 Calculate the average bond-dissociation energy of XeF_4. The standard enthalpy of formation of $XeF_4(g)$ is -215 kJ mol^{-1}.

19-130 Calculate the molecular mass of argon from the observation that 2111.0 mL weighed 2.5094 g at 0°C and 506.67 torr.
[G. P. Baxter and W. W. Starkweather, *Proc. Natl. Acad. Sci. U.S.*, *14*, 60 (1928)]

19-131 A solution was described as being 3.85×10^{-3} M in a xenon(VI) species. If the species was XeO_4^{2-}, what was the concentration in grams per milliliter?

19-132 A fluoride of xenon was synthesized at the University of Washington. In one experiment 0.409 g of xenon produced 0.758 g of the fluoride. Calculate the empirical formula of the fluoride synthesized.
[F. B. Dudley, G. Gard, and G. H. Cady, *Inorg. Chem.*, *2*, 229 (1963)]

19-133 What was the simplest formula of the fluoride from which 0.0327 g of xenon was recovered from a sample weighing 0.0607 g?
[F. B. Dudley, G. Gard, and G. H. Cady, *Inorg. Chem.*, *2*, 229 (1963)]

19-134 A 0.3934-g sample of the fluoride described in Problem 19-132 was treated with hydrogen. Xenon (0.2125 g) and hydrogen fluoride (0.1770 g) were formed. Do the results of this experiment support the formula calculated in Problem 19-132?
[J. G. Maim, I. Sheft, and C. L.. Chernick, *J. Am. Chem. Soc.*, *85*, 110—111 (1963)]

19-135

a. Write a Lewis structure for the compound in Problem 19-134.
b. Predict the shape of the molecules of the compound.

19-136 A white solid, formed by the reaction of aqueous barium hydroxide with aqueous XeO_3, contained 71.75% BaO, 20.60% Xe, and 7.55% O. What is the formula of the compound?
[A. D. Hirschenbaum and A. V. Grosse, *Science*, *142*, 580 (1963)]

19-137 A sample of xenon hexafluoride weighing 0.0911 g reacted with mercury at 100°. Calculate the mass of mercury(II) fluoride formed.
[E. E. Weaver, B. Weinstock, and C. P. Knop, *J. Am. Chem. Soc.*, *85*, 111 (1963)]

19-138 A sample of a fluoride of xenon (0.0607 g) reacted with aqueous KI solution to yield 0.0327 g Xe. What is the formula of the xenon fluoride?
[F. B. Dudley, G. Gard, and G. H. Cady, *Inorg. Chem.*, *2*, 229 (1963)]

19-139 What is the shape of the molecules of XeF_2?

19-140 What is the structure of the molecules of XeO_3?

19-141 The first compound of xenon reported resulted from the reaction of the gas with PtF_6. In one experiment an initial pressure of PtF_6 in a silica bulb was 93.0 mm. Xe was added until the total pressure was 201 mm. After reaction, the residual pressure of xenon was 17.0 mm. Assuming that direct combination occurred, calculate the formula of the product.
[N. Bartlett, *Proc. Chem. Soc.*, *1962*, 218]

19-142 A sample of xenon hexafluoride weighing 0.3934 g was treated with hydrogen by J. G. Malin, I. Selig, and C. L. Chernick, who obtained 0.2125 g of Xe and 0.1864 g of HF as products. Compare the results with the theoretical yields of the products.
[*J. Am. Chem. Soc.*, *85*, 110 (1963)]

19-143 A sample of a xenon fluoride weighing 158.9 mg exerted a pressure of 4.61 mm at 24.8°C in a container the volume of which was 2609.5 cc. Calculate the molecular mass.
[B. Weinstock, E. E. Weaver, and C. P. Knop, *Inorg. Chem.*, 5, 2196 (1966)]

19-144 Calculate the average bond dissociation energy of XeF_6 from the heat of formation at 298.15 K, -294 kJ mol^{-1}.
[B. Weinstock, E. E. Weaver, and C. P. Knop, *Inorg. Chem.*, 5, 2202 (1966)]

19-145 The $\Delta H°_f$ of $KrF_2(g)$ was determined by S. R. Gunn as +57.3 kJ mol^{-1}, and the $\Delta H°_f$ of $KF_2(s)$ as +15 kJ mol^{-1}. Compute the molar enthalpy of sublimation of $KrF_2(s)$ at 25°C.
[*J. Am. Chem. Soc.*, *88*, 5924 (1966)]

COPPER, SILVER, AND GOLD

19-146 Calculate the mass of copper deposited by the passage of 2.02 amp for 2.25 hr through a solution of copper(II) sulfate.

19-147 Determine the potential of the cell $Fe|Fe^{2+}\|Cu^{2+}|Cu$.

19-148 What volume of nitric oxide can be produced by the reaction of 0.705 g of copper with concentrated nitric acid?

19-149 Calculate the silver ion concentration in a solution that is 0.100 M in NH_3 and in equilibrium with solid silver chloride.

19-150 Calculate the standard enthalpy of decomposition of copper(II) sulfate to copper(II) oxide and sulfur trioxide.

19-151 A compound was prepared that contained 21.63% K, 35.74% Te, 25.15% H_2O, and the remainder oxygen. A sample of the compound reacted with hot hydrochloric acid to form Cl_2 and $TeCl_4$. Write an equation for the reaction of the compound with hydrochloric acid.
[E. B. Hutchins, Jr., *J. Am. Chem. Soc.*, *27*, 1172 (1905)]

19-152 Calculate the potential of the cell H_2, Pt|HBr (0.97 m) AgBr(s)|Ag.
[W. J. Biermann and R. S. Yamasaki, *J. Am. Chem. Soc.*, *77*, 241 (1955)]

19-153 Find the potential of the cell Pb(Hg)|$PbCl_2$, Cl^-(1 m), AgCl|Ag.
[W. R. Carmody, *J. Am. Chem. Soc.*, *51*, 2909 (1929)]

19-154 Calculate the copper(II) ion concentration of a 0.100 M solution of $[Cu(NH_3)_4](NO_3)_2$.

19-155 How many grams of sulfur can be formed by the reaction of concentrated nitric acid with 0.125 mol of CuS?

ZINC, CADMIUM, AND MERCURY

19-156 A sample of mercury(II) chloride weighing 9.8500 g was dissolved in 32.75 g of ethanol. The boiling-point elevation was 1.275°. Calculate the molecular mass of the solute.
[E. Beckmann, *Z. Physik. Chem. (Leipzig)*, *6*, 453 (1890)]

19-157 A boiling-point-elevation experiment was run to determine the extent of dissociation of mercury(II) chloride in water. The boiling-point elevation of a solution containing 3.8113 g of the chloride in 43.69 g of water was 0.165°.

a. What species were present in the solution?
b. What is the type of bonding present in $HgCl_2$?
c. What is the structure of $HgCl_2$?

[E. Beckmann, *Z. Physik. Chem. (Leipzig)*, *6*, 453 (1890)]

19-158 The boiling-point elevation of an aqueous solution of 5.6977 g of cadmium iodide in 44.69 g of water was 0.181°. What solute species were present in the solution?
[E. Beckmann, *Z. Physik. Chem. (Leipzig)*, *6*, 453 (1890)]

19-159 The microwave spectrum of the volatile compound prepared by the action of chlorine gas on mercury(II) oxide was consistent with a Cl—O bond distance of 1.700 Å and a Cl—O—Cl angle of 110.96°. Do these results support the prediction of the shape of the molecules from the electron-pair repulsion principle?
[G. E. Herberich, R. H. Jackson, and D. J. Miller, *J. Chem. Soc.*, *1966*, 336]

19-160 Calculate the potential of the cell Zn|Zn^{2+}‖ Sn^{2+}|Sn.

19-161 What is the potential of the cell $Zn(Hg)|ZnSO_4(0.02\ m)$, $PbSO_4(s)|Pb(Hg)$?

[I. A. Cowperthwaite and V. K. Lamer, *J. Am. Chem. Soc.*, *53*, 4333 (1931)]

19-162 The E° of the cell $Cd(Hg)|CdCl_2$, $AgCl|Ag$ is 0.573 V. Calculate the standard potential E° of the half-cell $Cd(Hg)|Cd^{2+}$.

[H. S. Harned and M. E. Fitzgerald, *J. Am. Chem. Soc.*, *58*, 2624 (1936)]

19-163 Calculate a value of the atomic mass of mercury from the mercury(II) chloride-silver ratio, 1.25847.

[O. Honigschmid, L. Birkenbach, and M. Steinhell, *Chem. Ber.*, *56B*, 1212 (1923)]

19-164 Determine the mercury(I) ion concentration in a solution that is 0.0873 M in KCl.

19-165 Calculate the quantity of ammonium nitrate formed by the action of very dilute nitric acid on 0.261 g of zinc metal.

TRANSITION ELEMENTS

19-166 A 0.10 M Na_2HPO_4 solution was added dropwise to a 20% excess of 0.10 M $CrCl_3$, and the violet, crystalline precipitate that was formed contained 42.5% water. Further analysis showed the presence of 20.15% Cr and 12.26% P. What was the compound?

[A. T. Ness, R. E. Smith, and R. L. Evans, *J. Am. Chem. Soc.*, *74*, 4688 (1952)]

19-167 A black material containing 35.3% Cr was obtained by heating the product in Problem 19-166 to 800°C. What was the compound?

[A. T. Ness, E. E. Smith, and R. L. Evans, *J. Am. Chem. Soc.*, *74*, 4688 (1952)]

19-168 In a novel method of preparing metal chlorides, T. E. Austin and S. Y. Tyree, Jr. obtained a compound containing 34.52% Fe and 65.51% Cl. What was the compound?

[*J. Inorg. Chem.*, *14*, 141–142 (1960)]

19-169 A nickel ferrite was prepared by heating the compound $Ni_3Fe_6(CH_3CO_2)_{17}O_3OH$ to 1000°. The ferrite analysis was Ni, 25.03; Fe, 47.54. What is the formula of the ferrite?
[D. G. Wickham, E. R. Whipple, and E. G. Larson, *J. Inorg. Nucl. Chem.*, *14*, 217 (1960)]

19-170 What is the formula of the fluoride of niobium that melts at 78.9° and contains 49.7% Nb and 50.2% F?
[J. H. Junkins, R. L. Farrar, Jr., E. J. Barber, and H. A. Bernhardt, *J. Am. Chem. Soc.*, *74*, 3464 (1952)]

19-171 A 3.66098-g sample of titanium(IV) chloride was dissolved in nitric acid. A solution made by dissolving silver in nitric acid was added to determine the atomic mass of titanium. Using the accepted value of the atomic mass of titanium, calculate the mass of silver required to react with the titanium tetrachloride.
[G. P. Baxter and A. Q. Butler, *J. Am. Chem. Soc.*, *48*, 3117 (1926)]

19-172 Calculate the atomic mass of Ti from the titanium(IV) bromide-silver ratio, 0.851788.
[G. P. Baxter and A. Q. Butler, *J. Am. Chem. Soc.*, *50*, 408 (1928)]

19-173 What is the emf of the cell $Pt|\ Fe^{2+}(0.10\ m),\ Fe^{3+}(0.0010\ m)\|Ce^{3+}(0.0010\ m),\ Ce^{4+}(0.10\ m)|Pt$?

19-174 Pt reacted with fluorine and oxygen to give a product that contained 32.7% F, 10.4% O, and 57.5% Pt. What was the compound?
[N. Bartlett and D. H. Lohmann, *Proc. Chem. Soc.*, *1962*, 115]

19-175 Calculate E° for the cell $Pt,\ O_2|H_2O_2\|Mn^{2+},\ MnO_2(s)|Pt$.

19-176 Determine the standard emf of the following cell: $Pt|Fe(CN)_6{}^{4-},\ Fe(CN)_6{}^{3-}\|I^-,\ I_2(s)|Pt$.

19-177 Calculate the emf of the following cell: $Pt|Fe^{2+}(0.1\ m),\ Fe^{3+}(0.01\ m)\|Fe^{2+}(0.1\ m)|Fe$.

19-178 Balance the following equations.

a. $Cr_2O_7{}^{2-} + I^- + H^+ = I_2 + Cr^{3+} + H_2O$

b. $CrO_4^{2-} + SO_2 + H^+ = Cr^{3+} + SO_4^{2-} + H_2O$

c. $MnO_4^- + H_2S = MnO_2 + S + H_2O + OH^-$

d. $MnO_4^- + H_2S + H^+ = Mn^{2+} + S + H_2O$

e. $Fe^{2+} + O_2 + H_2O = Fe^{3+} + OH^-$

19-179 Calculate the solubility of silver chromate in a saturated solution $K_S = 9 \times 10^{-12}$.

19-180 Find the Fe^{2+} ion concentration in a 0.10 M H_2S solution that is also 0.30 M in HCl. Calculate the Fe^{2+} ion concentration in a solution in which the S^{2-} ion concentration is 0.010 M.

19-181 Calculate a value of the atomic mass of vanadium from the experimental observation that 4.8564 g $NaVO_3$ gave 2.3277 g NaCl when treated with hydrochloric acid.

[D. J. McAdam, Jr., *J. Am. Chem. Soc.*, *32*, 1603 (1910)]

19-182 A sample of a chloride of iron weighing 0.1404 g gave 0.0695 g of Fe_2O_3 and 0.3732 g of AgCl. Calculate the simplest formula of the chloride.

[W. Grunewald and V. Meyer, *Chem. Ber.*, *21*, 690 (1888)]

19-183 A chloride of iron contained 34.65% Fe and 65.76% Cl. A sample weighing 0.0864 g exerted a pressure of 742 mm in a bulb whose volume was 6.7 cc at a temperature of 29°C. What was the formula of the vapor?

[W. Grunewald and V. Meyer, *Chem. Ber.*, *21*, 690 (1888)]

19-184 The standard energy of combustion of NbC was found to be 1188.0 kJ mol^{-1}. Calculate the standard enthalpy of formation of NbC from this observation and the standard enthalpies of formation of Nb_2O_5 and CO_2.

[A. D. Mah and B. J. Boyle, *J. Am. Chem. Soc.*, *77*, 6513 (1955)]

19-185 Find the standard enthalpy of formation of $Mo(CO)_6(s)$ from the standard enthalpy of combustion, -2123.5 kJ mol^{-1}.

[F. A. Cotton, A. K. Fischer, and G. Wilkinson, *J. Am. Chem. Soc.*, *78*, 5169 (1956)]

19-186 Calculate the number of faradays of electricity required to deposit 100 g of chromium metal from a solution of $H_2Cr_2O_7$.

19-187 To 10.0 mL of a 0.100 M $K_2Cr_2O_7$ solution was added 10.0 mL of a 0.100 M $Pb(NO_3)_2$ solution. What is the mass of the precipitate that was formed?

19-188 An attempt was made to dissolve 0.100 mol of K_2MnO_4 in a dilute H_2SO_4 solution. Calculate the quantity of each of the products formed.

19-189 Find the volume of 0.100 M $KMnO_4$ solution required to oxidize the iron in 25.0 mL of 0.100 M $FeSO_4$ solution that also contains H_2SO_4.

19-190 Calculate the volume of chlorine (at STP) that will result from the action of excess hydrochloric acid on 0.268 g of MnO_2.

19-191 Balance the following equations.

a. $Co^{2+} + NH_3 + O_2 + H_2O = ? + OH^-$

b. $[Fe(CN)_6]^{4-} + Cl_2 = ? + Cl^-$

c. $Cr(OH)_3 + Na_2O_2 = ? + NaOH + H_2O$

19-192 Yttrium oxide was dissolved in concentrated hydrochloric acid, and the solution was evaporated until crystals formed. Analysis found: Y, 29.08; Cl, 34.75; H_2O, 36.17. Calculate the formula.

[H. J. Nalting, C. R. Simmons, and J. J. Klingenberg, *J. Inorg. Nucl. Chem.*, *14*, 208 (1960)]

19-193 Chromium(III) oxide and lampblack were heated in a hydrogen furnace at 1525°. The solid product contained 13.2% C and 86.2% Cr. What is the formula of the carbide?

[R. A. Oriani and W. K. Murphy, *J. Am. Chem. Soc.*, *76*, 3345 (1954)]

19-194 Calculate the time required to deposit 125 g of Ni by the passage of a current of 1.60 amp through a solution of nickel sulfate.

19-195 What volume of nickel tetracarbonyl, measured at 87°C and 730 torr, can be formed by action of excess carbon monoxide on 10.0 g of nickel?

19-196 A sample of an oxide or uranium that contained 88.12% uranium weighed 0.3363 g. It was heated in an atmosphere of oxygen, and the product weighed 0.3493 g. Write an equation for the reaction that occurred.
[E. F. Smith and J. M. Matthews, *J. Am. Chem. Soc.*, *17*, 686 (1895)]

19-197 Protactinium(V) oxide was mixed with carbon, and an excess of bromine vapor was passed over the mixture at 750° for several days. In addition to carbon monoxide, two products were formed. One contained 36.4% Pa and 63.9% bromine and sublimed at 300--350° to leave the other as a yellow-green solid. The second contained 47.8% Pa and 49.07% Br. Write an equation for the reaction of Br_2 and C with Pa_2O_5.
[D. Brown and P. J. Jones, *J. Chem. Soc.*, *1966*, 262]

19-198 Calculate the enthalpy of formation of NdOCl from the equation $NdCl_3(s) + H_2O(g) \rightleftharpoons NdOCl(s) + 2HCl(g)$. The experimental value of the heat of hydrolysis was 1014 kJ mol^{-1}.
[*J. Am. Chem. Soc.*, *76*, 1473 (1954)]

19-199 Calculate the formula of the volatile compound containing 66.5% Pu and 30.9% F.
[B. Weinstock and J. G. Malm, *J. Inorg. Nucl. Chem.*, *2*, 380 (1956)]

19-200 Calculate the volume of hydrogen, measured at standard conditions, that can be formed by the reaction of 1.39 g of lanthanum with water.

Appendix A
LOGARITHMS

Logarithms are the exponents to which a base, commonly 10, is raised to equal a number in question. The logarithm of 100 is 2, since $10^2 = 100$; log 1000 = 3, since $10^3 = 1000$. Although $10^{0.3010} = 2.000$ is not so obvious, log 2.000 = 0.3010.

The values of logarithms of numbers are found in tables such as Table D-2, which supplies four-place logarithms. Use of the table is illustrated in the examples that follow.

The log of a number has two parts, called the *mantissa* and the *characteristic*. The purpose of the *characteristic* is to locate the decimal point of the number. The *mantissa* establishes the identity of the digits of the number. The value of the mantissa of a number is found in a table of logarithms.

The characteristic is determined by the application of two rules. For a number greater than 1 the characteristic is one less than the *number of digits* to the *left* of the decimal point.

For numbers less than 1 the characteristic is negative and is *one more than the number of zeros* between the decimal point and the first digit to the right of the decimal point. Negative characteristics are written in either of two ways. In the first way, the negative sign is placed *above* the characteristic, $\bar{1}$, $\bar{2}$, $\bar{3}$, etc. The other way is to represent the negative characteristic by 9, *mantissa* -10, 8, *mantissa* -10, 7, *mantissa* -10, respectively, and so on (see examples below).

Examples

Number	Characteristic	Mantissa	Logarithm
2	0	.3010	0.3010
20	1	.3010	1.3010
200	2	.3010	2.3010
27	1	.4314	1.4314
270	2	.4314	2.4314
2.7	0	.4314	0.4314
2700	3	.4314	3.4314
832	2	.9201	2.9201
8.32	0	.9201	0.9201
83.2	1	.9201	1.9201
8320	3	.9201	3.9201

Examples

Number	Characteristic	Mantissa	Logarithm
0.2	-1	.3010	$\bar{1}.3010$ or 9.3010 - 10
0.02	-2	.3010	$\bar{2}.3010$ or 8.3010 - 10
0.27	-1	.4314	$\bar{1}.4314$ or 9.4314 - 10
0.027	-2	.4314	$\bar{2}.4314$ or 8.4314 - 10
0.00832	-3	.9201	$\bar{3}.920$, or 7.9201 - 10

PROPORTIONAL PARTS

The mantissas in the examples given thus far can be found directly in the table. A number containing four significant figures requires the use of the table of proportional parts. The logarithm of 2764 has a characteristic of 3. The nearest mantissas to be found in the table are for 2.760 and 2.770, or .4409 and .4425. Under 4 in the table of proportional parts is the number 6, which is added to .4409 to give the mantissa for 2764, or .4415. The log of 2674 is then 3.4415.

Examples

Find the logarithms of the following numbers:

log 2.816 = 0.4496

log 471.2 = 2.6732

log 0.8138 = $\bar{1}$.9105 or 9.9105 - 10

log 0.006418 = $\bar{3}$.8074 or 7.8074 - 10

log 76.44 = 1.8833

The procedure for reversing the process and finding the number corresponding to a logarithm is as follows:

1. log x = 1.6018

2. Because the characteristic is 1, there will be two digits to the left of the decimal point.

3. The mantissa .6018 lies between .6010 and .6021, which are mantissas for 39.90 and 40.00. The first three digits of the number are therefore 39.9.

4. The difference between .6010 and .6018 is 8. In the table of proportional parts the 8 is found under 7, which is added to 39.9 to produce the number 39.97.

USE OF LOGARITHMS IN MULTIPLICATION AND DIVISION

The use of logs for multiplying and dividing is based on the relationships

$$\log ab = \log a + \log b$$

and

$$\log \frac{a}{b} = \log a - \log b$$

Examples

Use logs to compute the following:

a. $236.0 \times 4.61 \times 21.2 = x$

$$\begin{aligned} \log 236.0 &= 2.3729 \\ \log 4.610 &= 0.6637 \\ \log 21.20 &= \underline{1.3263} \\ \log x &= 4.3629 \\ x &= 23{,}070 \end{aligned}$$

b. $2.42 \times 0.3160 \times 0.01750 = x$

$$\begin{aligned} \log 2.420 &= 0.3838 && 0.3838 \\ \log 0.3160 &= \bar{1}.4997 \text{ or } && 9.4997 - 10 \\ \log 0.01750 &= \underline{\bar{2}.2430} \text{ or } && \underline{8.2430 - 10} \\ \log x &= \bar{2}.1265 && 18.1265 - 20 \\ x &= 1.338 \times 10^{-2} \text{ or } 0.01338 \end{aligned}$$

c. $\dfrac{16.82}{0.8436} = x$

$$\begin{aligned} \log 16.82 &= 1.2258 \text{ or } && 11.2258 - 10 \\ -\log 0.8436 &= \underline{\bar{1}.9261} \text{ or } && \underline{9.9261 - 10} \\ \log x &= 1.2997 && 1.2997 \\ x &= 19.94 \end{aligned}$$

An important point to remember is that a logarithm, as usually written, is the *sum* of two parts, the characteristic and the mantissa.

POWERS AND ROOTS

The use of logs gives a convenient means of determining *nth* powers and *nth* roots of numbers. The relationships are

$$n \log a = \log a^n$$

and

$$\frac{\log a}{n} = \log \sqrt[n]{a}$$

Example

Find the values of x by means of logs.

a. $x = 2.416^2$

$\log 2.416 = 0.3831$

$2 \log 2.416 = 0.7662 = \log x$

$x = 5.847$

b. $x = 21.61^3$

$\log 21.61 = 1.3347$

$3 \log 21.61 = 4.0041 = \log x$

$x = 10{,}100$

c. $x = 0.003170^3$

$\log 0.003170 = \bar{3}.5011 \text{ or } 7.5011 - 10$

$3 \log 0.003170 = \bar{8}.5033 \text{ or } 22.5033 - 30$

$x = 3.187 \times 10^{-8}$

d. $x = \sqrt{61.42}$

$\log 61.42 = 1.7883$

$\frac{1}{2} \log 61.42 = 0.8942 = \log \sqrt{61.42}$

$x = 7.838$

e. $x = 481.7^{1/3}$

$\log 481.7 = 2.6827$

$\frac{1}{3} \log 481.7 = 0.8942 = \log x$

$x = 7.838$

f. $x = 0.003605^{1/3}$

$\log 0.003605 = \bar{3}.5569$ or $6.5569 - 9$

$\frac{1}{3} \log 0.003605 = \bar{1}.1856$ or $2.1856 - 3$

$x = 0.1533$

EXPONENTIAL NUMBERS

Very large and very small numbers are often best expressed as powers of 10. For example, the number 2,610,000 is 2.61×10^6 with one digit to the left of the decimal point. The exponent 6 is found by counting the number of places the decimal point must be moved to the left in order to leave one digit to the left of the decimal point.

The number 0.00375 is written 3.75×10^{-3}; the exponent is the number of places the decimal point must be moved to the right in order to leave one digit to the left of the decimal point.

Examples

Express the following numbers in exponential form.

a. $128{,}000 = 1.28 \times 10^5$

b. $7860 = 7.86 \times 10^3$

c. $0.000654 = 6.54 \times 10^{-4}$

d. $0.00101 = 1.01 \times 10^{-3}$

e. $10{,}500 = 1.05 \times 10^4$

When numbers raised to powers are multiplied, the exponents are added; when divided, the exponents are subtracted.

Examples

Carry out the following multiplications by using the exponential forms of the numbers.

a. $2.00 \times 10^{-4} \times 3.01 \times 10^{23} = 6.02 \times 10^{19}$

b. $4.00 \times 10^{-2} \times 1.50 \times 10^8 = 6.00 \times 10^6$

c. $0.0025 \times 0.040 = 1.0 \times 10^{-4}$

d. $0.0103 \times 2.00 \times 10^8 = 2.06 \times 10^6$

e. $86{,}000 \times 4000 = 3.44 \times 10^8$

Examples

Use exponential forms to compute the following:

a. $\dfrac{0.00480}{1.20 \times 10^{-3}} = \dfrac{4.80 \times 10^{-3}}{1.20 \times 10^{-3}} = 4.00 \times 10^0 = 4.00$

since $10^0 = 1$

b. $\dfrac{0.000648}{324} = \dfrac{6.48 \times 10^{-4}}{3.24 \times 10^2} = 2.00 \times 10^{-6}$

c. $\dfrac{197}{6.02 \times 10^{23}} = 32.7 \times 10^{-23} = 3.27 \times 10^{-22}$

When a number written in exponential form is raised to a power, the exponent is multiplied by the power. For example,

$(2.00 \times 10^4)^2 = 4.00 \times 10^8$

$(3.00 \times 10^{-4})^3 = (27.0 \times 10^{-12})$, $(8.00 \times 10^{-6})^{1/3} = 2.00 \times 10^{-2}$

When numbers are written in exponential form, the characteristics of their logarithms are equal to the exponents. The log of 3.000×10^6 is 6.4771; the log of 3.000×10^{-6} is $\bar{6}.4771$, or 4.4711 - 10.

Looked at from another point of view, the log of a number written in exponential form is the sum of two logs. The log 3.000×10^6 is the log of 3.000 plus the log of 10^6, 0.4771 + 6.0000.

$\log 3.000 \times 10^{-6} = 0.4771 - 6.000$

Examples

Write the logarithms of the following:

a. $\log 2.000 \times 10^4 = 4.3010$

b. $\log 2.000 \times 10^{-3} = \bar{3}.3010$

c. $\log 4.710 \times 10^6 = 6.6730$

d. $\log 8.565 \times 10^{-4} = \bar{4}.9323$

e. $\log 28.74 \times 10^{-2} = 1.4585 + \bar{2}.000$

PROBLEMS

A-1 Use logs to compute the following:

a. $\dfrac{6.28}{44.0}$ b. $\dfrac{16.0}{6.02}$ c. $\dfrac{0.133}{117}$

d. $\dfrac{49.7}{92.9}$ e. $\dfrac{50.2}{19.0}$ f. $\dfrac{23.1}{39.1}$

g. $\dfrac{14.2}{47.9}$ h. $\dfrac{62.7}{35.5}$ i. $\dfrac{22.36}{14.01}$

A-2 Use logs to calculate the following:

a. $\dfrac{22.36}{14.01}$ b. $\dfrac{40.20}{95.94}$ c. $\dfrac{58.98}{35.45}$

d. $\dfrac{39.46}{69.72}$ e. $\dfrac{60.46}{35.45}$

A-3 Solve the following with logs:

a. $\dfrac{20.82}{34.83} \times 100$ b. $\dfrac{14.01}{34.83} \times 100$ c. $\dfrac{28.21}{205.7} \times 100$

d. $\dfrac{42.03}{205.7} \times 100$ e. $\dfrac{35.45}{205.7} \times 100$

A-4 Solve the following with logs:

a. $268 \times \dfrac{17.9}{760} \times \dfrac{273}{342}$

b $\dfrac{0.0516 \times 82.0 \times 342 \times 760}{17.9 \times 268}$

c. $\dfrac{2.66 \times 0.0820 \times 293 \times 760}{749}$

d. $\dfrac{736 \times 200 \times 39.9}{760 \times 82.0 \times 300}$

e. $\dfrac{0.2182 \times 82.05 \times 414.3 \times 760}{45.00 \times 668.5}$

f. $\dfrac{6.517 \times 82.05 \times 284.2 \times 760}{776.5 \times 135.1}$

A-5 Solve the following with logs:

a. 26.2^2 b. 2.713^3

c. 43.68^3 d. 0.1641^4

e. 0.2424^2 f. 81.4^5

g. 653.4^2 h. $\sqrt[3]{4.452}$

i. $\sqrt[3]{0.08610}$ j. $\sqrt{0.5432}$

A-6 Express the following numbers in exponential form:

a. 156,000 b. 8420

c. 0.000419 d. 0.00502

e. 63,080,000 f. 0.0005004

A-7 Compute the following:

a. $6.40 \times 10^{-4} \times 2.00 \times 10^8$

b. $3.00 \times 10^{-2} \times 480$

c. $0.00286 \times 3{,}000$

d. $0.00194 \times 2.00 \times 10^{-7}$

e. $680{,}000 \times 0.00400$

A-8 Solve the following:

a. $(2.00 \times 10^{-9})^3$

b. $(4.00 \times 10^{-8})^{1/2}$

c. $(27.0 \times 10^{-15})^{1/3}$

d. $(25 \times 10^4)^2$

e. $(270 \times 10^{-7})^{1/3}$

A-9 What are the logs of the following?

a. 3.00×10^{5} b. 6.250×10^{-4}

c. 87.61×10^{-3} d. 1.63×10^{-9}

e. 4.07×10^{-11}

A-10 What are the numbers that correspond to the following logs?

a. 3.7125 b. $\bar{4}.8274$

c. 4.8725 - 10 d. $\bar{3}.1486$

e. 8.1942 - 10

Appendix B
NATURAL LOGS, CALCULATORS, AND COMPUTERS

NATURAL LOGARITHMS

In addition to the common logs discussed in Appendix A, another type of log is often encountered in science. Since it appears as a natural function in higher mathematics, it is known as a natural or Naperian logarithm. For instance, the integral of dx/x is the natural log of x and not the common log of x. The symbol for a Naperian function is e.

Whereas common logs have 10 as their base, natural logs have e as their base. The natural log is expressed as $\log_e x$ or usually as ln x. The base e can be represented by the following series:

$$e = \frac{1}{0!} + \frac{1}{1!} + \frac{1}{2!} + \frac{1}{3!} + \frac{1}{4!} + \frac{1}{5!} + \frac{1}{6!} + \cdots \frac{1}{n!}$$

When evaluated, the value of e to six digits is

$$e = 2.71828$$

Just as a logarithm in base 10 is the power to which 10 has to be raised in order to equal a given number, a logarithm in base e is the power to which e has to be raised in order to equal the number.

Examples

$2 = e^{0.693} = (2.71828)^{0.693}$ $\qquad \log_e 2 = 0.693$

$4 = e^{1.386}$ $\qquad \log_e 4 = 1.386$

$2.71828 = e^{1.000}$ $\qquad \log_e 2.71828 = 1.000$

The formula for the half-life of a first-order reaction, as given in Section 17-5, is

$$t_{1/2} = \frac{0.693}{k}$$

The 0.693 here comes from the $\log_e 2$.

When using a computer in the language BASIC, the LOG function available is the $\log_e$. In most forms of BASIC, in order to find the $\log_{10}$ of a number, you divide the $\log_e$ by the $\log_e 10$ or 2.303 (which is $\log_e 10$).

$$\log_{10} x = \frac{\log_e x}{2.303}$$

USING CALCULATORS AND COMPUTERS

A calculator or computer knows nothing about significant figures. Most calculators store eight digits, and computers store six or more digits, depending on the type of variable used. For this reason, some of the answers given in the book may differ from your answer in the third or fourth digit. This all depends on how much you round off while using the calculator or computer.

Some of the least expensive calculators have an x^y or y^x button and not an e^x button. With these calculators, the x^y button can be used as an e^x button by using the value of e given in the first section of this appendix.

With calculators that use algebraic notation the following procedure can be used:

To find $e^{1.3}$, enter 1.3; press the x^y button; enter 2.71828; and then press the equal key. The result should be 3.6693. With calculators that use RPN arithmetic, press 1.3; enter 2.71828; and then press the x^y button.

Particular care must be taken to use the proper number of significant figures in going from an exponential to a number, or in taking the logarithm of a number. Let us examine a few cases of e^x.

x	e^x	
1.00	2.71828	Here a change of 1% in x leads to a change of 1% in e^x.
1.01	2.7456	

5.00	148.413	Here a change of 1% in x leads to a change of 5% in e^x.
5.05	156.022	
10.0	22026	Here a change of 1% in x leads to a change of 10% in e^x.
10.1	24343	

In each of these cases, the value of x used in e^x differed by 1% but the uncertainty in the result increased as x got larger. At the largest value of x we lost one significant figure as we went from x to e^x. The value of x has three significant figures whereas that of e^x differs in the second significant figure. This is true whether x is positive or negative since e^{-x} is $1/e^x$. In general, keep an extra digit in making these transformations, and round off the final result.

In solving these problems on a computer, there are various useful techniques (called algorithms) that can be used over and over again. The following three programs can be useful. If one learns to write and use programs such as these, problems in advanced chemistry courses can be handled in a shorter time than without the computer.

Iterations

Sometimes mathematical relationships are encountered that cannot be solved exactly or easily. For instance:

$$x = 5 - 5e^x$$

has no exact solution. It can be solved by repeated substitutions of x. The following program illustrates how:

```
100 rem    A program to solve a function by iteration.
110 PRINT "Guess a starting value of x."
120 INPUT X
130 X1 = 5 - 5*EXP(X)
140 X = X1
150 IF ABS(X-X1) > 0.0001 THEN GOTO 130
160 PRINT "X = ";X
200 END
```

This program loops through lines 130–150 until x is accurate to four digits to the right of the decimal point.

The Van der Waals equation discussed in Chapter 5 is a cubic equation with respect to the volume. Though cubic equations have an exact solution similar to quadratic equations, most people are not aware of the formula. This program can be used to find the volume of a Van der Waals gas. Find the volume of 1 mol of ammonia gas at 27° and 1 atm pressure using this program.

Quadratic Equations

```
50   REM  A program to solve the equation AX^2+BX +C = 0
100  A = 1
110  B = 10
120  C = 2
130  Q = B^2 = 4*A*C
140  IF Q < 0 THEN GOTO 210
150  PRINT "FIRST ROOT = "; (-B+SQR(Q))/(2*A)
160  PRINT "SECOND ROOT = "; (-B-SQR(Q))/(2*A)
200  END
210  PRINT "ROOTS ARE IMAGINARY"
220  Q = -Q
230  PRINT "FIRST ROOT = "; (-B/(2*A));"+i";SQR(Q)/(2*A)
240  PRINT "SECOND ROOT = "; (-B/(2*A));"-i";SQR(Q)/(2*A)
```

This program can be used to solve the equilibrium problems in Chapters 11, 12, and 13. Change the variables in lines 100–120 to fit your equation.

Linear Least-Squares

One of the techniques used to find the best fit of a set of data to a straight line is known as linear least-squares. The following program fits the data from Example 17-2 to a straight line using this technique. The program is known as a nonweighted least-squares calculation.

```
10  REM this program calculates a least-square fit to the equation
20  REM y = mx + b for up to 50 sets of data points
30  REM Using the values of m and b from the ℓs equation it then
40  REM calculates a y value, the deviation, and the standard deviation
50  REM N is the number of pairs of data points X,Y
90  DIM X(50),Y(50),YC(50),D(50)
100  READ N
110  FOR I = 1 TO N
```

```
120 READ X(I),Y(I)
130 NEXT I
140 FOR I = 1 TO N
145 Y(I) = LOG(Y(I))/2.303
150 X1 = X1 + X(I)
160 Y1 = Y1 + Y(I)
170 X2 = X2 + X(I)^2
180 XY = XY + X(I)*Y(I)
190 NEXT I
200 M = (N*XY-X1*Y1)/(N*X2-X1^2)
210 B = (X2*Y1-X1*XY)/(N*X2-X1^2)
220 FOR I =1 TO N
230 YC(I) = M*X(I) + B
240 NEXT I
250 FOR I = 1 TO N
260 D(I) = YC(I)-Y(I)
270 NEXT I
280 FOR I = 1 TO N
290 D1 = D1 + D(I)
300 D2 = D2 + D(I)^2
310 NEXT I
320 SD = SQR(D2/(N-2))
330 PRINT "FOR A SET OF ";N;" DATA POINTS"
340 PRINT "Y = ";M;"*X +";B
350 PRINT "THE SD = ";SD
390 END
400 DATA 6
410 DATA 0,.0121,10,.0104,20,.00880,33,.00706,46,.00576
420 DATA 65,.00434
```

To use this program with other data, lines 400 on down have to be changed. Line 400 tells the computer how many pairs of x and y values are to be used. Lines 410 and after are the x and y values in the order x1,y1,x2,y2,x3,y3, etc. Line 145 may be deleted. This makes the y axis a $\log_{10}$ axis.

Appendix C
ANSWERS TO PROBLEMS

1-1
a. 3 b. 3
c. 6 d. 3
e. 2 f. ?
g. 2 h. 4
i. 2 j. 4

1-2
a. 98.5 g
b. 99.9%
c. 4.47×10^{-2} mg
d. 77.8%
e. 14.33 g

1-3
a. 6.15 mL
b. 23.8 L
c. 223.6 g
d. 0.0159 g

1-4
a. 2.809 g
b. 2.615 L
c. 2.45×10^{5}
d. 0.119

1-5
a. 1.85×10^{5}
b. 24.4
c. 24.47
d. 24.47
e. 6.067×10^{23}

1-6
a. 2.732
b. 1.077
c. 1.27×10^{-1}
d. 2.54
e. 3.20×10^{-4}

1-7
a. 7.4×10^{-2} nm
b. 7.4×10^{-5} μm
c. 7.4×10^{-8} mm

1-8
a. 6.939 mg 0.006939 kg

1-9 1000 γ

1-10 50 cc

1-11 2.580 g/L

1-12 0.463 g

1-13 272 mL

1-14 \$12.30/100 g

1-15 C.Al

1-16 10×10^{-3} msec; 10×10^{-6} μsec

1-17 yes; 4 oz = 113.5 g

1-18 9.52×10^{-2} mm

1-19 475 g

1-20 128 g

1-21 yes; 383 g = 0.845lb

1-22 484 cc

1-23 348 g

1-24 a. 1776 g b. 965 g

1-25 42.0°F

1-26 a. 98.6°F
b. body temperature

1-27 2.90×10^3 J

1-28 25.4°

1-29 36.1°C

1-30 a. 10°C
b. 1°C
c. 18°F; 1.8°F

1-31 a. 1.10×10^{-1} nm
b. 1.10×10^{-4} μm
c. 1.10×10^{-7} mm

1-32 a. 1.32×10^{-7} mm
b. 1.32×10^{-4} μm
c. 1.32×10^{-1} nm

1-33 200,590 mg 0.20059 kg

1-34 10 mL

1-35 250 mL

1-36 11,508 cc

1-37 5.155 g

1-38 1.51 lb

1-39 5.5 lb

1-40 $14.00/L

1-41 61.5 cc

1-42 188 mL

1-43 11.2 kg

1-44 205

1-45 0.003 sec 3000 μsec

1-46 0.0545 g

1-47 113.5 g

1-48 880 g

1-49 3.77 ℓ 3,770 mL

1-50 472 mL 0.472 L

1-51 22.7 g 431 g

1-52 66.4 g

1-53 35.6 %

1-54 43.9°F

1-55 15°C

1-56 0.171 cal

1-57 21.2°

2-1	61.32% Cd 38.66% Cl	2-2	1.585
2-3	0.92188	2-4	47.970% Zn 52.028% Cl
2-5	1.60 g	2-6	0.07148 mol
2-7	3.82×10^{-23} g 3.82×10^{-20} mg	2-8	1.97×10^{-22} g
2-9	4.77×10^{23} atoms	2-10	0.525 mol
2-11	1.15×10^{23} atoms	2-12	12.84 mol 7.73×10^{24} atoms
2-13	5.33×10^{-11} g	2-14	3.48×10^{8} atoms
2-15	3.28×10^{24} atoms	2-16	1.46×10^{24} atoms
2-17	107.87	2-18	35.46
2-19	69.72	2-20	83.80
2-21	112.4	2-22	14.007
2-23	24.39	2-24	1.0076
2-25	1.0025	2-26	63.90 g
2-27	63.80 g	2-28	3.01×10^{-19} mg
2-29	104 g	2-30	3.56 mol
2-31	2.14×10^{22} atoms/g 2.14×10^{19} atoms/mg 2.14×10^{16} atoms/μg 2.14×10^{16} atoms/γ	2-32	6.03×10^{10} atoms
2-33	5.19×10^{20} atoms	2-34	2.69×10^{-6} mol
2-35	1.30×10^{-19} mol	2-36	4.42×10^{16} atoms
2-37	0.338 mol	2-38	2.42 mol

2-39 1.14×10^{23} atoms

2-40 69.723

2-41 79.89

2-42 65.366

2-43 127.58

2-44 87.617

2-45 15.999

3-1 304 g

3-2 55.5 mol

3-3 1.03×10^{23} P atoms
4.12×10^{23} O atoms

3-4 3.39×10^{22} molecules
6.78×10^{22} O atoms

3-5 1.34×10^{-22} g

3-6 a. 160.2 g
b. 288 g
c. 1.15×10^{-5} g
d. 169 g
e. 1000 g

3-7 1.27×10^{24} Ca^{2+}
2.54×10^{24} Cl^-

3-8 3.55×10^{-2} g

3-9 1.10

3-10 0.644

3-11 32.8% As, 14.9% Si

3-12 67.4% Co; 32.6% S

3-13 40.36% Me; 59.65% Cl

3-14 AgF_3

3-15 BNH_4

3-16 Si_2NH_5

3-17 Tl_2TiCl_6

3-18 TiF_3

3-19 $CeCl_3 \cdot 7H_2O$

3-20 $Co(NH_4)_2Cl_4 \cdot 2H_2O$

3-21 B_6H_{12}

3-22 $MoCl_2$

3-23 $B_3N_3H_3F_3$

3-24 BF_3NH_3

3-25 $P_2F_2Cl_2O_3$

3-26 71.0 g

3-27 0.0859 mol

3-28 1.03×10^{22} molecules

3-29	a. 257 g b. 128 g c. 64.2 g d. 77.0 g e. 2.07×10^{-2} g	3-30	a. 591 g b. 86.6 c. 0.806 g d. 1540 g e. 61.3 g
3-31	7.60×10^{20} Mg^{2+} 1.52×10^{21} Br^-	3-32	1.12
3-33	0.6891	3-34	1.933
3-35	41.1% V; 46.0% F	3-36	19.3% NH_3; 36.1% Cu, 43.2% F
3-37	BI_3	3-38	$ZrBr_4$
3-39	WCl_6	3-40	$CrOF_3$
3-41	ThF_4	3-42	ReO_3F
3-43	K_2ReCl_6	3-44	PSI_2
3-45	ICl_3	3-46	TiF_4
3-47	$ZrBr_3$	3-48	K_2TeI_6
3-49	WBr_3	3-50	RhF_6
3-51	B_2O_2N BN	3-52	CdP_4
3-53	HgNCl	3-54	K_2AsO_3F
3-55	$LiBCH_6$	3-56	$AgNH_2$
3-57	$Ag_2TeO_4 \cdot 2H_2O$	3-58	$CaCl_2 \cdot 6H_2O$
3-59	Sc_2S_3	3-60	S_3Cl_2
3-61	$P_3N_3F_6$ $P_4N_4F_8$	3-62	CO_2F_2

4-1 a. 3 4 1 4 b. 2 2 1 2 c. 1 2 1
d. 1 2 1 2 e. 2 2 1 f. 2 2 1
g. 2 10 6 1 h. 2 2 2 1 i. 8 4 7 2 8 8
j. 3 1 2 1 k. 1 1 3 1 l. 1 3 1 2
m. 1 12 4 3 n. 1 6 2 2 3 o. 1 3 2 3
p. 2 5 2 2 4 q. 2 1 2 1 r. 1 3 1 1
s. 1 2 1 1 t. 1 2 1 2 u. 2 3 2 2
v. 2 1 2 1 w. 1 2 3 1 x. 1 1 2 2
y. 1 1 2 z. 1 1 2

4-2 a. 2 9 1 6 b. 4 11 4 3 c. 6 2 6 6 1
d. 1 12 3 4 e. 3 4 2 6 1 f. 6 3 2 8
g. 1 6 2 4 1 h. 2 2 2 2 1 i. 1 3 1 2
j. 1 2 1 2

4-3 6 and 12 mol

4-4 24 mol

4-5 1.84×10^{-3} g

4-6 0.146 g

4-7 0.647 g

4-8 2.62 g

4-9 0.0458 g

4-10 14.1 g

4-11 1.81 mmol

4-12 $WCl_4 + 2H_2 \rightarrow W + 4HCl$
$HCl + NaOH \rightarrow NaCl + H_2O$

4-13 $TlVO_3$
$Tl_2CO_3 + V_2O_5 \rightarrow 2TlVO_3 + CO_2$

4-14 $2HF \rightarrow H_2 + F_2$

4-15 $2K_2MnF_6 + 4SbF_5 \rightarrow F_2 + 4KSbF_6 + 2MnF_3$

4-16 $2KMnO_4 + 2KF + 10HF + 3H_2O_2 \rightarrow 2K_2MnF_6 + 8H_2O + 3O_2$

4-17 a. $NF_3 + F_2 + SbF_5 \rightarrow NF_4SbF_6$ b. SbF_5 c. 5.26 g

4-18 a. $NF_3 + BF_3 + F_2 \rightarrow NF_4BF_4$ b. F_2 c. 62.9%

4-19 $6CCl_4 + 9NH_4Cl \rightarrow C_6N_9H_3 + 33HCl$

4-20 58.8% $WOCl_4$, 41.2% WOF_4 based on W
$2WO_2 + 2CCl_2F_2 \rightarrow WOF_4 + WOCl_4 + 2CO$

4-21 1.88 g $InAlCl_4$

4-22 3.25×10^{-7} mol
7.52×10^{-5} g

4-23 1.06×10^{-3} g

4-24 88.7%

4-25 39.55%

4-26 99.530%

4-27 Hg_3N_2

4-28 18.89 g

4-29 1.01 g

4-30 99.84%

4-31 14.00

4-32 Mn^{+3} reduced
Cr^{+3} oxidized

4-33 a. Al_2O_3
b. 97.4%

4-34 Cl reduced
O oxidized

4-35 a. 4 5 4 6 b. 3 1 2 1 c. 4 4 1 2
d. 1 2 1 4 e. 1 3 5 3 5 1 f. 1 3 1 3
g. 1 6 2 3 h. 3 1 3 i. 1 1 2 1
j. 4 4 1 2 k. 1 1 1 1 l. 3 2 1 3
m. 1 1 2 n. 6 1 2 o. 1 6 4
p. 1 4 1 2 1 q. 1 3 1 3 1 r. 1 3 3 2
s. 2 3 1 t. 2 3 3 3 1 u. 1 1 1 2
v. 1 4 1 4 w. 2 1 1 1 x. 2 2 4 1
y. 1 6 3 2 z. 1 10 2 10 4

4-36 20.4 mmol

4-37 190 g

4-38 86.3%

4-39 $Zr + 2Cl_2 \rightarrow ZrCl_4$
$3ZrCl_4 + Zr \rightarrow 4ZrCl_3$

4-40 67.3 g

4-41 $3WCl_4 \rightarrow 2WCl_5 + WCl_2$

4-42 $SnS + 5F_2 \rightarrow SnF_4 + SF_6$

4-43 $Gd_2O_3 + 3CCl_4 \rightarrow 3COCl_2 + 2GdCl_3$

4-44 99.970%

4-45 a. $NaBH_4 + 4I_2 \rightarrow BI_3 + NaI + 4HI$
b. I_2
c. 3.69%

4-46 98.52%

4-47 88.9%

4-48 33.0%

4-49 71.2%

4-50 0.01979 g

4-51 18.15 g

4-52 11.1% (theoretical loss)

4-53 S oxidized; Cl reduced

4-54 284 g

4-55 94%

4-56 0.48 mM HF

4-57 a. $Mg + 2SiH_3Br \rightarrow SiH_4 + H_2 + Si + 2MgBr$
b. Mg oxidized; H reduced
c. 7.94×10^{-3} g

4-58 Cl reduced; S oxidized

4-59 1.81 g

4-60 76.68%

4-61 94.45%

4-62 41%

4-63 1.289

4-64 9.013

4-65 107.87

4-66 55.845

4-67 40.08

4-68 50.96

5-1 1.37

5-2 219 mL

5-3 542 mL

5-4 69.7 g/mol

5-5 4.45 dm^3

5-6 a. 3168 g/m^3
b. 3.168 g/dm^3

5-7 a. 39.92 g/mol
b. 3.998 g/mol
c. 2.013 g/mol
d. 83.06 g/mol
e. 20.16 g/mol
f. 28.01 g/mol
g. 32.01 g/mol
h. 218 g/mol
i. 131.2 g/mol

5-8	127.4 g/mol	5-9	a. 5.92×10^{-3} atm to 6.91×10^{-3} atm b. 4.50 torr to 5.25 torr
5-10	8.80 atm	5-11	303.5 g/mol
5-12	a. 0.332 mmol b. 0.658 mmol c. CH_3F and CO_2	5-13	133.5 g/mol $C_7H_{14}O_2$
5-14	B_4Cl_4	5-15	9.674×10^{-2} g
5-16	46.3 cc	5-17	17.2%
5-18	14.7 mm	5-19	101.3 g/mol
5-20	VCl_4	5-21	871 mL
5-22	298 mL	5-23	39.3 g/mol
5-24	44.54 g/mol	5-25	540 mL
5-26	100%	5-27	B_2Cl_4
5-28	Ga_2Cl_6	5-29	22.1-22.6 L
5-30	0.225 L/min	5-31	1.08/L
5-32	0.683 atm	5-33	33.74 atm
5-34	0.00364	5-35	210 cc
5-36	210 mL	5-37	67.3 mL
5-38	8.319	5-39	1.97 g/L
5-40	a. 17.26 g/mol c. 39.92 g/mol e. 67.2 g/mol g. 36.74 g/mol		b. 127.3 g/mol d. 78.5 g/mol f. 81.60 g/mol h. 38.3 g/mol
5-41	0.198 mol	5-42	$B_3N_3Cl_3$

5-43 3.27 g/L

5-44 $B_5H_9 + 15H_2O \rightarrow 5H_3BO_3 + 12H_2$

5-45 96.7% 92% 95.3%

5-46 49.25 vol% 57.8 wt%

5-47 $SO_3 + H_2O$ molecules in vapor

5-48 No 43.2 g/mol

5-49 30.01 g/mol

5-50 28.00 g/mol

5-51 68.2 g/mol

5-52 B_2H_6

5-53 55.9%

5-54 $Ga_2Cl_6 + Ga \rightarrow 3GaCl_2$

5-55 4.95 g 1.53 g

5-56 7.72%

5-57 10/1

5-58 0.0987 g

5-59 0.335 g

5-60 1 min 43 sec

5-61 5.97×10^{-5} mL/sec

5-62 1.005/L

5-63 94.5 atm

6-1 a. 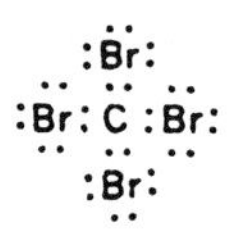

b. 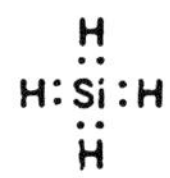

c. 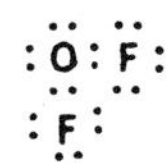

d. 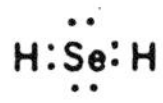

e. 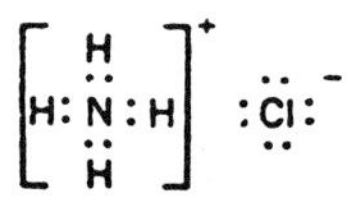

f. 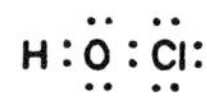

g. 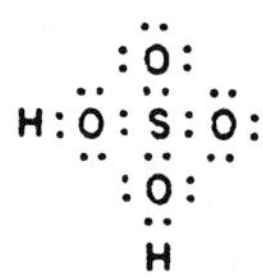

h. 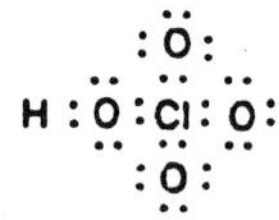

i. 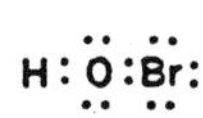

j. 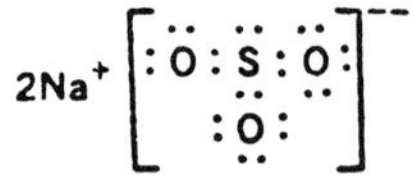

k. 

l. 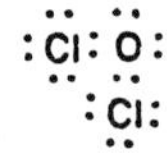

m. 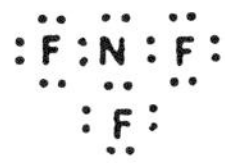

n. 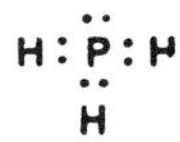

o. 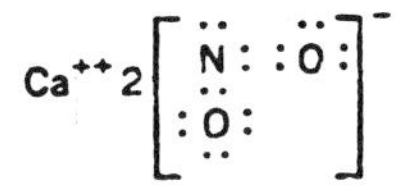

6-2 a. 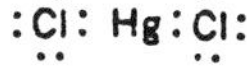

b. 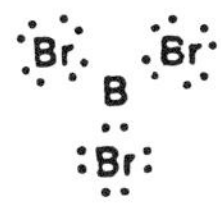

c. 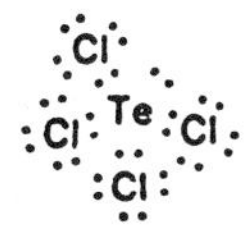

d. 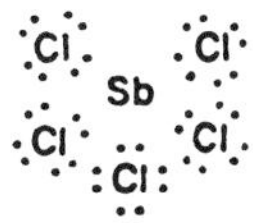

e. 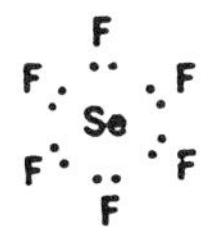

6-3 a. 

b. 

c. 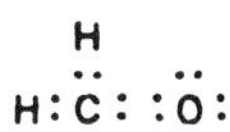

d. 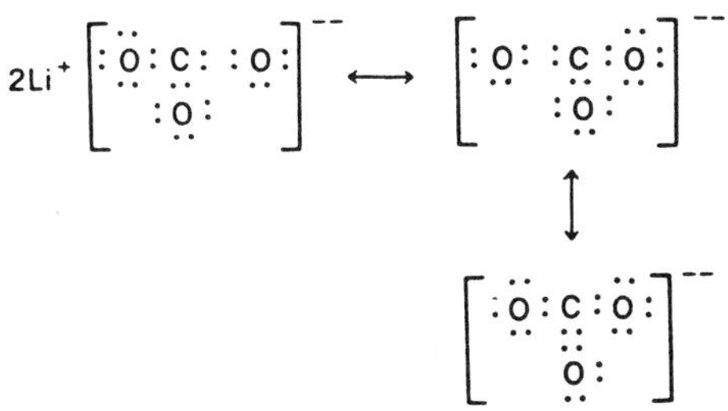

e. 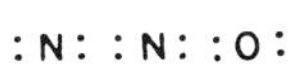

f. 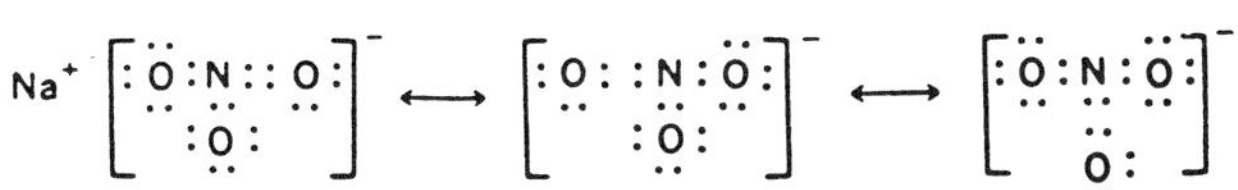

g. 

h. 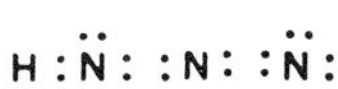

i. 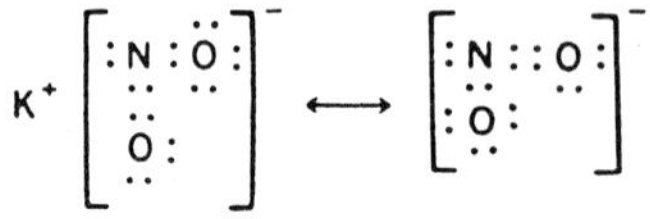

j. 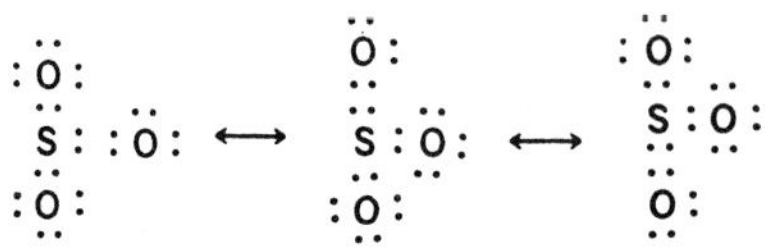

6-4 a. 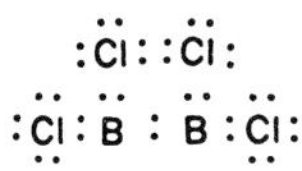

b. 

c. 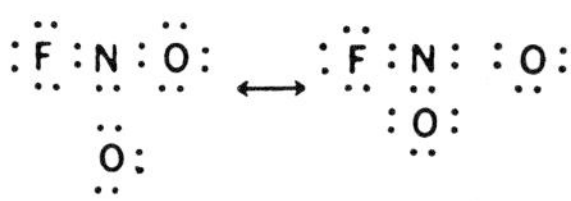

d. 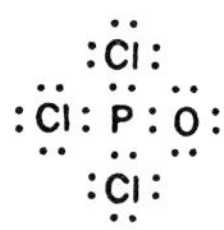

e. 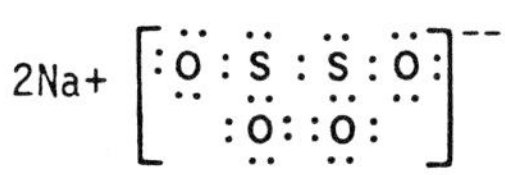

f. 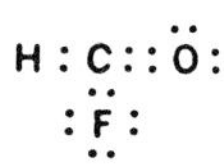

6-5 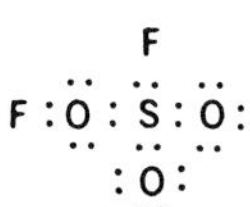

6-6 a. 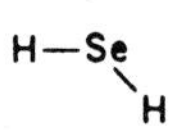

b. 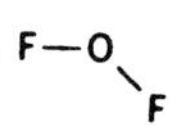

c. 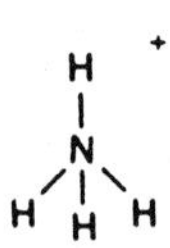

d. 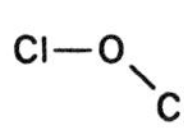

e. 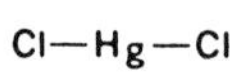

f. 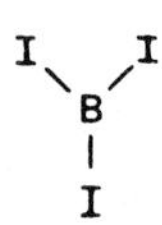

g. 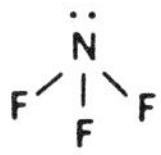

h. 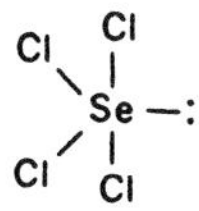

i. 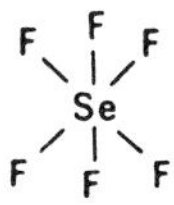

j. 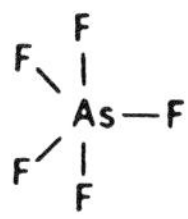

6-7 a. V shape
b. V shape
c. tetrahedron
d. tetrahedron
e. trigonal pyramid
f. tetrahedron
g. plane triangle
h. trigonal bipyramid
i. T shape
j. octahedron

6-8 a. H—C≡N

b. 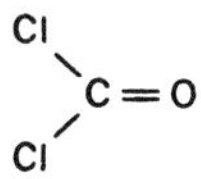

c. 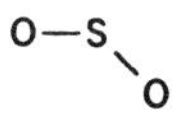

d. 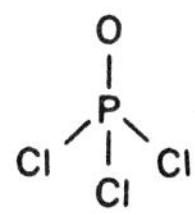

e. 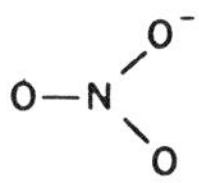

f. 

g. 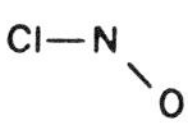

h. 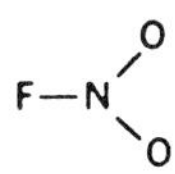

6-9 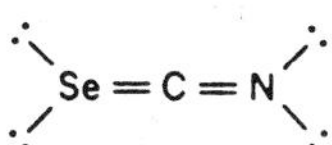

linear

6-10 

6-11 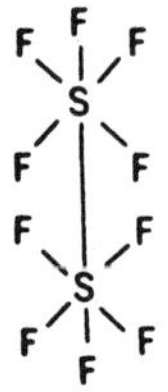

6-12 a. B(SCN)$_3$

b. BCl_3 slight excess

c. 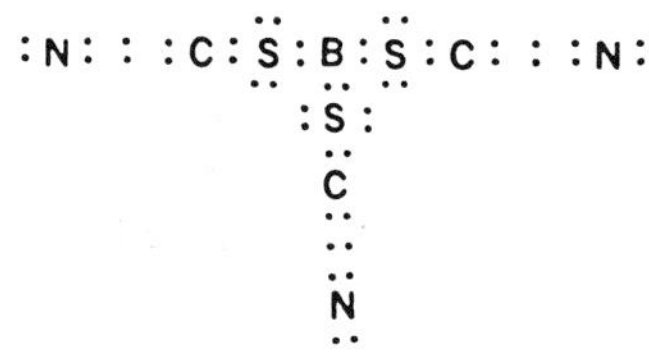

6-13 a. 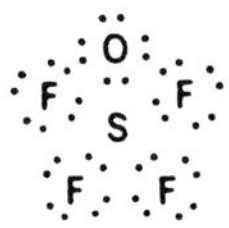

b. 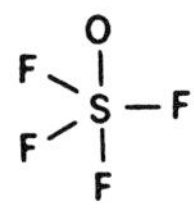

6-14 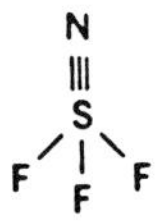

6-15 a. 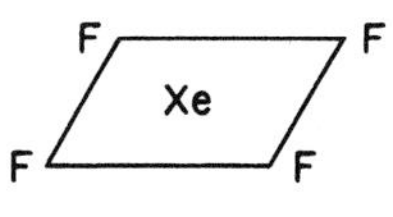

b. F—Xe—F

c. 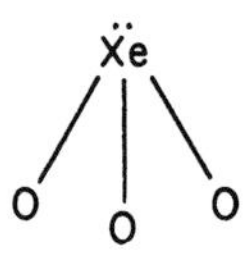

d. F—Kr—F

e. 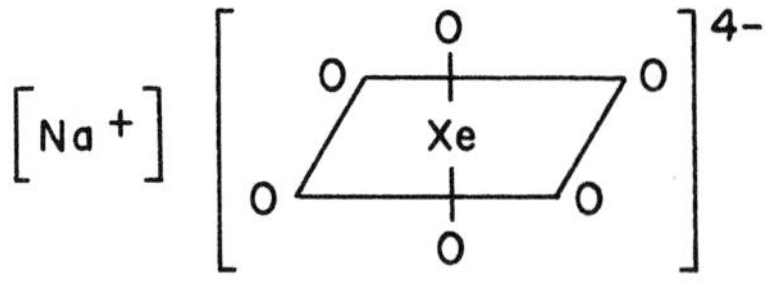

6-16 a. 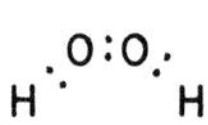

b. F:O:O:F

c. 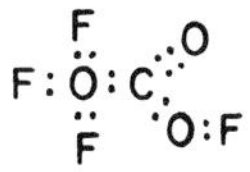

d. 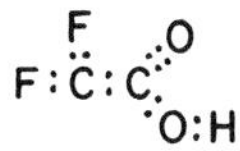

e. 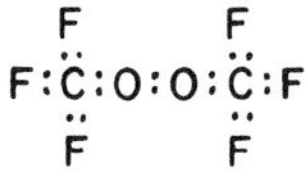

6-17 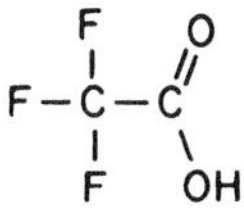

6-18 a. 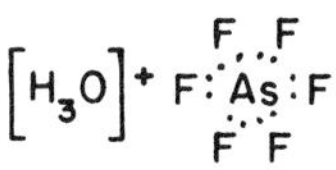

b. 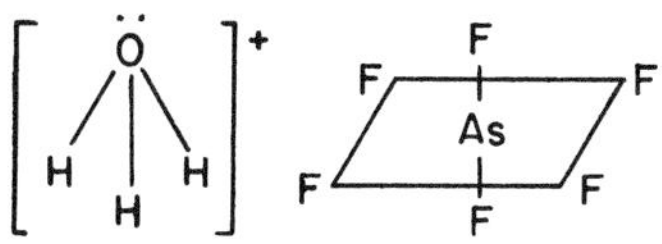

6-19 a. 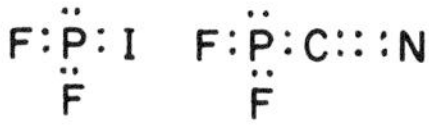

b. 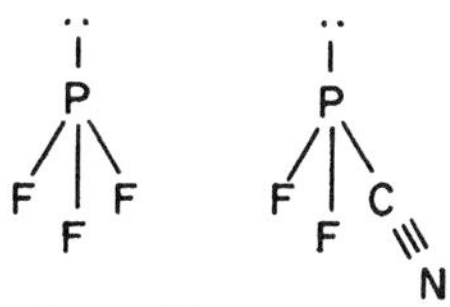

6-20 a. 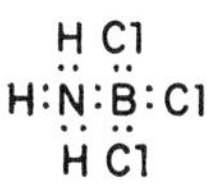

b. 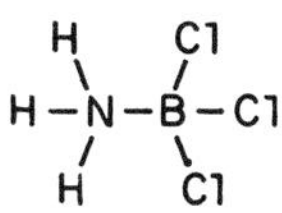

6-21 a. 

b. 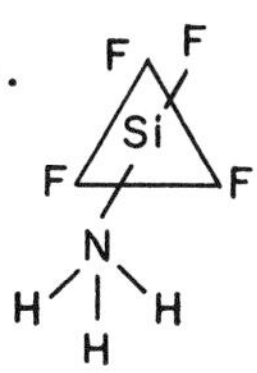

Note: 2:1 adduct is also well known

6-22 a. 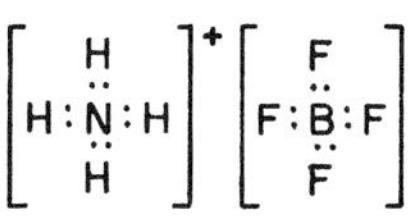

b. 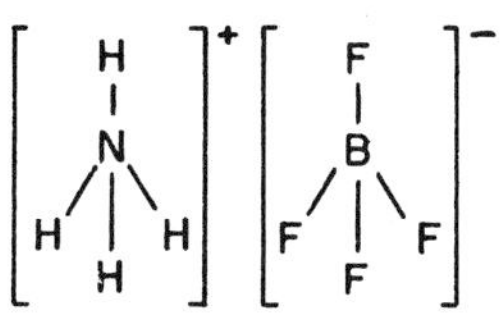

6-23 a. 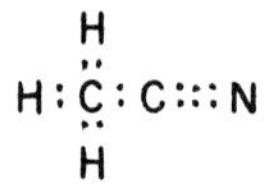 b. 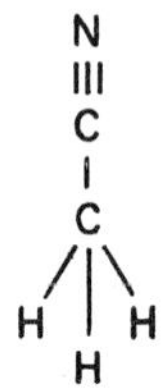

6-24
a. bent
b. bent
c. tetrahedral
d. bent
e. linear
f. equilateral triangle
g. (trigonal) pyramid
h. distorted tetrahedron
i. octahedron
j. trigonal (triangular) bipyramid

6-25
a. sp^3
b. sp^3
c. sp^3
d. sp^3
e. sp^3
f. sp^3
g. sp^2
h. sp^3d
i. sp^3d
j. sp^3d^2

7-1 10 ppm

7-2 100 ppm

7-3 2000 ppm

7-4 a. 50.4 g b. 78.62 g c. 7.5 g
d. 7.24 g e. 174 g

7-5 a. 42.4 g b. 412.5 g c. 0.3935 g
d. 1.215×10^{-2} g e. 9.75 g

7-6 a. 8.415 g b. 124.8 g c. 0.7988 g
d. 58.2 g e. 8.75×10^{-4} g

7-7 a. 1.88 M b. 2.87 M c. 1.30 M

7-8 a. 24.5 mL b. 100 mL c. 0.781 mL
d. 0.00830 mL e. 0.0329 mL

7-9 a. 0.144 M b. 0.8 M c. 4 ppm
d. 1.92 M

7-10 6725 mL

7-11 12.4 M

7-12 16.09 M

7-13 12.1 M 16.0 M

7-14 7.576 M

7-15 15.6 m

7-16 0.0922 m

7-17 1.362 M 1.500 m

7-18 a. 7.059 m
b. HF(aq)
c. 19.2 g

7-19 3.14 mL

7-20 0.977 solvent
0.023 solute

7-21 0.0158

7-22 5798 ppm

7-23 a. 66.46 g
b. 1.25 g
c. 6.044 g
d. 0.605 g
e. 2.411 g
f. 157.6 g
g. 39.8 g
h. 2.64 g
i. 3.87×10^{-3} g
j. 0.401 g

7-24 1.2489 g

7-25 a. 100 ppm
b. 1) 5 ppm
2) 10 ppm
3) 15 ppm

7-26 a. 2.0 g
b. 1.7 g
c. 2.12 g
d. 0.0602 g
e. 0.0917 g

7-27 5.54×10^{-3} g

7-28 a. 0.0603 M
b. 5.03×10^{-4} M
c. 0.0151 M

7-29 a. 0.5770 M
b. 0.212 M
c. 0.0194 M
d. 0.00579 M
e. 0.00845 M

7-30 1481 mL

7-31 4094 mL

7-32	8.71 M	7-33	a. 25.0 mL b. 1.67 mL c. 3.47 mL
7-34	23.0 M	7-35	11.7 M
7-36	3.99 m	7-37	0.495 m
7-38	4.3110 m	7-39	2.68 m
7-40	16.0 L	7-41	85.2 g

8-1 17.35 mm

8-2 634.4 mm

8-3 35.5 mm

8-4 From Raoult's Law

106.5 mm	146.0 mm	250.4 mm
346.7 mm	446.1 mm	552.9 mm

8-5	0.240°	8-6	40.6°
8-7	-6.42°C	8-8	0.362°
8-9	2.40 atm	8-10	182 g/mol
8-11	128.7 g/mol	8-12	463 g/mol
8-13	128 g/mol	8-14	258 g/mol S_8
8-15	P_4	8-16	5.08 °C kg/mol
8-17	2860 g/mol	8-18	28.2 g/mol
8-19	-7.30°	8-20	49 g/mol
8-21	1.75×10^6 g/mol	8-22	$S_4N_4F_4$
8-23	613.9 mm	8-24	0.156 mm

8-25 32.18 mm

8-26 From Raoult's Law

46.10 mm	176.9 mm	243.6 mm
286.9 mm	309.4 mm	356.8 mm

8-27	71.8 mm	8-28	0.170°
8-29	46.5°C	8-30	-8.80°C
8-31	-0.0374°	8-32	-0.1485°
8-33	-4.10°	8-34	0.518 g
8-35	2.40×10^{-3} atm	8-36	129 g/mol
8-37	53.8 g/mol	8-38	258.6 g/mol
8-39	185 g/mol	8-40	a. 0.278 deg b. 0.408 deg
8-41	P_2I_4	8-42	S_8
8-43	P_4	8-44	341 g/mol
8-45	186 g/mol	8-46	258 g/mol
8-47	2.32×10^5 g/mol		

9-1	43.1 J	9-2	10.86 kJ
9-3	2.68 kJ	9-4	B_2H_6
9-5	a. -5443.0 kJ/mol b. -1380 kJ/mol	9-6	-0.8351 kJ/mol
9-7	-1682 kJ/mol	9-8	-1076 kJ/mol
9-9	-285.98 kJ/mol	9-10	550.2 kJ/mol
9-11	a. -495.4 kJ/mol b. -497.5 kJ/mol	9-12	-780.7 kJ/mol

9-13 533.0 kJ/mol

9-14 -1302.9 kJ/mol

9-15 -66.1 kJ/mol

9-16 -94.5 kJ/mol

9-17 -383.7 kJ/mol

9-18 -184.0 kJ/mol

9-19 a. -1233 kJ/mol
b. 387 kJ/mol

9-20 -898.3 kJ/mol

9-21 -184.9 kJ/mol

9-22 -1055 kJ/mol

9-23 460.8 kJ/mol
110.1 kcal/mol

9-24 13.414 kJ/°c

9-25 158.2 kJ/mol

9-26 463.2 kJ/mol

9-27 457.7 kJ/mol

9-28 383.8 kJ/mol

9-29 431.4 kJ/mol first bond
391 kJ/mol ave

9-30 366 kJ/mol

9-31 -50 kJ

9-32 121.1 kJ

9-33 -18.4 kJ/mol

9-34 -1084 kJ/mol

9-35 -1037 kJ/mol

9-36 -63.22 kJ/mol

9-37 -237.6 kJ/mol

9-38 -40.3 kJ/mol

9-39 -1385.8 kJ/mol
24.06 kJ/mol

9-40 -925.9 kJ/mol

9-41 -255.2 kJ/mol

9-42 -607.9 kJ/mol

9-43 -338 kJ/mol

9-44 -225 kJ/mol

9-45 899.6 kJ/mol

9-46 -910.9 kJ/mol

9-47 377 kJ/mol

9-48 -271.6 kJ/mol

9-49 -1718 kJ/mol

9-50 -89.29 kJ/mol

9-51 -133.4 kJ/mol

9-52 -461.5 kJ/mol

9-53	140.4 kJ/mol	9-54	-926.8 kJ/mol
9-55	3212.6 g	9-56	-3026.6 kJ/mol
9-57	430.9 kJ/mol	9-58	322 kJ/mol
9-59	248 kJ/mol first bond 279 kJ/mol ave	9-60	366 kJ/mol
9-61	326 kJ/mol	9-62	50.33 kJ/mol
9-63	594.3 kJ/mol	9-64	523 kJ/mol

10-1 a. 7.16 b. 9.06
c. 2.92 d. 1.95
e. 3.43

10-2 2.42 3.17 3.75 3.27 2.01

10-3 a. 1.12×10^{-10} mol/L b. 1.07×10^{-2} mol/L
c. 1.78×10^{-6} mol/L d. 8.32×10^{-11} mol/L
e. 3.80×10^{-3} mol/L

10-4	a. 1.74 b. 1.94 c. 2.96	10-5	2.32
10-6	1.16	10-7	9.46×10^{-3} mL
10-8	8.46×10^{-2} mL	10-9	2.15×10^{-2} mol/L
10-10	0.244 mol/L	10-11	24.5 g
10-12	27.7 g	10-13	0.0130 g
10-14	0.226 mm	10-15	0.0810°
10-16	0.0165°	10-17	0.0934°

10-18 0.216° ions 0.108° no ions

10-19 From Raoult's Law
17.41 mm 16.91 mm
17.28 mm 16.79 mm
17.16 mm 16.61 mm
17.04 mm 16.32 mm
16.97 mm

10-20 361

10-21 57.2

10-22 0.0188 0.1179
0.0358 0.2164
0.0613 0.4344

10-23 0.101 M 0.202 M

10-24 0.108 M

10-25 2.42

10-26 3.828 g/500 mL

10-27 3.431 g/500 mL

10-28 0.0219 M

10-29 pH = 2.67

10-30 KPF_6

10-31 1.442 g/100 mL

11-1 a. $K_c = \frac{[N_2][O_2]}{[NO]^2}$ b. $K_c = \frac{[NH_3]^2}{[N_2][H_2]^3}$

c. $K_c = \frac{[NH_3]^2[H_2O]^4}{[NO_2]^2[H_2]^7}$ d. $K_c = \frac{[CO]^2}{[O_2]}$

e. $K_c = \frac{[H_2O]^4}{[H_2]^4}$ f. $K_c = \frac{[CO][H_2O]}{[CO_2][H_2]}$

g. $K_c = [O_2]^{1/2}$ h. $K_c = \frac{[NO_2][NO]}{[N_2O_3]}$

11-2 a. $K_p = \frac{P_{N_2} \cdot P_{O_2}}{P^2_{NO}}$ b. $K_p = \frac{P^2_{NH_3}}{P_{N_2} \cdot P^3_{H_2}}$

c. $K_p = \frac{P^2_{NH_3} \cdot P^4_{H_2O}}{P^2_{NO_2} \cdot P^7_{H_2}}$ d. $K_p = \frac{P^2_{CO}}{P_{O_2}}$

e. $K_p = \frac{P^4_{H_2O}}{P^4_{H_2}}$

f. $K_p = \frac{P_{CO} P_{H_2O}}{P_{CO_2} P_{H_2}}$

g. $K_p = P_{O_2}^{1/2}$

h. $K_p = \frac{P_{NO_2} P_{NO}}{P_{N_2O_3}}$

11-3 0.144

11-4 9.64×10^{-4}

11-5 45.4

11-6 0.0218

11-7 8.16×10^{-3}

11-8 6.20

11-9 0.49

11-10 144 atm

11-11 2.95 atm

11-12 3.29×10^{-4}

11-13 3.13×10^{-2}

11-14 a. $K_p = \frac{P^2_{AgBr} P_{OPF}}{P_{P(O)F\ Br_2}}$

b. 1.39×10^{-9} bar^2

11-15 0.298 mol/L

11-16 0.176

11-17 a. 0.992 atm
b. 0.984 atm

11-18 158 mm

11-19 26.7

11-20 85.0 mm

11-21 2.76×10^{-4}

11-22 0.00478

11-23 2.90×10^{-4} 2.87×10^{-4} 3.05×10^{-4} 2.66×10^{-4}
K_p(ave) = 2.87×10^{-4}

11-24 4.65×10^{-7} mol/L

11-25 1.57

11-26 1.36

11-27 3.87×10^{-3} mol/L

11-28 1.17×10^{-3} mol/L

11-29 0.56 mol 0.44 mol

11-30 0.661 mol/L

11-31 a. $K_p = \frac{P^2_{Se_2Cl_2}}{P_{SeOCl_2}}$ b. $K_p = \frac{P^2_{SeCl_2}}{P_{SeOCl_2}}$

11-32 1.46×10^{-5}

11-33 a. 0.790 $atm^{1/2}$
b. 1.6 atm^{-1}

11-34 6.85×10^4

11-35 85.6 mm 0.113 atm

11-36 0.0375 mm^{-1}

11-37 4.32×10^{-3}

11-38 3.27×10^{-5} atm

11-39 2.20×10^{-5}

11-40 4.15×10^{11}

11-41 1.91 atm

11-42 31.6 mm

11-43 1.59

11-44 0.597 mm

11-45 0.787

11-46 1.71×10^{-11}

11-47 $K_c = 169$; $K_p = 5.82$

11-48 9.56×10^{-3}

11-49 4.48×10^{-10} mm^{-2}

11-50 1.79×10^{-10}

11-51 27.8

11-52 94.44 mm

11-53 0.118 atm NO_2

11-54 insufficient data

11-55 0.165 atm NH_3
4.918 atm N_2
4.752 atm H_2

11-56 0.0770 atm

11-57 0.714

11-58 0.428 mol/L

11-59 0.84%

11-60 90.8%

11-61 21.4 mol%

11-62 1.41×10^{-3} mol/L

11-63 1.86×10^{-3} mol/L

11-64 1.76×10^{-2} mol/L for monomer

12-1	1.84×10^{-5}	12-2	1.51×10^{-5}
12-3	6.74×10^{-5}	12-4	6.5×10^{-4} mol/L
12-5	2.73	12-6	0.925 M
12-7	2.97×10^{-3} mol/L	12-8	7.75×10^{-3} mol/L 6.78×10^{-3} mol/L 5.66×10^{-3} mol/L
12-9	2.94×10^{-6} mol/L	12-10	10.39
12-11	3.09	12-12	1.51×10^{-6}
12-13	3.64×10^{-9}	12-14	2.21
12-15	2.60	12-16	11.50
12-17	1.56	12-18	1.05×10^{-5} mol/L
12-19	3.88	12-20	0.455 g
12-21	11.3/L	12-22	4.54
12-23	0.075	12-24	a. 0.055/L b. 17.8/L c. 2/L
12-25	1.88×10^{-5}	12-26	4.80
12-27	1.21×10^{-22} mol/L	12-28	6.23×10^{-21} mol/L
12-29	5.52	12-30	9.84×10^{-7} mol/L pH = 7.99
12-31	12.7	12-32	5.98
12-33	4.21	12-34	1.86×10^{-5} 4.73
12-35	3.24	12-36	10.92
12-37	2.86	12-38	10.86

12-39	1.16% 3.66% 11.6%	12-40	2.86
12-41	6.26×10^{-5} mol/L	12-42	4.30
12-43	2.26	12-44	1.98
12-45	2.47	12-46	2.2 g
12-47	3.46	12-48	0.077 M
12-49	1.40	12-50	3.97
12-51	1.16	12-52	5.07
12-53	2.21	12-54	2.60
12-55	1.48×10^{-3} mol/L	12-56	0.109 M
12-57	3.50×10^{-6} mol/L	12-58	1.12×10^{-3} mol/L
12-59	0.0501 mol/L		

13-1	3.400×10^{-6}	13-2	7.02×10^{-7}
13-3	4.15×10^{-18}	13-4	5.67×10^{-8}
13-5	3.24×10^{-5}	13-6	3.48×10^{-5}
13-7	7.67×10^{-5}	13-8	$K_S = 2.13 \times 10^{-11}$
13-9	4.44×10^{-11}	13-10	1.09×10^{-8} mol/L
13-11	3.74×10^{-4} mol/L	13-12	1.09×10^{-3} mol/L
13-13	a. 5.01×10^{-16} b. 6.56×10^{-5} m	13-14	1.59×10^{-2} mol/L
13-15	$[Ag^+] \times 10^6$ a. 31.9 b. 12.1 c. 2.91 d. 0.60 e. 0.47 f. 0.07	13-16	before 2.30×10^{-4} mol/L after 4.9×10^{-9} mol/L

13-17	a. 1.6×10^{-3} m b. 1.6×10^{-8} m	13-18	1.52×10^{-7} mol/L
13-19	0.0565 mol/L	13-20	3.85×10^{-5} mol/L
13-21	4.31×10^{19}	13-22	9.23
13-23	1.18×10^{-8} mol/L	13-24	9.24
13-25	4.61	13-26	2.75×10^{-5}
13-27	9.44	13-28	8.73
13-29	5.21	13-30	8.90
13-31	1.69×10^{-11}	13-32	1.53×10^{-4}
13-33	7.53×10^{-8}	13-34	3.40×10^{-6}
13-35	3.08×10^{-11}	13-36	3.77×10^{-10}
13-37	4.48×10^{-17}	13-38	7.99×10^{-8}
13-39	6.22×10^{-5}	13-40	2.64×10^{-3} mol/L
13-41	6.5×10^{-5} mol/L	13-42	7.99×10^{-4} mol/L
13-43	1.37×10^{-6} mol/L	13-44	2.64×10^{-3} mol/L
13-45	1.4×10^{-5} mol/L	13-46	7.51×10^{-20} mol/L
13-47	9.04×10^{-4} mol/L	13-48	6.0×10^{-9} mol/L
13-49	7.02×10^{-16} mol/L	13-50	2×10^{-10} mol/L
13-51	1.73×10^{-20} mol/L	13-52	3×10^{-6} mol/L
13-53	0.286 mol/L	13-54	1.008 mol/L
13-55	5.13 g/100 mL	13-56	2.48×10^{-5}
13-57	1.02%	13-58	8.41
13-59	5.36	13-60	8.71

14-1

	Oxidized	Reduced
a.	H	Fe
b.	Cl	Mn
c.	S	Cr
d.	C	P
e.	None	
f.	O	Cl
g.	None	
h.	N	N
i.	None	
j.	Br	Br

14-2

a. $NO_2 \rightarrow NO_3^-$ (ox)
$HClO \rightarrow Cl^-$ (red)

b. $Mg \rightarrow Mg^{2+}$ (ox)
$NO_3^- \rightarrow N_2$ (red)

c. $H_2S \rightarrow SO_2$ (ox)
$IO_3^- \rightarrow I_2$ (red)

d. $HNO_2 \rightarrow NO_3^-$ (ox)
$CrO_4^{2-} \rightarrow Cr^{3+}$ (red)

e. $Cl^- \rightarrow Cl_2$ (ox)
$MnO_2 \rightarrow Mn^{2+}$ (red)

14-3

a. $I_2 + 2S_2O_3^{2-} \rightarrow 2I^- + S_4O_6^{2-}$

b. $6H^+ + 5Br^- + BrO_3^- \rightarrow 3Br_2 + 3H_2O$

c. $64H^+ + 16NO_3^- + 24CuS \rightarrow 24Cu^{2+} + 3S_8 + 16NO + 32H_2O$

d. $16H^+ + 2MnO_4^- + 10Cl^- \rightarrow 2Mn^{2+} + 5Cl_2 + 8H_2O$

e. $4Zn + 10H^+ + NO_3^- \rightarrow 4Zn^{2+} + NH_4^+ + 3H_2O$

f. $3Cu + 8H^+ + 2NO_3^- \rightarrow 3Cu^{2+} + 2NO + 4H_2O$

g. $Cu + 3H^+ + HSO_4^- \rightarrow Cu^{2+} + SO_2 + 2H_2O$

h. $64H^+ + 24H_2S + 8Cr_2O_7^{2-} \rightarrow 3S_8 + 16Cr^{3+} + 56H_2O$

i. $2Cu^{2+} + 4I^- \rightarrow 2CuI + I$

j. $H^+ + 2ClO_3^- + SO_2 \rightarrow 2ClO_2 + HSO_4^-$

k. $3H_2O + 5ICl \rightarrow 2I_2 + IO_3^- + 5Cl^- + 6H^+$

l. $3H_2SO_3 + 5H^+ + Cr_2O_7^{2-} \rightarrow 2Cr^{3+} + 3HSO_4^- + 4H_2O$

m. $2Mn^{2+} + 5S_2O_8^{2-} + 8H_2O \rightarrow 2MnO_4^- + 10HSO_4^- + 6H^+$

n. $6H^+ + 2MnO_4^- + 5H_2O_2 \rightarrow 2Mn^{2+} + 5O_2 + 8H_2O$

o. $2H^+ + Ag + NO_3^- \rightarrow Ag^+ + NO_2 + H_2O$

14-4

a. $2OH^- + 2H_2O + SbO_2^- + 2ClO_2 \rightarrow 2ClO_2^- + Sb(OH)_6^-$

b. $Cl_2 + 2OH^- + IO_3^- \rightarrow 2Cl^- + IO_4^- + H_2O$

c. $3I_2 + 6OH^- \rightarrow 5I^- + IO_3^- + 3H_2O$

d. $H_2O + 6Fe_3O_4 + 2MnO_4^- \rightarrow 9Fe_2O_3 + 2MnO_2 + 2OH^-$

e. $2OH^- + 2MnO_2 + 3H_2O_2 \rightarrow 2MnO_4^- + 4H_2O$

14-5

a.	3	2	8	3	2	2	11
b.	1	40	40	12	2	5	
c.	2	27	64	6	2	54	32
d.	3	14	18	14	6	9	

14-6 2.54 g

14-7 4372 mL

14-8 83.92 g 100.7 g

14-9 151 g

14-10 11.2 g

14-11 S excess 96.2%

14-12 307 g 25.8% 6340 mL

14-13 8.69%

14-14 19.0 g

14-15 81.1%

14-16 0.9192 g

14-17 57.5 mL

14-18 $2NH_3 + Hg_2Cl_2 \rightarrow Hg + Hg(NH_2)Cl + NH_4Cl$

14-19 $4CoCl_2 + 4NH_4Cl + 20NH_3 + O_2 \rightarrow 4[CO(NH_3)_6]Cl_3 + 2H_2O$

14-20 a. $3NO_2(g) + H_2O(\ell) \rightarrow NO(g) + 2HNO_3(aq)$
b. N
c. N

15-1 3671.8 C 4.8288 g

15-2 0.83046 g

15-3 3880 sec

15-4 4.99 g

15-5 0.680 g

15-6 1.6021×10^{-19} C

15-7 a. 1.159 volts
b. 1.041 volts
c. 1.218 volts

15-8 17.5 mL

15-9 a. 0.224 L
b. 0.112 L

15-10 Cu 94.1%
Te 78.4%

15-11 96,490 C

15-12 4.1042 g

15-13 0.010447 mg/C
0.082909 mg/C
0.3043 mg/C

15-14 a. +2.1223 volts
+0.6251 volt
-1.523 volts
+0.3588 volt
-1.1373 volts
b. $Zn + Cl_2 \rightarrow Zn^{2+} + 2Cl^-$
$V + Cu^{2+} \rightarrow V^{2+} + Cu$
$2Ag + Cl_2 \rightarrow 2AgCl$
$Cd + 2AgCl \rightarrow Cd^{2+} + 2Ag + 2Cl^-$
$Pb + 2H_3O^+ + SO_4^{2-} \rightarrow PbSO_4 + H_2 + 2H_2O$

15-15 0.958 volt

15-16 0.5738 volt

15-17 a. $Al|Al^{3+}\|Fe^{2+}|Fe$
$Al + 3Fe^{2+} \rightarrow Al^{3+} + 3Fe$
b. $Al|Al^{3+}\| Ag^{+}|Ag$
$Al + 3Ag^{+} \rightarrow Al^{3+} + 3Ag$

15-18 a. $Al + 3Ag^{+} \rightarrow Al^{3+} + 3Ag$
b. yes

5-19 a. +0.56 volt, yes
b. +0.71 volt, yes
c. +0.82 volt, yes
d. -0.486 volt, no
e. -0.935 volt, no

15-20 0.731 volt

15-21 -0.414 volt

15-22 5.0×10^{71} 1.3×10^{21}
3.5×10^{-52} 1.3×10^{12}
3.8×10^{-39}

15-23 a. $Ag|Ag_2SO_4\|PbSO_4|Pb$
b. -1.1579 volts

15-24 a. $1/2S_2(g, 1\ atm) + 2e^- \rightleftharpoons S^{2-}$
b. $2Ag^\circ \rightleftharpoons 2Ag^+ + 2e^-$
c. $2Ag(s) + 1/2S_2(g, 1\ atm) \rightleftharpoons Ag_2S$
d. $E = E^\circ - \frac{RT}{nF} \ln 1/p_{S_2}^{1/2}$
e. at 490°C slope = 0.0378

15-25 a. $2Ag^+ + O^{2-} + 1/2O_2 + 2Ag$
b. $E = E^\circ - \frac{RT}{2F} \ln \frac{p_{O_2}^{1/2}}{[O^{2-}]}$
c. yes; slope = -0.0289 V
d. 0.6167 V

15-26 4.8274 g

15-27 0.02000 F

15-28 96,475 C

15-29 1990 sec

15-30 0.603 L

15-31 2.27 g

15-32 3,6665 C

15-33 3,745.0 C

15-34 229 C

15-35 CuTe; 76.5%

15-36 0.0454 volt

15-37 0.115 volt

15-38 a. $Al|Al^{3+}\|Fe^{2+}|Fe$
b. $2Al + 3Fe^{2+} \rightarrow 2Al^{3+} + 3Fe$

15-39 a. no, $2CuF \rightleftharpoons CuF_2 + Cu$, + 0.368 volt
b. no, $2FeI_3 \rightleftharpoons 2FeI_2 + I_2$, +0.235 volt
c. no, $TlI_3 \rightleftharpoons TlI + I_2$, +0.71 volt
d. yes, $2AgF_2 \rightleftharpoons 2AgF + F_2$, -0.89 volt

15-40 -0.391 volt

15-41 0.2214 volt

15-42 0.269 volt

15-43 a. 5.7×10^{56} b. 9.2×10^{59} c. 1.3×10^{83}
d. 3.8×10^{-17} e. 2.6×10^{-32}

15-44 a. $2Ag^+ + O^{2-} \rightleftharpoons 2Ag + 1/2O_2$
b. $E = E° - \frac{RT}{2F} \ln \frac{P^{1/2}_{O_2}}{[O^{2-}]}$
c. yes, slope = 0.0583
d. 0.6192 V

15-45 a. 1.581 volts
b. $PbO_2 + Pb + 4H^+ \rightarrow 2Pb^{2+} + 2H_2O$
c. $PbO_2 + Pb + 2H_2SO_4 \rightarrow 2PbSO_4 + 2H_2O$

15-46 a. $Cd + 2OH^- \rightarrow Cd(OH)_2 + 2e^-$
b. $NiO_2 + 2H_2O + 2e^- \rightarrow Ni(OH)_2 + 2OH^-$
c. 1.299 volts

15-47 a. 1.345 volts
b. activity of all chemicals are 1; log term in Nernst equation is 0

16-1 1.034 J K^{-1}

16-2 79.18 J K^{-1}

16-3 326 J K^{-1}

16-4 2296 J K^{-1}

16-5 $-7.66\ J\ K^{-1}$

16-6 $1.79\ J\ K^{-1}$

16-7 $2.17\ J\ K^{-1}$

16-8 $1.44\ J\ K^{-1}$

16-9 $5.76\ J\ K^{-1}$

16-10 $3.37\ J\ K^{-1}$

16-11 1

16-12 a. $2.93\ J\ K^{-1}\ mol^{-1}$ C
b. $19.8\ J\ K^{-1}\ mol^{-1}\ H_2$
c. $-86.78\ J\ K^{-1}\ mol^{-1}$ CO
d. $-83.18\ J\ K^{-1}\ mol^{-1}$ S
e. $-94.5\ J\ K^{-1}\ mol^{-1}\ SO_2$
f. $-44.35\ J\ K^{-1}\ mol^{-1}\ H_2$

16-13 $71.3\ J\ K^{-1}\ mol^{-1}$

16-14 a. -1184 kJ spontaneous
b. -1314 kJ spontaneous
c. -1331 kJ spontaneous
d. -953.5 kJ spontaneous
e. -966 kJ spontaneous
f. +217 kJ nonspontaneous

16-15 $-1093\ kJ\ mol^{-1}$

16-16 $-94.98\ kJ\ mol^{-1}$

16-17 -39.12 kJ

16-18 14.4 kJ

16-19 -173.7 kJ

16-20 $-59.24\ kJ\ mol^{-1}$

16-21 1.03 V

16-22 $1.279\ J\ g^{-1}$

16-23 $41.30\ J\ mol^{-1}$

16-24 $279.3\ J\ K^{-1}\ mol^{-1}$

16-25 $332\ J\ mol^{-1}$

16-26 $245.1\ J\ K^{-1}\ mol^{-1}$

16-27 35.5 J

16-28 $\Delta S = 84.1$ J
$\Delta S_{surr.} = 0$
$\Delta S_{univ.} = 84.1$ J

16-29 0

16-30 2

16-31 $-189\ kJ\ mol^{-1}$

16-32 $+32\ kJ\ mol^{-1}$

16-33	-42.6 kJ mol^{-1}	16-34	-36.7 kJ
16-35	-1.11 kJ mol^{-1}	16-36	-9730 J
16-37	-94.440 kJ	16-38	-75.880 kJ
16-39	-245.650 kJ		
17-1	0.268 min^{-1}	17-2	0.271 min^{-1}
17-3	0.0152 min^{-1}	17-4	0.08145 min^{-1}
17-5	24.7×10^{-4} sec^{-1}	17-6	1.7×10^{-4} sec^{-1} (from graph)
17-7	2.16×10^{3} sec	17-8	2.18×10^{-4} sec^{-1} 3.18×10^{3} sec
17-9	first order 6.0×10^{-4} sec^{-1} (from graph)	17-10	1165 sec 11.7%
17-11	1.32×10^{-3} mol/L	17-12	1.50×10^{-2} mol/L
17-13	0.10 sec^{-1}	17-14	$\frac{1}{1.8 \times 10^{-4}\ a}$
17-15	0.30 M^{-1} min^{-1}	17-16	second order
17-17	6.61×10^{-2} M^{-1} sec^{-1}	17-18	3.06 M^{-1} sec^{-1}

17-19 a. 2.90×10^{-3} min^{-1}
b. 0.140 M^{-1} min^{-1} (divide a by 0.02 M to get b)

17-20	a. second order b. second order c. fourth order	17-21	first order NaOCl first order MbN second order overall
17-22	first order [A] 3/2 order [B] 5/2 order overall	17-23	zero order
17-24	8.34 kcal mol^{-1}	17-25	5.69×10^{-4} sec^{-1}

17-26 $0.2688\ min^{-1}$

17-27 $1.44 \times 10^{-4}\ sec^{-1}$

17-28 $3.30 \times 10^{-4}\ sec^{-1}$

17-29 $5.93 \times 10^{-4}\ sec^{-1}$

17-30 $1.65 \times 10^{-4}\ sec^{-1}$

17-31 $6.93 \times 10^{-3}\ min^{-1}$
$6.22 \times 10^{-3}\ min^{-1}$

17-32 $10.5\ sec^{-1}$

17-33 2.33×10^{-5} mol/g

17-34 3466 sec 78.7%

17-35 4.90 mol/L

17-36 7.14×10^{-3} mol/L

17-37 rate = $k[ClO_2^-][HSO_5^-]$

17-38 0.0493 0.0481

17-39 $11.7\ M^{-1}\ min^{-1}$
second order

17-40 1.24×10^{-3} mol/L; I^-

17-41 8.25 sec

17-42 if zero order $k = 2.75 \times 10^{-4}\ mol\ min^{-1}$
if first order $k = 2.66 \times 10^{-4}\ mol\ min^{-1}$

17-43 second order [A]
zero order [B]
second order overall

17-44 a. zero order
b. $2.5 \times 10^{-2}\ sec^{-1}$

17-45 $12.9\ kJ\ mol^{-1}$

17-46 $21.38\ kcal\ mol^{-1}$

18-1
a. tetraamminecopper(II) ion
b. potassium hexafluoronickelate(II)
c. tetrathiocyanatocolbaltate(II) ion
d. tetraisothiocyanatocolbaltate(II) ion
e. tetracarbonylnickel(0)
f. tris(ethylenediamine)cobalt(III) sulfate
g. potassium trichlorotrifluoroplatinate(IV)
h. triamminetrichlorocobalt(III)

18-2
a. $[Cd(NH_3)_4](NO_3)_2$
b. $[Ni(NH_3)_6]Cl_2$
c. $Cr(CO)_6$
d. $[Co(NH_3)_4Cl_2]NO_3$
e. Na $[PtCl_2F_2]$
f. $[Rh(en)_3]Cl_3$
g. $[Cr(NH_3)_6][Cu(Cl)_5]$

18-3 a. trans-dibromodichloroplatinate(II) ion
b. trans-triamminetrichlorocobalt(III)
c. trans-diamminebis(ethylenediamine)cobalt(III) ion

18-4

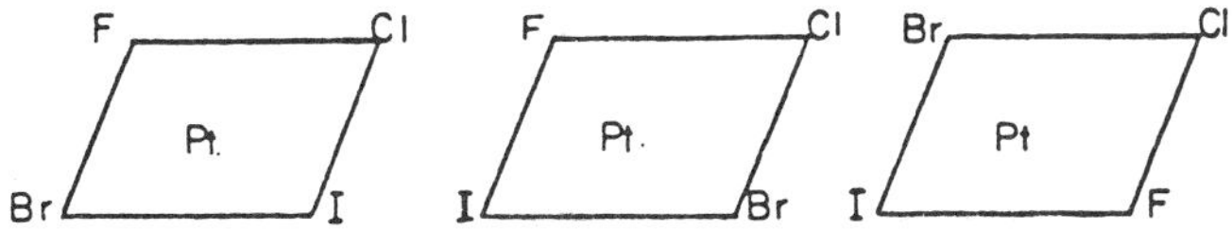

18-5 a. hexaamminechromium(III) pentachlorocuprate(II)
b. hexamminechromium(III) pentacyanonickelate(II)
c. hexaaminecobalt(III) hexachloroferrate(III)

18-6 potassium hexathiocyanatoniobate(V)
$K[Nb(SCN)_6]$

18-7 diamminecobalt(II) nitrate
$[Co(NH_3)_2](NO_3)_2$

18-8 90%

18-9 42.9%

18-10 $K[Nb(SCN)_6]$ potassium hexathiocyanatoniobate(V)

18-11 ↑↓ ↑↓ ↑ |_ _ _ _ _ _| d^2sp^3 octahedron

18-12 a. low spin b. low spin c. low spin
d. low spin e. high spin

18-13 a. ↑ _ _ |_ _ _ _ _ _| d^2sp^3 octahedron

b. e_g _ _

t_{2g} ↑ _ _ = 20,300 cm^{-1}

c. λ (absorbed) = 4926 Å

18-14 $CoCl_2^{2-}$ > $CoBr_4^{2-}$ > CoI_4^{2-}

18-15 a. tetraamminecadmium(II) ion
b. dicyanoargentate(I) ion
c. diamminetetrachlorocobaltate(III) ion

d. potassium octacyanomolybdate(IV)
e. pentacarbonyliron(0)
f. pentaammineaquocobalt(III) perrhenate
g. pentaamminechloroiridium(III) chloride
h. bis(ethylenediamine)dichlororuthenium(III) chloride
i. tris(ethylenediamine)chromium(III) pentacyanonickelate(II)

18-16 a. $[Co(NH_3)_4ClNO_2]Cl$
b. $[Co(NH_3)_4Cl_2]NO_2$
c. $Co(CO)_3NO$
d. $K_2Zn(CN)_4$
e. $[Ru(NH_3)_5Cl]Cl_2$
f. $[Cr(en)_3]_2(SO_4)_3$
g. $K_4[Mo(CN)_8]$

18-17 a. cis-dibromodichloroplatinate(IV) ion
b. cis-triamminetrichlorocobalt(III)
c. cis-diamminebis(ethylenediamine)cobalt(III) ion

18-18 a. pentaammineaquocobalt(III) trioxalatocobaltate(III)
b. pentaammineaquocobalt(II) diaamminetetrathiocyanatochromate(II)
c. pentaammineaquocobalt(II) diaamminetetranitrocobaltate(II) or pentaammineaquocobalt(II) diaamminetetranitritocobaltate(II)

18-19 $CoC_6H_{20}N_7O_7$

18-20 $K[Ta(SCN)_6]$ potassium hexathiocyanatotantalate(V)

18-21 a. $C_{12}H_{24}CoN_6S_4$
b. tetramethylammonium tetrathiocyanatocobaltate(II) $[(CH_3)_4N]_2Co(SCN)_4$
c. 87.5%

18-22 70.4%

18-23 92.6

18-24 4 high spin 0 low spin

18-25 hexaaquoruthenium(II) ion
tris(ethylenediamine)ruthenium(II) ion
bis(propylenediamine)ruthenium(II) ion
bis(diethylenediamine)ruthenium(II) ion

hexaammineruthenium(II) ion
$[Ru(dien)_2]^{2+} > [Ru(pn)_3]^{2+} > [Ru(en)_3]^{2+} >$
$[Ru(NH_3)]^{2+} > [Ru(H_2O)_6]^{2+}$

18-26 yes (theoretical yield 0.3620 g)

18-27 $[Co(NH_3)_5Cl]Cl_2$
pentaamminechlorocobalt(III) chloride

18-28 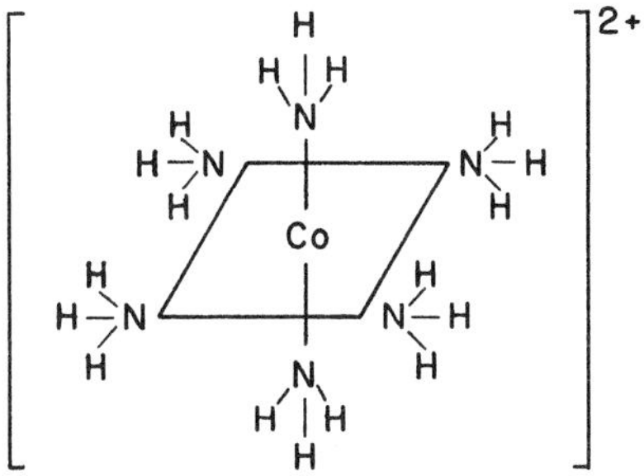

18-29 a. $[Cu(NH_3)_4]SO_4 \cdot 4H_2O$
b. 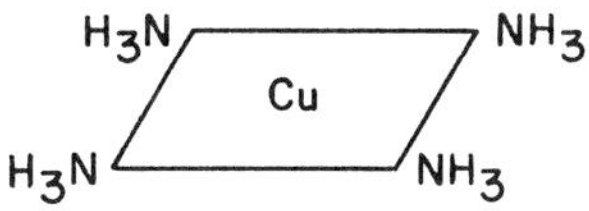

18-30 0.664 g

18-31 $Cu(NH_3)_4{}^{2+}$

18-32 hexaamminecolbalt(III) chloride

19-1 1.0080

19-2 8.62×10^6 C

19-3 0.556 atm CO and H_2O
0.444 atm CO_2 and H_2

19-4 2.62

19-5 9820 mL

19-6 318 kJ mol^{-1}

19-7 30.3439 g

19-8 464 kJ mol^{-1}
498 kJ mol^{-1}
427 kJ mol^{-1}

19-9	96.4	19-10	0.0364 M 1.44
19-11	165.5	19-12	a. +0.2223 volt b. 0.3936 V
19-13	2.01	19-14	0.511 V
19-15	2.96×10^{-2} atm	19-16	15.9994
19-17	1.60×10^{-3} mol Cu 7.95×10^{-4} mol O_2	19-18	70.8%
19-19	3.92×10^{-6} atm 1.96×10^{-6} atm	19-20	$CaCr_2O_7 \cdot 3H_2O$
19-21	-46.96	19-22	142 kJ mol^{-1}
19-23	-228 kJ	19-24	0.458 volt
19-25	1.04 L	19-26	0.75 atm
19-27	+0.82 volt	19-28	1.000 g
19-29	WO_3	19-30	0.2888 g N_2 0.3298 g O_2
19-31	6.939	19-32	$(C_2H_5Li)_6$
19-33	10.646 g	19-34	-56.90 kJ mol^{-1}
19-35	-89.29 kJ mol^{-1}	19-36	40.09
19-37	320 mL	19-38	1.3×10^{-9} mol/L
19-39	-117 kJ	19-40	0.727 L
19-41	10.80	19-42	BCl_3
19-43	250	19-44	28.1
19-45	$B_{18}H_{22}$	19-46	BN
19-47	67.2	19-48	Plane triangle

19-49 $BF_3 \cdot NH_3$

19-50 B_2Cl_4

19-51 13.6 cc H_2

19-52 $2Ga + 3I_2 \rightarrow 2GaI_3$
$2GaI_3 + O_2 \rightarrow 2GaOI + 2I$
$GaOI + 3F_2 \rightarrow GaOF + IF_5$

19-53 -493.3 kJ mol^{-1} AlF_3

19-54 0.936 atm due to dimer

19-55 Yes

19-56 SiS_2

19-57 SiI_4 $GeCl_4$ tetrahedra

19-58 a. NaF
b. 81.4%
c. 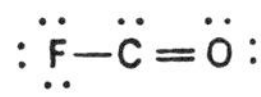

d. 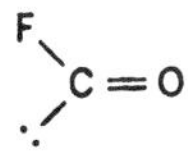

19-59 O=C=C=C=O

19-60 Plane triangle
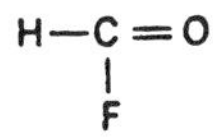

Plane triangle

19-61 62.5

19-62 114 mL

19-63 2.20 g

19-64 Plane triangle
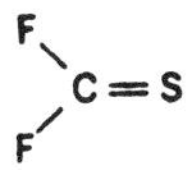

Plane triangle

19-65 28.3 g

19-66 1.88 kJ mol^{-1}

19-67 0.473 volt

19-68 Tetrahedron, octahedron

19-69 118.69

19-70 1.78×10^{-4} M

19-71 14.007

19-72 124.1

19-73 78.4

19-74 Trigonal pyramid

19-75 $3NH_2OH \rightarrow NH_3 + N_2 + 3H_2O$

19-76 $N_3H_{10}Cl_3$

19-77 a. $2NF_3 + Cu \rightarrow CuF_2 + N_2F_4$
b. 10.2 g

19-78 Yes

19-79 NF_3

19-80 a. 71.3
b. NF_3

19-81 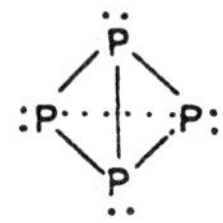

19-82 PIF_2 Trigonal pyramid

19-83 a. P_2F_4
b. 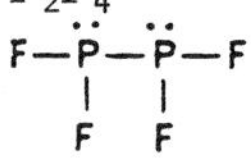
c. two tetrahedra
d. $2PIF_2 + Hg \rightarrow HgI_2 + P_2F_4$

19-84 69%

19-85 75.5% CaF_2

19-86 Octahedron

19-87 Electron pair repulsion of basal electron pairs and apical electron pairs.

19-88 $AsOF_3$

19-89 126

19-90 14.053

19-91 420 mm

19-92 14.4 M 0.523

19-93 SF_2

19-94 V shape

19-95 a. SeF_5OF
b. Octahedron

19-96 P_4S_3

19-97 32.07

19-98 1.21×10^{-22} M

19-99 Cu^{2+} 6.9×10^{-23} M
Zn^{2+} 0.1 M

19-100 SO_3F_2

19-101 CF_3SF_5

19-102 S_4N_3Cl

19-103 $2P_4S_3 + 6S \rightarrow P_4S_5 + P_4S_7$

19-104 S_8

19-105 TeO_2

19-106 79.926

19-107 2.58 g

19-108 2.16×10^{-3} M

19-109 0.15 volt

19-110 0.19 volt

19-111 1240 mm

19-112 1.303×10^{-3}

19-113 9.33×10^{-2} M

19-114 T shape

19-115

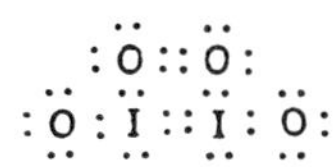

19-116 Square-based pyramid
Pentagonal bipyramid

19-117 1600 mm

19-118 94.7 mm

19-119 a. Square pyramid
b. Repulsion of unshared pair

19-120 562.1 mm

19-121 IF_5

19-122 0.0547 M

19-123 V shape

19-124 -4711 J

19-125 1552 J

19-126 XeF_4

19-127 Yes

19-128 Square plane

19-129 133 kJ mol^{-1}

19-130 39.964

19-131 7.52×10^{-4} g/mL

19-132 XeF_6

19-133 XeF_6

19-134 Yes

19-135 b. Distorted octahedron

19-136 Ba_3XeO_6

19-137 0.266 g

19-138 XeF_6

19-139 V shape

19-140 Trigonal pyramid

19-141 $XePtF_6$

19-142 0.2106 g 0.1924 g

19-143 245.4

19-144 128 kJ mol^{-1}

19-145 41.8 kJ mol^{-1}

19-146 5.40 g

19-147 0.7772 volt

19-148 166 mL

19-149 3.6×10^{-9} mol/L

19-150 217 kJ mol^{-1}

19-151 $8HCl + K_2TeO_4 \rightarrow Cl_2 + TeCl_4 + 4H_2O + 2KCl$

19-152 0.073 volt

19-153 +0.490 volt

19-154 2.85×10^{-4} M

19-155 4.01 g

19-156 288

19-157 a. $HgCl_2$ molecule
b. Covalent
c. Linear

19-158 CdI_2 molecules

19-159 V shape

19-160 0.627

19-161 0.513 volts

19-162 -0.351 volt

19-163 200.60

19-164 1.70×10^{-16} M

19-165 0.756 g

19-166 $CrPO_4 \cdot 6H_2O$

19-167 $CrPO_4$

19-168 $FeCl_3$

19-169 $NiFe_2O_4$

19-170 NbF_5

19-171 8.3265 g

19-172 47.89

19-173 0.909 volt

19-174 O_2PtF_6

19-175 0.55 volt

19-176 -0.18

19-177 -1.182 volt

19-178 a. $14H^+ + Cr_2O_7^{2-} + 6I^- \rightarrow 2Cr^{3+} + 3I_2 + 7H_2O$
b. $4H^+ + 2CrO_4^{2-} + 3SO_2 \rightarrow 2Cr^{3+} + 3SO_4^{2-} + 2H_2O$
c. $2MnO_4^- + 3H_2S \rightarrow 2MnO_2 + 3S + 2H_2O + 2OH^-$
d. $6H^+ + 2MnO_4^- + 5H_2S \rightarrow 2Mn^{2+} + 5S + 8H_2O$
e. $4Fe^{2+} + 2H_2O + O_2 \rightarrow 4Fe^{3+} + 4OH^-$

19-179 1.31×10^{-4} M

19-180 3.06×10^{3} M
3.7×10^{-17} M

19-181 50.942

19-182 $FeCl_3$

19-183 Fe_2Cl_6

19-184 -154 kJ mol^{-1}

19-185 -989.9 kJ mol^{-1}

19-186 11.54 F

19-187 0.423 g

19-188 0.1 mol $KMnO_4$
0.1 mol $KHSO_4$
0.5 mol H_2

19-189 5.0 mL

19-190 69.1 mL

19-191 a. $4Co^{2+} + 24NH_3 + O_2 + 2H_2O \rightarrow 4Co(NH_3)_6^{3+} + 4OH^-$
b. $2[Fe(CN)_6]^{4-} + Cl_2 \rightarrow 2[Fe(CN)_6]^{3-} + 2Cl^-$
c $2Cr(OH)_3 + 3Na_2O_2 \rightarrow 2Na_2CrO_4 + 2H_2O + 2NaOH$

19-192 $YCl_3 \cdot 6H_2O$

19-193 Cr_3C_2

19-194 71 hr, 24 min

19-195 5.23 L

19-196 $3UO_2 + O_2 \rightarrow U_3O_8$

19-197 $Pa_2O_5 + 5C + 4Br_2 \rightarrow PaBr_5 + PaBr_3 + 5CO$

19-198 -104.48 kJ mol^{-1}

19-199 PuF_6

19-200 0.336 L

Appendix D
TABLES

TABLE D-1A General Physical Constants[a]

Constant	Value
Avogadro's number	$6.022 \times 10^{23}\ mol^{-1}$
Boltzmann constant	$1.38 \times 10^{23}\ J\ K^{-1}$
Charge on electron	$1.60 \times 10^{-19}\ C$
Faraday	$96{,}485.309\ C\ mol^{-1}$
Gas constant	$8.314\ J\ K^{-1}\ mol^{-1}$
Planck's constant	$6.626 \times 10^{-34}\ J\ S$
Speed of light in vacuum	$3.00 \times 10^{8}\ ms^{-1}$

[a]Most are taken from NBS reprint in *J. Chem. Ed.*, *48*, 569 (1971).

TABLE D-1B Vapor Pressure of Water

Temp (°C)	Vapor pressure (torr)	Vapor pressure (Pa)
0	4.579	610.48
5	4.750	633.28
10	9.209	1227.77
15	12.788	1704.93
20	17.535	2337.81
25	23.756	3167.21
30	31.824	4242.85

TABLE D-2 Four-Place Common Logarithms

Natural numbers	0	1	2	3	4	5	6	7	8	9	Proportional parts								
											1	2	3	4	5	6	7	8	9
10	0000	0043	0086	0128	0170	0212	0253	0294	0334	0374	4	8	12	17	21	25	29	33	37
11	0414	0453	0492	0531	0569	0607	0645	0682	0719	0755	4	8	11	15	19	23	26	30	34
12	0792	0828	0864	0899	0934	0969	1004	1038	1072	1106	3	7	10	14	17	21	24	28	31
13	1139	1173	1206	1239	1271	1303	1335	1367	1399	1430	3	6	10	13	16	19	23	26	29
14	1461	1492	1523	1553	1584	1614	1644	1673	1703	1732	3	6	9	12	15	18	21	24	27
15	1761	1790	1818	1847	1875	1903	1931	1959	1987	2014	3	6	8	11	14	17	20	22	25
16	2041	2068	2095	2122	2148	2175	2201	2227	2253	2279	3	5	8	11	13	16	18	21	24
17	2304	2330	2355	2380	2405	2430	2455	2480	2504	2529	2	5	7	10	12	15	17	20	22
18	2553	2577	2601	2625	2648	2672	2695	2718	2742	2765	2	5	7	9	12	14	16	19	21
19	2788	2810	2833	2856	2878	2900	2923	2945	2967	2989	2	4	7	9	11	13	16	18	20
20	3010	3032	3054	3075	3096	3118	3139	3160	3181	3201	2	4	6	8	11	13	15	17	19
21	3222	3243	3263	3284	3304	3324	3345	3365	3385	3404	2	4	6	8	10	12	14	16	18
22	3424	3444	3464	3483	3502	3522	3541	3560	3579	3598	2	4	6	8	10	12	14	15	17
23	3617	3636	3655	3674	3692	3711	3729	3747	3766	3784	2	4	6	7	9	11	13	15	17
24	3802	3820	3838	3856	3874	3892	3909	3927	3945	3962	2	4	5	7	9	11	12	14	16
25	3979	3997	4014	4031	4048	4065	4082	4099	4116	4133	2	3	5	7	9	10	12	14	15
26	4150	4166	4183	4200	4216	4232	4249	4265	4281	4298	2	3	5	7	8	10	11	13	15
27	4314	4330	4346	4362	4378	4393	4409	4425	4440	4456	2	3	5	6	8	9	11	13	14
28	4472	4487	4502	4518	4533	4548	4564	4579	4594	4609	2	3	5	6	8	9	11	12	14
29	4624	4639	4654	4669	4683	4698	4713	4728	4742	4757	1	3	4	6	7	9	10	12	13
30	4771	4786	4800	4814	4829	4843	4857	4871	4886	4900	1	3	4	6	7	9	10	11	13
31	4914	4928	4942	4955	4969	4983	4997	5011	5024	5038	1	3	4	6	7	8	10	11	12
32	5051	5065	5079	5092	5105	5119	5132	5145	5159	5172	1	3	4	5	7	8	9	11	12
33	5185	5198	5211	5224	5237	5250	5263	5276	5289	5302	1	3	4	5	6	8	9	10	12
34	5315	5328	5340	5353	5366	5378	5391	5403	5416	5428	1	3	4	5	6	8	9	10	11
35	5441	5453	5465	5478	5490	5502	5514	5527	5539	5551	1	2	4	5	6	7	9	10	11
36	5563	5575	5587	5599	5611	5623	5635	5647	5658	5670	1	2	4	5	6	7	8	10	11
37	5682	5694	5705	5717	5729	5740	5752	5763	5775	5786	1	2	3	5	6	7	8	9	10
38	5798	5809	5821	5832	5843	5855	5866	5877	5888	5899	1	2	3	5	6	7	8	9	10
39	5911	5922	5933	5944	5955	5966	5977	5988	5999	6010	1	2	3	4	5	7	8	9	10
40	6021	6031	6042	6053	6064	6075	6085	6096	6107	6117	1	2	3	4	5	6	8	9	10
41	6128	6138	6149	6160	6170	6180	6191	6201	6212	6222	1	2	3	4	5	6	7	8	9
42	6232	6243	6253	6263	6274	6284	6294	6304	6314	6325	1	2	3	4	5	6	7	8	9
43	6335	6345	6355	6365	6375	6385	6395	6405	6415	6425	1	2	3	4	5	6	7	8	9
44	6435	6444	6454	6464	6474	6484	6493	6503	6513	6522	1	2	3	4	5	6	7	8	9
45	6532	6542	6551	6561	6571	6580	6590	6599	6609	6618	1	2	3	4	5	6	7	8	9
46	6628	6637	6646	6656	6665	6675	6684	6693	6702	6712	1	2	3	4	5	6	7	7	8
47	6721	6730	6739	6749	6758	6767	6776	6785	6794	6803	1	2	3	4	5	5	6	7	8
48	6812	6821	6830	6839	6848	6857	6866	6875	6884	6893	1	2	3	4	4	5	6	7	8
49	6902	6911	6920	6928	6937	6946	6955	6964	6972	6981	1	2	3	4	4	5	6	7	8
50	6990	6998	7007	7016	7024	7033	7042	7050	7059	7067	1	2	3	3	4	5	6	7	8
51	7076	7084	7093	7101	7110	7118	7126	7135	7143	7152	1	2	3	3	4	5	6	7	8
52	7160	7168	7177	7185	7193	7202	7210	7218	7226	7235	1	2	2	3	4	5	6	7	7
53	7243	7251	7259	7267	7275	7284	7292	7300	7308	7316	1	2	2	3	4	5	6	6	7
54	7324	7332	7340	7348	7356	7364	7372	7380	7388	7396	1	2	2	3	4	5	6	6	7

(continued)

TABLE D-2 (continued)

Natural numbers	0	1	2	3	4	5	6	7	8	9	Porportional parts 1	2	3	4	5	6	7	8	9
55	7404	7412	7419	7427	7435	7443	7451	7459	7466	7474	1	2	2	3	4	5	5	6	7
56	7482	7490	7497	7505	7513	7520	7528	7536	7543	7551	1	2	2	3	4	5	5	6	7
57	7559	7566	7574	7582	7589	7597	7604	7612	7619	7627	1	2	2	3	4	5	5	6	7
58	7634	7642	7649	7657	7664	7672	7679	7686	7694	7701	1	1	2	3	4	4	5	6	7
59	7709	7716	7723	7731	7738	7745	7752	7760	7767	7774	1	1	2	3	4	4	5	6	7
60	7782	7789	7796	7803	7810	7818	7825	7832	7839	7846	1	1	2	3	4	4	5	6	6
61	7853	7860	7868	7875	7882	7889	7896	7903	7910	7917	1	1	2	3	4	4	5	6	6
62	7924	7931	7938	7945	7952	7959	7966	7973	7980	7987	1	1	2	3	3	4	5	6	6
63	7993	8000	8007	8014	8021	8028	8035	8041	8048	8055	1	1	2	3	3	4	5	5	6
64	8062	8069	8075	8082	8089	8096	8102	8109	8116	8122	1	1	2	3	3	4	5	5	6
65	8129	8136	8142	8149	8156	8162	8169	8176	8182	8189	1	1	2	3	3	4	5	5	6
66	8195	8202	8209	8215	8222	8228	8235	8241	8248	8254	1	1	2	3	3	4	5	5	6
67	8261	8267	8274	8280	8287	8293	8299	8306	8312	8319	1	1	2	3	3	4	5	5	6
68	8325	8331	8338	8344	8351	8357	8363	8370	8376	8382	1	1	2	3	3	4	4	5	6
69	8388	8395	8401	8407	8414	8420	8426	8432	8439	8445	1	1	2	2	3	4	4	5	6
70	8451	8457	8463	8470	8476	8482	8488	8494	8500	8506	1	1	2	2	3	4	4	5	5
71	8513	8519	8525	8531	8537	8543	8549	8555	8561	8567	1	1	2	2	3	4	4	5	5
72	8573	8579	8585	8591	8597	8603	8609	8615	8621	8627	1	1	2	2	3	4	4	5	5
73	8633	8639	8645	8651	8657	8663	8669	8675	8681	8686	1	1	2	2	3	4	4	5	5
74	8692	8698	8704	8710	8716	8722	8727	8733	8739	8745	1	1	2	2	3	4	4	5	5
75	8751	8756	8762	8768	8774	8779	8785	8791	8797	8802	1	1	2	2	3	3	4	5	5
76	8808	8814	8820	8825	8831	8837	8842	8848	8854	8859	1	1	2	2	3	3	4	5	5
77	8865	8871	8876	8882	8887	8893	8899	8904	8910	8915	1	1	2	2	3	3	4	4	5
78	8921	8927	8932	8938	8943	8949	8954	8960	8965	8971	1	1	2	2	3	3	4	4	5
79	8976	8982	8987	8993	8998	9004	9009	9015	9020	9026	1	1	2	2	3	3	4	4	5
80	9031	9036	9042	9047	9053	9058	9063	9069	9074	9079	1	1	2	2	3	3	4	4	5
81	9085	9090	9096	9101	9106	9112	9117	9122	9128	9133	1	1	2	2	3	3	4	4	5
82	9138	9143	9149	9154	9159	9165	9170	9175	9180	9186	1	1	2	2	3	3	4	4	5
83	9191	9196	9201	9206	9212	9217	9222	9227	9232	9238	1	1	2	2	3	3	4	4	5
84	9243	9248	9253	9258	9263	9269	9274	9279	9284	9289	1	1	2	2	3	3	4	4	5
85	9294	9299	9304	9309	9315	9320	9325	9330	9335	9340	1	1	2	2	3	3	4	4	5
86	9345	9350	9355	9360	9365	9370	9375	9380	9385	9390	1	1	2	2	3	3	4	4	5
87	9395	9400	9405	9410	9415	9420	9425	9430	9435	9440	0	1	1	2	2	3	3	4	4
88	9445	9450	9455	9460	9465	9469	9474	9479	9484	9489	0	1	1	2	2	3	3	4	4
89	9494	9499	9504	9509	9513	9518	9523	9528	9533	9538	0	1	1	2	2	3	3	4	4
90	9542	9547	9552	9557	9562	9566	9571	9576	9581	9586	0	1	1	2	2	3	3	4	4
91	9590	9595	9600	9605	9609	9614	9619	9624	9628	9633	0	1	1	2	2	3	3	4	4
92	9638	9643	9647	9652	9657	9661	9666	9671	9675	9680	0	1	1	2	2	3	3	4	4
93	9685	9689	9694	9699	9703	9708	9713	9717	9722	9727	0	1	1	2	2	3	3	4	4
94	9731	9736	9741	9745	9750	9754	9759	9763	9768	9773	0	1	1	2	2	3	3	4	4
95	9777	9782	9786	9791	9795	9800	9805	9809	9814	9818	0	1	1	2	2	3	3	4	4
96	9823	9827	9832	9836	9841	9845	9850	9854	9859	9863	0	1	1	2	2	3	3	4	4
97	9868	9872	9877	9881	9886	9890	9894	9899	9903	9908	0	1	1	2	2	3	3	4	4
98	9912	9917	9921	9926	9930	9934	9939	9943	9948	9952	0	1	1	2	2	3	3	4	4
99	9956	9961	9965	9969	9974	9978	9983	9987	9991	9996	0	1	1	2	2	3	3	3	4

TABLE D-3 Table of Atomic Masses, 1983 (Based on the Atomic Mass of ^{12}C = 12)

Name	Symbol	Atomic number	Atomic masses
Actinium	Ac	89	(227)
Aluminum	Al	13	26.98154
Americium	Am	95	(243)
Antimony	Sb	51	121.75
Argon	Ar	18	39.948
Arsenic	As	33	74.9216
Astatine	At	85	(210)
Barium	Ba	56	137.33
Berkelium	Bk	97	(247)
Beryllium	Be	4	9.01218
Bismuth	Bi	83	208.9804
Boron	B	5	10.811
Bromine	Br	35	79.904
Cadmium	Cd	48	122.41
Calcium	Ca	20	40.08
Californium	Cf	98	(251)
Carbon	C	6	12.01115
Cerium	Ce	58	140.12
Cesium	Cs	55	132.9054
Chlorine	Cl	17	35.453
Chromium	Cr	24	51.996
Cobalt	Co	27	58.9332
Copper	Cu	29	63.546
Curium	Cm	96	(247)
Dysprosium	Dy	66	162.50
Einsteinium	Es	99	(252)
Erbium	Er	68	167.26
Europium	Eu	63	151.96
Fermium	Fm	100	(257)
Fluorine	F	9	18.998403
Francium	Fr	87	(223)
Gadolinium	Gd	64	157.25
Gallium	Ga	31	69.72
Germanium	Ge	32	72.59
Gold	Au	79	196.9665

(continued)

TABLE D-3 (Continued)

Name	Symbol	Atomic number	Atomic masses
Hafnium	Hf	72	178.49
Helium	He	2	4.00260
Holmium	Ho	67	164.9304
Hydrogen	H	1	1.00797
Indium	In	49	114.82
Iodine	I	53	126.9045
Iridium	Ir	77	192.22
Iron	Fe	26	55.847
Krypton	Kr	36	83.80
Lanthanum	La	57	138.9055
Lawrencium	Lr	103	(260)
Lead	Pb	82	207.19
Lithium	Li	3	6.941
Lutetium	Lu	71	174.967
Magnesium	Mg	12	24.305
Manganese	Mn	25	54.9380
Mendelevium	Md	101	(258)
Mercury	Hg	80	200.59
Molybdenum	Mo	42	95.04
Neodymium	Nd	60	144.24
Neon	Ne	10	20.179
Neptunium	Np	93	(237)
Nickel	Ni	28	58.70
Niobium	Nb	41	92.9064
Nitrogen	N	7	14.0067
Nobelium	No	102	(259)
Osmium	Os	76	190.2
Oxygen	O	8	15.9994
Palladium	Pd	46	106.4
Phosphorus	P	15	30.97376
Platinum	Pt	78	195.09
Plutonium	Pu	94	(244)
Polonium	Po	84	(209)
Potassium	K	19	39.0983
Praseodymium	Pr	59	140.9077
Promethium	Pm	61	(145)

(continued)

TABLE D-3 (Continued)

Name	Symbol	Atomic number	Atomic masses
Protactinium	Pa	91	(231)
Radium	Ra	88	226.0254
Radon	Rn	86	(222)
Rhenium	Re	75	186.207
Rhodium	Rh	45	102.9055
Rubidium	Rb	37	85.4678
Ruthenium	Ru	44	101.07
Samarium	Sm	62	150.54
Scandium	Sc	21	44.9559
Selenium	Se	34	78.96
Silicon	Si	14	28.0855
Silver	Ag	47	107.868
Sodium	Na	11	22.98977
Strontium	Sr	38	87.62
Sulfur	S	16	32.064
Tantalum	Ta	73	180.9479
Technetium	Tc	43	(98)
Tellurium	Te	52	127.60
Terbium	Tb	65	158.9254
Thallium	Tl	81	204.37
Thorium	Th	90	232.0381
Thulium	Tm	69	168.9342
Tin	Sn	50	118.69
Titanium	Ti	22	47.90
Tungsten	W	74	183.85
Uranium	U	92	238.0829
Vanadium	V	23	50.9415
Xenon	Xe	54	131.30
Ytterbium	Yb	70	173.04
Yttrium	Y	39	88.9059
Zinc	Zn	30	65.38
Zirconium	Zr	40	91.22

A mass in parentheses represents the longest-lived isotope.

TABLE D-4 Periodic Table

1 IA	2 IIA	3 IIIB	4 IVB	5 VB	6 VIB	7 VIIB	8 VIII	9 VIII	10 VIII	11 IB	12 IIB	13 IIIA	14 IVA	15 VA	16 VIA	17 VIIA	18
1 H																	2 He
3 Li	4 Be											5 B	6 C	7 N	8 O	9 F	10 Ne
11 Na	12 Mg											13 Al	14 Si	15 P	16 S	17 Cl	18 Ar
19 K	20 Ca	21 Sc	22 Ti	23 V	24 Cr	25 Mn	26 Fe	27 Co	28 Ni	29 Cu	30 Zn	31 Ga	32 Ge	33 As	34 Se	35 Br	36 Kr
37 Rb	38 Sr	39 Y	40 Zr	41 Nb	42 Mo	43 Tc	44 Ru	45 Rh	46 Pd	47 Ag	48 Cd	49 In	50 Sn	51 Sb	52 Te	53 I	54 Xe
55 Cs	56 Ba	57 La	72 Hf	73 Ta	74 W	75 Re	76 Os	77 Ir	78 Pt	79 Au	80 Hg	81 Tl	82 Pb	83 Bi	84 Po	85 At	86 Rn
87 Fr	88 Ra	89 Ac															
				58 Ce	59 Pr	60 Nd	61 Pm	62 Sm	63 Eu	64 Gd	65 Tb	66 Dy	67 Ho	68 Er	69 Tm	70 Yb	71 Lu
				90 Th	91 Pa	92 U	93 Np	94 Pu	95 Am	96 Cm	97 Bk	98 Cf	99 Es	100 Fm	101 Md	102 No	103 Lw

INDEX

C

D

E

F